RENORMALIZATION

This is a volume in
PURE AND APPLIED MATHEMATICS

A Series of Monographs and Textbooks

Editors: SAMUEL EILENBERG AND HYMAN BASS

A list of recent titles in this series appears at the end of this volume.

RENORMALIZATION

Edward B. Manoukian

ROYAL MILITARY COLLEGE OF CANADA
KINGSTON, ONTARIO, CANADA

 1983

ACADEMIC PRESS

A Subsidiary of Harcourt Brace Jovanovich, Publishers

New York   London
Paris   San Diego   San Francisco   São Paulo   Sydney   Tokyo   Toronto

ACADEMIC PRESS, INC.
111 Fifth Avenue, New York, New York 10003

United Kingdom Edition published by
ACADEMIC PRESS, INC. (LONDON) LTD.
24/28 Oval Road, London NW1   7DX

Library of Congress Cataloging in Publication Data

Manoukian, Edward B.
   Renormalization.

   (Pure and applied mathematics)
   Includes index.
   1. Renormalization (Physics) 2. Feynman integrals.
3. Quantum field theory. I. Title. II. Series: Pure
and applied mathematics (Academic Press)
QA3.P8 [QC20.7.R43]   510s [530.1'5]   82-8772
ISBN 0–12–469450–0   AACR2

To Tanya

CONTENTS

Chapter 4 $\varepsilon \to +0$ **Limit of Feynman Integrals and Lorentz Covariance**

Chapter 5   **The Subtraction Formalism**

Chapter 6   **Asymptotic Behavior in Quantum Field Theory**

Appendix   **Subtractions versus Counterterms**

References

PREFACE

Renormalization theory is still with us and very much alive since its birth over three decades ago. It has reached such a high level of sophistication that any book on the subject has to be mathematically rigorous to do any justice to it. Since the early classic works on quantum electrodynamics, it has been studied systematically and has become the method for computations in relativistic quantum field theory. The success of renormalizaztion has been recorded at least twice in the history of physics through Nobel prizes, once in 1965 for the work on quantum electrodynamics, and again in 1979 for the work on the unification of the electromagnetic and weak interactions. Quantum electrodynamics is in excellent agreement with experiments, and the unified field theories seem to be quite promising candidates for a more complete theory. In spite of the importance of renormalization theory in physics, very few field theorists seem to know the intricate details of this subject. This is essentially due to the complex nature of the subject. Accordingly a book on renormalization theory would be quite justifiable at this stage, if not a matter of urgency. We define two major lines of work on the subject. The first is to develop the subtraction formalism, provide its convergence proof, and extract general and valuable information from the subtractions, all done in a model-independent way, and is mathematical in nature. The other deals with model building, "modified" Feynman rules, symmetry principles and spontaneous symmetry breaking, renormalization group equations, and operator product expansions, to mention just a few problems. This second line of work is a rapidly developing branch of research, and there is still much to be done. The first line of work is well established, and this book is devoted exclusively to it and deals with the basic facts of renormalization theory. Perhaps another suitable title for the book would have been "The Core of Renormalization." The subtraction scheme we use has a very simple structure; we were inspired by the ingenious and classic work of Salam in its formulation. It is carried out in momentum space and applied directly to the Feynman integrand without ultraviolet cutoffs. A unifying theorem of renormalization is given that brings us into contact with other standard approaches of subtractions. In particular, the latter establishes the long-standing problem of essentially the equivalence of the paths taken in the ingenious approaches of Salam and Bogoliubov.

The book deals with the basics of renormalization theory presented in a mathematically rigorous manner. It presents the subject in a unified manner and model-independent way. It is primarily designed for graduate students and researchers in quantum field theory, mathematical physics, and elementary particles. It may be used as a text in renormalization theory, or its subject matter may be incorporated in the second half of a serious course in quantum field theory. It may be also used as a reference book and for individual study. Although the textbook is mathematical in nature, one does not need mathematics beyond what one learns in a conventional undergraduate honors physics program to read it. Some familiarity with Feynman rules (at least with the structure of a Feynman integral and a Feynman diagram) is, however, essential. Excellent books and monographs exist in the literature on the latter, and they may be read simultaneously with this book. Most of the material given in the text has already appeared in the published literature. The first chapter deals with basics of mathematical analysis and, in particular, with the theory of multiple integrals and Fubini's theorem that are so essential in this subject. The property of Feynman integrands as belonging to a very special class of functions, called the class B_n functions, is given in Chapter 2. In Chapter 3 we deal with the classic power-counting theorem that provides a convergence criterion for Feynman integrals and also provides a method of studying their asymptotic behavior. The $\varepsilon \rightarrow +0$ limit of Feynman integrals is dealt with in Chapter 4, where we establish their Lorentz covariance in this limit. In Chapter 5 we discuss the subtraction formalism and its convergence. Chapter 6 is devoted to the study of the asymptotic behavior of subtracted-out Feynman amplitudes in the Euclidean region. In an appendix the equivalence of the subtraction and the counterterm formalisms is given.

ACKNOWLEDGMENTS

Most of the book was written at the Royal Military College, and I could have found no more congenial atmosphere for completing this pleasant task than the one existing here. I should like to thank Professor S. Weinberg for the interest he has shown in my work on the subtraction formalism at its earlier stages. Professor W. Zimmermann's analysis of his Bogoliubov scheme has been extremely useful in formulating our analysis of subtractions. Thanks are due to many of my colleagues who have repeatedly encouraged me to write a book on the subject. My thanks to C. Chamberlain, L. Craven, E. Engelhardt, B. Ison, I. Kennedy, and J. Morin for all participating in typing the difficult manuscript. The book would never have been completed without the patience and understanding of my wife. I am deeply grateful to her.

INTRODUCTION

The years 1965 and 1979 marked, through the Nobel prizes,[1] the unique role renormalization theory had played in physics. Renormalization theory is undoubtedly one of the most important developments ever in elementary particle physics. It has become the method for computations in relativistic quantum field theory and thus been able to confront theory with experiments, which, after all, is the ultimate goal of physics. The success of quantum electrodynamics is undisputably recognized, and the numerical predictions of this theory are in excellent agreement with experiments.[2] The unified field theories of the weak and electromagnetic interactions,[3] with Yang–Mills–Shaw[4] fields, and in the same spirit the currently unified field theories of the strong, weak, and electromagnetic interactions,[5] seem to be promising candidates for a more complete theory. There have been even many interesting attempts to unite gravity[6] with the other interactions and develop a formalism

---

[1] Schwinger (1972), Feynman (1972), Tomonaga (1972), and Salam (1980), Weinberg (1980), Glashow (1980).

[2] Cf. Brodsky and Drell (1970). See also Lautrup et al. (1972); Hänsch et al. (1979).

[3] Salam (1968, 1980), Weinberg (1967, 1970, 1974, 1980), Glashow (1961, 1980), and some reviews by Abers and Lee (1973), Bernstein (1974), Bég and Sirlin (1974), Taylor (1976). See also Schwinger (1957), Glashow (1959).

[4] Yang and Mills (1954), Shaw (1955). See also Utiyama (1956).

[5] Cf. Bjorken (1972), Pati and Salam (1973), Georgi and Glashow (1974), Georgi et al. (1974), Buras et al. (1978).

[6] Cf. Zumino (1975), Arnowitt et al. (1975), Akulov et al. (1975), Freedman et al. (1976), Deser and Zumino (1976), Wess and Zumino (1977), Brink et al. (1978). See also Salam and Strathdee (1978).

such that gravity may be treated by currently available methods of renormalized quantum field theory.[7] In short, renormalization theory is very much alive and will undoubtedly stay with us for a very long time.

Almost immediately after the initial stages in the development[8] of quantum field theory, the so-called ultraviolet divergence problem was encountered in various computations.[9] As early as 1947, Bethe successfully carried out a mass renormalization[10] in his classic computation of the Lamb shift,[11] and this year may be said to have marked the beginning for the need of a careful study of renormalization in field theory.[12] The first covariant formulation of quantum field theory, in a form quite suitable for practical computations, was due to Schwinger, Feynman, and Tomonaga,[13] and the concept of renormalization was particularly touched upon in the classic papers of Schwinger (1948a, 1949a,b). Dyson (1949a,b) unified the works of Schwinger, Feynman, and Tomonaga and developed the initial stages of renormalization in quantum electrodynamics to arbitrary orders in the fine-structure constant.

The first systematic study of renormalization historically was carried out by Salam in 1951 in a classic paper (Salam, 1951b; see also 1951a) where the subtraction scheme of renormalization, in a general form, was formally sketched. Surprisingly, this classic paper was not carefully reexamined until much later. In 1960 Weinberg established and proved one of the *most* important theorems in field theory. This theorem, popularly known as the "power-counting theorem," embodied a power-counting criterion to establish the absolute convergence of Feynman integrals. Salam's work was first reexamined and brought to a mathematically consistent form in Manoukian (1976). In this latter paper, inspired by the classic paper of Salam, a subtraction scheme was developed and spelled out in momentum space with the subtractions, carried over subdiagrams, applied directly to the Feynman integrand with no ultraviolet cutoffs and taking into account all divergences. The absolute convergence of the corresponding renormalized Feynman amplitudes was then proved by the author (Manoukian, 1982a; see also 1977) by explicitly verifying in the process that the power-counting criterion of

---

[7] Cf. Weinberg (1979).

[8] Born *et al.* (1926), Dirac (1927). See also Heisenberg (1938) for later work.

[9] Oppenheimer (1930), Waller (1930). See also Heisenberg and Pauli (1929, 1930), Heisenberg (1938), Weisskopf and Wigner (1930a,b), Weisskopf (1934a,b, 1936, 1939).

[10] Apparently H. A. Kramer (cf. Schweber, 1961; Weinberg, 1980) first emphasized the idea of mass renormalization.

[11] Lamb and Retherford (1947).

[12] For a fairly detailed historical review of the early days of quantum field theory see, e.g., Schwinger (1958), Peierls (1973), Wentzel (1973).

[13] See Schwinger (1948a,b, 1949a,b, 1951a,b), Feynman (1949a,b, 1950), Tomonaga (1946), and also Koba *et al.* (1947), Kanazawa and Tomonaga (1948), Koba and Tomonaga (1948). See also Schwinger (1958).

Weinberg was satisfied, thus completing the Dyson–Salam (DS) program. Shortly after the appearance of Salam's work, Bogoliubov, together with Parasiuk (1957),[14] in a classic paper developed a subtraction scheme and outlined a proof of its convergence. In 1966 Hepp gave a convergence proof of the Bogoliubov–Parasiuk scheme by using in the intermediate stages ultraviolet cutoffs, and in 1969 Zimmermann formulated the Bogoliubov scheme in momentum space with no ultraviolet cutoffs and gave a convergence proof of this subtraction scheme. Thus these two latter authors completed the Bogoliubov–Parasiuk (BP) program. Finally, the equivalence of the Bogoliubov scheme, in the Zimmermann form, and our scheme was then proved, after some systematic cancellations in the subtractions, by the author (Manoukian, 1976) in a theorem that we have called the "unifying theorem of renormalization." Since the Zimmermann form grew out of Bogoliubov's work and our form grew out of Salam's work, this theorem establishes the long-standing problem of essentially the equivalence of the paths taken in the ingenious approaches of Salam and Bogoliubov (in momentum space). The situation may be summarized by the following diagram:

DS: Dyson → Salam → Weinberg → Author
(program)   (completion)

                  ↘

                   Author
                 (equivalence)

                  ↗

BP: Bogoliubov–Parasiuk → Hepp–Zimmermann
(program)   (completion)

The book deals with a mathematically rigorous formulation of renormalization presented in a unified manner and a model independent way. The subtraction scheme is introduced and its structure is studied, its convergence proof is provided, and finally valuable information from the so-called renormalized Feynman amplitudes is extracted. There are, however, many other interesting and important problems directly or indirectly related to renormalization that the reader may wish to read about. Some of these problems are gauge invariance,[15] "modified" Feynman rules,[16] anomalies[17]

---

[14] See also Bogoliubov and Shirkov (1959) and Parasiuk (1960).

[15] Cf. Schwinger (1951a), Zumino (1960), Bialynicki-Birula (1962), Lukierski (1963), Taylor (1971), Slavnov (1972), Lee (1973), Fradkin and Tyutin (1974).

[16] Cf. Matthews (1949), Lee and Yang (1962), Feynman (1963), deWitt (1964, 1967), Weinberg (1964a,b, 1969), Lam (1965), Fadeev and Popov (1967), Mandelstam (1968a,b), Fradkin and Tyutin (1969, 1970), Salam and Strathdee (1972), Bernard and Duncan (1975).

[17] Schwinger (1951a), Adler (1969), Bell and Jackiw (1969).

versus renormalizability,[18] symmetry breaking[19] versus renormalizability,[20] operator product expansions,[21] renormalization group equations,[22] and the important concept of asymptotic freedom,[23] and there is still much to be done. As a rule of thumb, a theory is renormalizable if no coupling parameters appear in the Lagrangian that have the dimensions of some negative powers of mass. This in turn is related to the important fact that in a renormalizable theory the numerical values of only a finite number of parameters (such as masses and coupling constants) are "adjusted" or "fixed" experimentally.[24] On the other hand, in a nonrenormalizable theory, generally speaking, an "infinite" number of parameters are to be "adjusted" or "fixed" experimentally and such theories lose their predictive powers.[24] The mathematical structure of the subtractions we have developed, however, is general to apply to nonrenormalizable theories as well. All the subtractions of renormalization are carried out at the origin of momentum space.[25] The following (partial) list of standard books on quantum field theory will be useful for the reader to consult when reading the present one so that he or she may benefit as much as possible from it: Bjorken and Drell (1965), Schweber (1961), Jauch and Rohrlich (1976), and Nishijima (1969).

---

[18] Cf. Gross and Jackiw (1972), Bouchiat *et al.* (1972).

[19] Cf. Higgs (1964, 1966), Kibble (1967), Englert and Brout (1964), Guranlik *et al.* (1964). See also Goldstone (1961), Goldstone *et al.* (1962), Anderson (1963).

[20] 't Hooft (1971a,b), 't Hooft and Veltman (1972a,b), Lee (1972), Lee and Zinn-Justin (1972a,b,c), Becchi *et al.* (1975, 1976). See also Fradkin and Tyutin (1974), Vaĭnshtein and Khriplovich (1974), Costa and Tonin (1975), Slavnov (1975).

[21] Cf. Wilson (1969), Frishman (1970), Brandt and Preparata (1971), Zimmerman (1973a,b).

[22] Stückelberg and Petermann (1953), Gell-Mann and Low (1954), Bogoliubov and Shirkov (1959), Callan (1970), Symanzik (1970), 't Hooft (1973), Weinberg (1973).

[23] Politzer (1973), Gross and Wilczek (1973). See also Politzer (1974), Buras (1980), Marciano and Pagels (1978).

[24] Cf. Matthews and Salam (1954), Salam (1952), Bogoliubov and Shirkov (1959) for such details.

[25] A renormalization with subtractions performed at the origin of momentum space is called an intermediate renormalization; cf. Bjorken and Drell (1965).

Chapter 1 / BASIC ANALYSIS

1.1 SETS AND THE HEINE–BOREL THEOREM

Consider two points $x = (x_1, \ldots, x_k)$ and $y = (y_1, \ldots, y_k)$ in the k-dimensional Euclidean space $\mathbb{R}^k$. Define the distance between x and y by $|x - y| = (\sum_{i=1}^{k}(x_i - y_i)^2)^{1/2}$. A neighborhood of a point $x \in \mathbb{R}^k$ is a set $N_\varepsilon(x)$ defined by $N_\varepsilon(x) = \{y : |x - y| < \varepsilon\}$, where $\varepsilon > 0$ is called the radius of $N_\varepsilon(x)$. A point x is a limit point of a set $S \subset \mathbb{R}^k$ if every neighborhood of x contains a point of S different from x.[1] This leads to a definition of a closed set as a set containing all of its limit points. A point x is an interior point of a set $S \subset \mathbb{R}^k$ if there is a neighborhood $N_\varepsilon(x) \subset S$. This leads to a definition of an open set as a set such that every point in it is an interior point. An elementary result in set theory relating open and closed sets is the following: A set is open if and only if its complement is closed. Let S^c be the complement of S in $\mathbb{R}^k$, i.e., $S^c = \mathbb{R}^k - S$, which denotes the set of elements in $\mathbb{R}^k$ but not in S. Suppose S^c is closed and let $x \in S$ (i.e., $x \notin S^c$). Then by definition of a closed set, x cannot be a limit point of S^c as the latter contains all of its limit points. This means that there exists a neighborhood $N_\varepsilon(x)$ of x such that $S^c \cap N_\varepsilon(x) = \varnothing$ and hence $N_\varepsilon(x) \subset S$. Thus x is an an interior point of S and S is open. Conversely, suppose that S is open and let x be a limit point of S^c. Then for every $\varepsilon > 0$, $N_\varepsilon(x)$ contains a point of S^c different from x [i.e., the $N_\varepsilon(x)$, for all $\varepsilon > 0$, are not subsets of S], and so x is not an interior point of S. Since S is open, it

---

[1] The symbol $\subset$ stands for containment. $A \subset B$ means A is a subset of B; it may also imply that A equals B. To exclude equality we write $\not\subset$.

follows that $x \in S^c$, S^c contains all of its limit points, and hence it is closed. Other useful results related to open and closed sets are the following: The union of an infinite number of open sets is open; the intersection of an infinite number of closed sets is closed; the union of a finite number of closed sets is closed; and the intersection of a finite number of open sets is open.

We discuss the important concept of a cover of a set. We say that a family of sets $\{S_\rho : \rho \in V\}$ covers a set S if $\bigcup_{\rho \in V} S_\rho \supset S$, where V is some set and need not be countable. For any set $S \subset \mathbb{R}^k$, the family of neighborhoods $\{N_\varepsilon(x) : x \in S\}$ obviously covers S. Again the set S need not be countable. Another, though trivial, example of a cover of a set $S \subset \mathbb{R}^k$ is the family $\{\mathbb{R}^k\}$.

The cover $\{N_\varepsilon(x) : x \in S\}$ of S provides an example of an open cover as every set $N_\varepsilon(x)$ in this family is open. A basic property of an open cover is the following: Every open cover of a set $S \subset \mathbb{R}^k$ has a countable subcover. The demonstration of this result is as follows.

Let $\{S_\rho : \rho \in V\}$ be an open cover of a set S. For any $x \in S$, there exists at least one $\rho \in V$ such that $x \in S_\rho$. Consider a neighborhood $N_r(x)$ of x such that $N_r(x) \subset S_\rho$. Let $y = (y_1, \ldots, y_k) \in \mathbb{R}^k$ such that the y_i are rational and $|y - x| < r/2$. Since between any two real numbers we may find a rational number, we may let t denote a rational number such that $|y - x| < t < r/2$. The latter implies that $x \in N_t(y)$, $N_t(y) \subset N_r(x) \subset S_\rho$. The family of all neighborhoods with rational radii about points $y = (y_1, \ldots, y_k) \in \mathbb{R}^k$ with the y_i varying over the set of rational numbers is obviously a countable set since the set of rational numbers is countable. We may then introduce a family of neighborhoods $\{N^1, N^2, \ldots\}$, where for some n, $N^n = N_t(y)$ with $x \in N^n$ and $N^n \subset S_\rho$. Accordingly we may introduce indices ρ_n such that $x \in N^n \subset S_{\rho_n}$. Now for any $x \in S$, we may find a neighborhood N^n such that $x \in N^n \subset S_{\rho_n}$; hence the family $\{S_{\rho_1}, S_{\rho_2}, \ldots\}$ provides a countable subcover of S by open sets.

We introduce the concept of a box in $\mathbb{R}^k$. Let $x = (x_1, \ldots, x_k)$, $m = (m_1, \ldots, m_k)$ and let I denote the set of integers and N the set of natural numbers. Consider the set

$$B_n(m) = \{x : m_j/2^n \leq x_j < (m_j + 1)/2^n, j = 1, \ldots, k\} \qquad (1.1.1)$$

for $n \in N$ and $m_j \in I$. Let

$$\mathscr{B}_n = \{B_n(m) : \text{all } m_j \in I\}. \qquad (1.1.2)$$

The latter defines a family of boxes with each box having the volume 2^{-kn}. Obviously for any boxes $B_n \in \mathscr{B}_n$ and $B_{n'} \in \mathscr{B}_{n'}$, with $n > n'$, we can either have $B_n \subset B_{n'}$ or $B_n \cap B_{n'} = \varnothing$. Let S be any nonempty open set in $\mathbb{R}^k$ and let $x \in S$. For a fixed $n \in N$, x is contained in one and only one of the boxes in $\mathscr{B}_n$. Since S is open, we may find, for n sufficiently large, a $B_n \in \mathscr{B}_n$ such that $x \in B_n$ and $B_n \subset S$. Accordingly, we may write S as the *union of* those elements in $\mathscr{B}_1$ contained in S, those elements in $\mathscr{B}_2$ contained in S but not lying in the boxes in $\mathcal{B}_1$, those elements in $\mathcal{B}_3$ contained in S but lying neither in the boxes in $\mathcal{B}_1$ nor in the boxes in $\mathcal{B}_2$, and so on. That is, S may be written as the countable union of disjoint boxes in $\mathcal{B}_1 \cup \mathcal{B}_2 \cup \mathcal{B}_3 \cup \cdots$. This construction is quite useful for defining the concept of a "volume" for the open set S as the sum of volumes of these disjoint boxes, just mentioned, in $\mathcal{B}_1 \cup \mathcal{B}_2 \cup \mathcal{B}_3 \cup \cdots$. The volume of a box in $\mathcal{B}_n$, as we have seen above, is 2^{-kn}. The following construction is of particular interest. Let $n \in N$ be fixed. The family $\mathcal{B}_n$ in (1.1.2) is countable; i.e., we may write $\mathcal{B}_n = \{B^1, B^2, \ldots\}$, where $B^1, B^2, \ldots$ are pairwise disjoint boxes, each box of volume 2^{-kn}, and

$$\mathbb{R}^k = \bigcup_{i=1}^{\infty} B^i. \tag{1.1.3}$$

Consider a set of real numbers E, i.e., $E \subset \mathbb{R}^1$. If for all $a \in E$, there is a number $b \in \mathbb{R}^1$ such that $a < b$, then the set E is said to be bounded above and b is called an upper bound of E. If b is an upper bound of E and if for every $c < b$, c is not an upper bound of E, then b is called the least upper bound (l.u.b.) of the set E. Similarly, we define a set E of real numbers to be bounded below if for all $a \in E$, there is a number $l \in \mathbb{R}^1$ such that $l \leq a$, then l is called a lower bound of E. If l is a lower bound of E and if for every $d > l$, d is not a lower bound of E, then l is called the greatest lower bound (g.l.b.) of the set E. Suppose E is a closed set and is bounded above. Let $b = \text{l.u.b. } E$;[2] then it is easy to see that $b \in E$. To show this, suppose that the contrary is true, i.e., $b \notin E$. Then for every $\varepsilon > 0$, there is an $a \in E$ such that $b - \varepsilon \leq a \leq b$; otherwise (i.e., $a \leq b - \varepsilon$ for all $a \in E$) $b - \varepsilon$ would have been an upper bound of E, contradicting the fact that $b = \text{l.u.b. } E$. This means that every neighborhood of b contains a point of E different from b since we have assumed that $b \notin E$. It then follows that b is a limit point of E and hence it must be in E, as the latter is closed, thus leading to a contradiction with the hypothesis that $b \notin E$.

Consider a set $S \subset \mathbb{R}^k$. The set S is said to be bounded if there is a point $y \in \mathbb{R}^k$ and a positive constant $0 < C < \infty$ such that $|x - y| < C$ for all $x \in S$. A sequence may be defined as a function whose domain $\mathcal{D}$ is a subset of (or coincides with) the set of natural numbers N: $\{a_n : n \in \mathcal{D} \subset N\}$.

1.1.1 The Heine–Borel Theorem

We show that each open cover of a closed and bounded set S contains a finite open cover. We have already shown that every open cover of a set contains a countable open cover. Accordingly suppose that $\{O_n\} = \{O_n : n = 1, 2, \ldots\}$ is a countable open cover of the set S: $\bigcup_n O_n \supset S$. Then we

---

[2] Here we are invoking the "completeness axiom" of real numbers, which states that the bounded set E has a l.u.b.

show that there exists a finite number of open sets $S_1, \ldots, S_M \in \{O_n\}$ that cover $S: S \subset \bigcup_{i=1}^{M} S_i$.

An example of a closed and bounded set is provided by a closed and bounded k-cell $\mathbb{R}^k$ defined by

$$W = \{x : a_j \leq x_j \leq b_j, j = 1, \ldots, k\}, \tag{1.1.4}$$

where $x = (x_1, \ldots, x_k) \in \mathbb{R}^k$, and

$$r = \left(\sum_{j=1}^{k} (b_j - a_j)^2 \right)^{1/2} < \infty. \tag{1.1.5}$$

Then we note that for all x and $y \in W$ in (1.1.4) we have $|x - y| \leq r$.

We consider the proof of the above statement first for a closed and bounded k-cell as defined in (1.1.4). Suppose there exists an open cover $\{O_n\}$ of W that contains no finite subcover. Let $d_j = (a_j + b_j)/2$, and divide the closed interval $[a_j, b_j]$ into two (closed) intervals $[a_j, d_j], [d_j, b_j]$ for $j = 1, 2, \ldots, k$. For $k = 2$, for example, we then generate 4 closed 2-cells: W^1, W^2, W^3, W^4 such that $\bigcup_{i=1}^{4} W^i = W$. In general we then generate 2^k closed k-cells: $W^1, W^2, \ldots, W^{2k}$ such that $\bigcup_{i=1}^{2k} W^i = W$. By hypothesis at least one of these 2^k closed k-cells, say, $W^j \equiv W_1$, cannot be covered by a finite number of open sets in $\{O_n\}$. We repeat the same construction, as done above for W, for the k-cell W_1 and thus generate 2^k closed k-cells. Again by hypothesis at least one of these latter k-cells, say, a k-cell W_2, cannot be covered by a finite number of open sets in $\{O_n\}$. Continuing this process, this generates a sequence of closed and bounded k-cells $W_1, W_2, \ldots$ such that $W \supset W_1 \supset W_2 \supset \cdots$, where by hypothesis the W_i cannot be covered by a finite number of open sets in $\{O_n\}$.

We first show that there is at least one element common to all the W_i, i.e., $\bigcap_{i=1}^{\infty} W_i$ is not empty. The W_i are of the form

$$W_i = \{x : a_j^i \leq x_j \leq b_j^i, j = 1, \ldots, k\}, \tag{1.1.6}$$

for $i = 1, 2, \ldots$. Let

$$c_j = \text{l.u.b.}\{a_j^i : i = 1, 2, \ldots\}, \tag{1.1.7}$$

and hence by definition $a_j^i \leq c_j$ for all $i = 1, 2, \ldots$. Because of the property of the k-cells $W_1 \supset W_2 \supset \cdots$, it is readily seen that the set $\{a_j^1, a_j^2, \ldots\}$, with j fixed, is bounded above by b_j^1. It is also clear that $c_j \leq b_j^i$, for all $i = 1, 2, \ldots$, otherwise we would arrive to the conclusion that c_j is not a least upper bound as defined in (1.1.7). Accordingly $a_j^i \leq c_j \leq b_j^i$ for all $i = 1, 2, \ldots$. Let $c = (c_1, c_2, \ldots, c_k)$. Then $c \in W_i$ for all $i = 1, 2, \ldots$, i.e., $c \in \bigcap_{i=1}^{\infty} W_i$, and hence the latter is not empty.

Since $\{O_n\}$ covers W, it then follows that we may find a member O_m such that $c \in O_m$. Also since O_m is open, c is an interior point of O_m; i.e., we may find a neighborhood $N_\varepsilon(c) \subset O_m$. By construction, for any x and y in a set W_i, our subdivisions imply that $|x - y| \le 2^{-i}r$, and in particular, $|x - c| \le 2^{-i}r$ for all $x \in W_i$. Accordingly, by choosing i sufficiently large, we may make $2^{-i}r < \varepsilon$, and hence we arrive to the conclusion that, for the corresponding i, $W_i \subset O_m$. This leads to a contradiction to the hypothesis that W_i cannot be covered by a finite number of elements from $\{O_n\}$ as the open set O_m covers W_i. This completes the demonstration that the closed and bounded k-cells can be covered by a finite number of open sets.

We now generalize the above result to any closed and bounded set $S \subset \mathbb{R}^k$. Let W be a closed and bounded k-cell containing the set S: $W \supset S$. Consider the set $\{O_n\} \cup S^c$, where the family $\{O_n\}$ is an open cover of S, i.e., $S \subset \bigcup_n O_n$. Obviously $\{O_n\} \cup S^c$ is an open cover of W as the set S^c is open and $S \cup S^c = \mathbb{R}^k$. We have already seen that a cover $\{O_n\} \cup S^c$ contains a finite cover of W. If this finite cover of W does not include the set S^c, then we have demonstrated the validity of our statement since $W \supset S$. On the other hand, if the finite cover does include the set S^c, then the finite open cover is of the form $\bigcup_{i=1}^{M} S_i \cup S^c \supset W \supset S$, where the $S_i \in \{O_n\}$. But $S \cap S^c = \varnothing$, hence $S \subset \bigcup_{i=1}^{M} S_i$ and $\{S_1, \ldots, S_M\}$ is a finite subcollection from the family $\{O_n\}$ and covers the closed and bounded set S. This completes the proof of our statement (the Heine–Borel theorem), which may be summarized as follows:

Theorem 1.1.1: *Every open cover of a set contains a countable open cover, and if the set is closed and bounded, then the latter cover contains a finite open cover.*

A useful result that follows by the application of the Heine–Borel theorem is the following:

Lemma 1.1.1: *Suppose that the domain $\mathscr{D} \subset \mathbb{R}^k$ of a real function f is closed and bounded and f is continuous on $\mathscr{D}$. Then f is bounded, i.e., $|f(x)| \le M$ for all $x \in \mathscr{D}$. Also there is a point $x_0 \in \mathscr{D}$ such that $|f(x_0)| = \text{l.u.b.}\{|f(x)| : x \in \mathscr{D}\}$.*

Since f is continuous on $\mathscr{D}$, then for each $x \in \mathscr{D}$, and any given $\varepsilon > 0$, we may find a $\delta_x > 0$ such that $|f(y) - f(x)| < \varepsilon$, or $|f(y)| \le |f(x)| + \varepsilon$, for all $y \in N_{\delta_x}(x) \subset \mathscr{D}$. In particular, we may choose $\varepsilon = 1$. The set $\{N_{\delta_x}(x) : x \in \mathscr{D}\}$ provides an open cover of $\mathscr{D}$, and from Theorem 1.1.1 we may find a sub-collection, say, $\{N_{\delta_{x_1}}(x_1), \ldots, N_{\delta_{x_m}}(x_m)\}$, that covers $\mathscr{D}$. f is bounded in each $N_{\delta_{x_i}}(x_i)$ as $|f(y)| \le |f(x_i)| + 1$ for all $y \in N_{\delta_{x_i}}(x_i)$. Let $M = 1 + \max\{|f(x_1)|, \ldots, |f(x_m)|\}$. Since for any $y \in \mathscr{D}$ we may find an $N_{\delta_{x_i}}(x_i)$, with $i \in [1, \ldots, m]$, we then have for all $y \in \mathscr{D}$, $|f(y)| \le M$. Let $b = \text{l.u.b.}\{|f(x)| : x \in \mathscr{D}\}$. Suppose $|f(y)| < b$ for all $y \in \mathscr{D}$. From the continuity condition of f, we may find for each $x \in \mathscr{D}$ and $\varepsilon = a(b - |f(x)|) > 0$, where $0 < a < 1$, a $\rho_x > 0$ such that $|f(y)| \le (1 - a)|f(x)| + ab$ for all $y \in N_{\rho_x}(x)$. Let $\{N_{\rho_{x_1}}(x_1), \ldots, N_{\rho_{x_n}}(x_n)\}$ be a cover of $\mathscr{D}$. Let $M_0 = \max\{|f(x_1)|, \ldots, |f(x_n)|\}$. Then for all $y \in \mathscr{D}$ we have $|f(y)| \leq (1 - a)M_0 + ab$. Since, by hypothesis, $M_0 < b$, we have $(1 - a)M_0 + ab < b$, contradicting the fact that b is an l.u.b. of $\{|f(x)|: x \in \mathscr{D}\}$. Hence there is a point $x_0 \in \mathscr{D}$ such that $|f(x_0)| = b$.

1.1.2 Algebras and σ-Algebras

Consider an arbitrary set X. A nonempty class $\mathscr{C}$ of subsets of X is called an algebra of sets if (i) $S_1, S_2 \in \mathscr{C}$ imply that $S_1 \cup S_2 \in \mathscr{C}$, (ii) $S \in \mathscr{C}$ implies that $X - S = S^c \in \mathscr{C}$.[3]

Elementary consequences of this definition are that the empty set $\varnothing$ and the whole set X are in $\mathscr{C}$ and $S_1 \cap S_2, S_1 - S_2$ are as well in $\mathscr{C}$ if $S_1, S_2 \in \mathscr{C}$. That $X \in \mathscr{C}$ follows from the fact that if a set $S \in \mathscr{C}$, then $S^c \in \mathscr{C}$ and $S \cup S^c = X \in \mathscr{C}$. It then also follows that $X^c = \varnothing \in \mathscr{C}$. We finally note that if $S_1, S_2 \in \mathscr{C}$, then $S_1^c, S_2^c \in \mathscr{C}$ and hence $S_1^c \cup S_2^c \in \mathscr{C}$ and $(S_1^c \cup S_2^c)^c = S_1 \cap S_2 \in \mathscr{C}$. Similarly, it is easily seen that $S_1 - S_2 \in \mathscr{C}$.

A nonempty class $\mathscr{C}_\sigma$ of subsets of a set X is called a σ-algebra of sets if (i) $S_1, S_2, \ldots, \in \mathscr{C}_\sigma$ imply that $\bigcup_{i=1}^\infty S_i \in \mathscr{C}_\sigma$ and (ii) $S \in \mathscr{C}_\sigma$ implies that $S^c \in \mathscr{C}_\sigma$. In particular, this definition implies that if $S_1, S_2, \ldots \in \mathscr{C}_\sigma$, then $\bigcap_{i=1}^\infty S_i \in \mathscr{C}_\sigma$ as well.

Let $\mathscr{C}_0$ be any collection of subsets of X. By considering the intersection[4] of all algebras containing the family $\mathscr{C}_0$, we then generate the smallest algebra containing $\mathscr{C}_0$. This algebra will be denoted by $\mathscr{C}(\mathscr{C}_0)$ and will be called the algebra generated by $\mathscr{C}_0$. Similarly, by considering the intersection of all σ-algebras containing $\mathscr{C}_0$, we then generate the smallest σ-algebra containing $\mathscr{C}_0$, which will be denoted $\mathscr{C}_\sigma(\mathscr{C}_0)$ and called the σ-algebra generated by $\mathscr{C}_0$. We now introduce the concept of a monotone class of subsets of a set X. A set $\mathscr{M}_c$ of subsets of X is called a *monotone class* if for every monotone sequence $S_1, S_2, \ldots \in \mathscr{M}_c$, i.e., for which $S_1 \subset S_2 \subset \cdots$ or $S_1 \supset S_2 \supset \cdots$, $\bigcup_{i=1}^\infty S_i \in \mathscr{M}_c$ or $\bigcap_{i=1}^\infty S_i \in \mathscr{M}_c$, respectively. Let $\mathscr{C}_0$ be any collection of subsets of a set X; then by considering the intersection of all monotone classes containing the family $\mathscr{C}_0$, we generate the smallest monotone class containing $\mathscr{C}_0$, which will be denoted $\mathscr{M}_c(\mathscr{C}_0)$. Let $\mathscr{C}$ be an algebra of subsets of a set X (also called an algebra in X). Let S be any subset of X. We define $\mathscr{C} \cap S = \{A \cap S : A \in \mathscr{C}\}$. Then it is easy to see that $\mathscr{C} \cap S$ is an algebra in (i.e., with complements taken relative to the set S) S. Let $\hat{\mathscr{C}}_\sigma(\mathscr{C} \cap S)$ denote the smallest σ-algebra in S containing $\mathscr{C} \cap S$. Similarly let $\hat{\mathscr{M}}_c(\mathscr{C} \cap S)$ denote the smallest monotone class with subsets in S containing $\mathscr{C} \cap S$. We note that for $S = X$,

---

[3] The set X will be called a universal set.

[4] It is easy to see that the intersection of algebras of sets is an algebra of sets, and that the intersection of σ-algebras of sets is a σ-algebra of sets.

$\hat{\mathscr{C}}_\sigma(\mathscr{C} \cap S) = \mathscr{C}_\sigma(\mathscr{C})$ and $\hat{\mathscr{M}}_c(\mathscr{C} \cap S) = \mathscr{M}_c(\mathscr{C})$ since $X \in \mathscr{C}$. Then we have the following important lemma concerning algebras, σ-algebras, and monotone classes as direct consequences of the above definitions:

Lemma 1.1.2: *Let $\mathscr{C}$ be an algebra of subsets of a set X. Then*

(i) $\hat{\mathscr{C}}_\sigma(\mathscr{C} \cap S) = \mathscr{C}_\sigma(\mathscr{C}) \cap S,$

where S is any subset of X, and $\mathscr{C}_\sigma(\mathscr{C}) \cap S$ is defined by

$$\mathscr{C}_\sigma(\mathscr{C}) \cap S = \{A \cap S : A \in \mathscr{C}_\sigma(\mathscr{C})\},$$

and the complements are taken relative to the set S, i.e., $\mathscr{C}_\sigma(\mathscr{C}) \cap S$ coincides with the smallest σ-algebra in S containing $\mathscr{C} \cap S$;

(ii) $\mathscr{C}_\sigma(\mathscr{C}) = \mathscr{M}_c(\mathscr{C}), \qquad \hat{\mathscr{C}}_\sigma(\mathscr{C} \cap S) = \hat{\mathscr{M}}_c(\mathscr{C} \cap S).$

Let X and Y be two sets. We define the *Cartesian product $X \times Y$* as a set given by

$$X \times Y = \{(x, y) : x \in X, y \in Y\}.$$

Let $\mathscr{C}_\sigma^x$ and $\mathscr{C}_\sigma^y$ be two σ-algebras of subsets of X and Y, respectively. Consider the family $\mathscr{F}$ of subsets of $X \times Y$ in the form

$$\mathscr{F} = \left\{ F = \bigcup_{i=1}^n F_i : F_i \cap F_j = \varnothing \text{ for all } i \neq j \text{ and all finite } n \right\},$$

where the F_i are of the form $F_i = A_i \times B_i$ with $A_i \in \mathscr{C}_\sigma^x$, $B_i \in \mathscr{C}_\sigma^y$.

Let E be any subset of $X \times Y$, we define the *x-section* and the *y-section* of E by

$$E_x = \{y : (x, y) \in E\} \subset Y, \qquad E_y = \{x : (x, y) \in E\} \subset X.$$

We may then state the following lemma.

Lemma 1.1.3: (i) *$\mathscr{F}$ is an algebra.*

(ii) *For any two algebras $\mathscr{E}_1$ and $\mathscr{E}_2$ of subsets of X and Y, respectively, let $\mathscr{E} = \mathscr{E}_1 \times \mathscr{E}_2$ and $\mathscr{C}_\sigma^x \equiv \mathscr{C}_\sigma(\mathscr{E}_1)$ and $\mathscr{C}_\sigma^y \equiv \mathscr{C}_\sigma(\mathscr{E}_2)$; then*

$$\mathscr{C}_\sigma(\mathscr{E}) = \mathscr{C}_\sigma(\mathscr{F}),$$

and if $E \subset \mathscr{C}_\sigma(\mathscr{F})$, then $E_x \in \mathscr{C}_\sigma(\mathscr{E}_2)$, $E_y \in \mathscr{C}_\sigma(\mathscr{E}_1)$.

(Note that $\mathscr{E} = \{A \times B : A \in \mathscr{E}_1, B \in \mathscr{E}_2\}$).

1.2   MEASURE, INTEGRATION, AND FUBINI'S THEOREM

1.2.1   Measure Theory

We have already defined the concept of a closed k-cell in $\mathbb{R}^k$ by the set

$$W = \{x : a_j \le x_j \le b_j, j = 1, \ldots, k\}, \tag{1.2.1}$$

with $x = (x_1, \ldots, x_k) \in \mathbb{R}^k$. Here we do not put any restrictions on the real constants a_j, b_j. In general we define a k-cell as any set of the form (1.2.1) or any set in this form with some or all of the signs $\le$ replaced by $<$ or $=$ as well. Thus in particular the set $\{x : x_j = a_j\}$ is a k-cell.

We define the family $\mathscr{E}$ of *elementary sets* as the family containing the empty set $\varnothing$ and any set that is a finite union of k-cells. Thus any nonempty set E in $\mathscr{E}$ is of the form $E = W_1 \cup \cdots \cup W_n$, where the W_i are k-cells. It is not difficult to show that $\mathscr{E}$ is an algebra and any nonempty set in it may be written as the union of a finite number of disjoint k-cells.

The measure of a k-cell is defined as its volume and is denoted by $\mu(W) = \prod_{j=1}^{k}(b_j - a_j)$. For the empty set $\varnothing$ we associate the "volume" 0, i.e., $\mu(\varnothing) = 0$. We may define the measure of an elementary nonempty set as a sum over a finite number of volumes of pairwise disjoint k-cells. We may also generalize the concept of a measure to an open set, as a subset of $\mathbb{R}^k$, by recalling (see Section 1.1) that an open set may be written as the disjoint union of boxes [see Eqs. (1.1.1) and 1.1.2)], where the latter are particular examples of k-cells. Hence the measure of an open set in $\mathbb{R}^k$ may be defined as the sum of the measures (volumes) of these disjoint boxes. In general, the "measure" of a set, such as $\mathbb{R}^k$, need not be finite, as $\mu(\mathbb{R}^k) = \infty$. But $\mathbb{R}^k$ may be written as the union of disjoint boxes B^1, B^2, $\ldots$ [see Eq. (1.1.3)] each of finite volume 2^{-nk}, with nonnegative integers n, i.e., $\mu(B^i) = 2^{-nk} < \infty$. This particular property of a set as being written as the union over sets each of finite measure is called a σ-finiteness property.

The above analysis suggests, quite generally, introducing the concept of a measure as an extended (extended means that the value ∞ is permissible) real-valued nonnegative set function on a family D (containing in particular the elementary sets) consisting of a suitable class of subsets of $\mathbb{R}^k$ such that (i) $\mu(A) \ge 0$ for any set $A \in D$, (ii) $\varnothing \in D$ and $\mu(\varnothing) = 0$, (iii) if $A_1, A_2, \ldots \in D$ are pairwise disjoint sets, i.e., $A_i \cap A_j = \varnothing$, for $i \ne j$, such that $\bigcup_{i=1}^{\infty} A_i \in D$, then $\mu(\bigcup_{i=1}^{\infty} A_i) = \sum_{i=1}^{\infty} \mu(A_i)$. The last property (iii) is called the σ-additivity property of a measure. Here it is worth recalling that if a family of sets is a σ-algebra, then the union of any infinite number of sets in it also belongs to the family. We now study how a family D may be defined and the measures of the sets in D are constructed. We already know how to define the measure of the elementary sets in $\mathscr{E}$; we generalize this to a larger class of subsets in $\mathbb{R}^k$. With this in mind we define a new extended real-valued set function as follows.

For any set $S \subset \mathbb{R}^k$, we define the extended real-valued nonnegative set function

$$\mu^*(S) = \inf\left\{ \sum_{k=1}^{\infty} \mu(E_k) : S \subset \bigcup_{k=1}^{\infty} E_k, E_k \in \mathscr{E}, k = 1, 2, \ldots \right\}.^5 \quad (1.2.2)$$

The infimum (or g.l.b.) runs over all coverings $\{E_1, E_2, \ldots\}$ of the set S with $E_1, E_2, \ldots \in \mathscr{E}$. In particular, we note that if S is an elementary set, i.e., $S \in \mathscr{E}$, then $\mu^*(S) = \mu(S)$. It is also easy to see that if $S \supset S'$, then $\mu^*(S) \geq \mu^*(S')$. Some elementary properties of the set function μ^* are (i) $\mu^*(S) \geq 0$, for any $S \subset \mathbb{R}^k$, (ii) $\mu^*(\varnothing) = 0$, (iii) $\mu^*(\bigcup_{i=1}^{\infty} S_i) \leq \sum_{i=1}^{\infty} \mu^*(S_i)$ for all $S_1, S_2, \ldots$ as subsets of $\mathbb{R}^k$. The set function μ^*, however, does not necessarily satisfy the σ-additivity property of a measure. It becomes σ-additive by *reducing* the family of sets on which it is defined. This leads to the concept of a μ^*-measurable set.

A set A is called *μ^*-measurable* if for any $S \subset \mathbb{R}^k$,

$$\mu^*(S) = \mu^*(S \cap A) + \mu^*(S \cap A^c). \quad (1.2.3)$$

Intuitively speaking, a μ^*-measurable set A is one that does not cut an arbitrary set S into two parts, one part lying "outside" of A and a part "inside" of A, such that the values of the set function μ^* for the two parts do not add up correctly. By restricting the definition of μ^* to the μ^*-measurable sets, satisfying (1.2.3), we obtain a measure, i.e., an extended real-valued set function satisfying all the requirements for a measure, including the σ-additivity property. An important lemma concerning μ^*-measurability is the following:

Lemma 1.2.1:   (i)   *An elementary set is μ^*-measurable.*
   (ii)   *The family M of all μ^*-measurable sets is a σ-algebra.*
   (iii)   *The set function μ^* in (1.2.2) when restricted to M is a measure.*
   (iv)   *Any subset S with $\mu^*(S) = 0$, is μ^*-measurable, i.e., $S \in M$.*

Since M is a σ-algebra and contains the elements in $\mathscr{E}$, we then conclude that $M \supset \mathscr{C}_\sigma(\mathscr{E})$.

Now we prove the following uniqueness theorem.

Theorem 1.2.1:   *There is a unique measure μ_0 on the σ-algebra $\mathscr{C}_\sigma(\mathscr{E})$ generated by $\mathscr{E}$ such that $\mu_0(S) = \mu(S)$ for any $S \in \mathscr{E}$.*

---

[5] We note that by the definition of μ^* as an extended real-valued nonnegative set function, the value $+\infty$ is not excluded for the expression on the right-hand side of (1.2.2) in the so-called extended system $[0, +\infty]$.

Suppose that μ_1 and μ_2 are two measures on $\mathscr{C}_\sigma(\mathscr{E})$ such that $\mu_1(S) = \mu_2(S)$ for all $S \in \mathscr{E}$. We decompose $\mathbb{R}^k$ into disjoint boxes of finite measures [see (1.1.3)]: $\mathbb{R}^k = \bigcup_{i=1}^\infty B^i, \mu(B^i) < \infty, B^i \in \mathscr{E}$. Let A be any set in $\mathscr{C}_\sigma(\mathscr{E})$ and write $A = \bigcup_{i=1}^\infty (A \cap B^i)$. Then $\mu_j(A \cap B^i) \le \mu_j(B^i) = \mu(B^i) < \infty$, for $j = 1, 2$. This finiteness property will be used explicitly. Let $A^n = A \cap B^n$. Consider the σ-algebra $\mathscr{C}_\sigma(\mathscr{E}) \cap B^n$ with subsets of B^n, i.e., a σ-algebra in B^n (see Lemma 1.1.2). It is actually the smallest σ-algebra in B^n containing $\mathscr{E} \cap B^n$. Let $\mathscr{M}_c$ be the family of all sets in $\mathscr{C}_\sigma(\mathscr{E}) \cap B^n$ such that for any set $A \in \mathscr{M}_c, \mu_1(A) = \mu_2(A)$. We show that $\mathscr{M}_c$ is a monotone class.

Let $S_1, S_2, \ldots$ be a monotone sequence in $\mathscr{M}_c$, i.e., $\mu_1(S_m) = \mu_2(S_m)$, $m = 1, 2, \ldots$, such that $S_{m+1} \supset S_m$ or $S_m \supset S_{m+1}$. Consider first the case with $S_{m+1} \supset S_m$. Let $R_1 = S_1$, and for $i \ge 2$ let $R_i = S_i - S_{i-1}$. Then $R_i \cap R_j = \varnothing$, for $i \ne j$, and $\bigcup_{i=1}^\infty R_i = \bigcup_{i=1}^\infty S_i$. Also $S_n = \bigcup_{i=1}^n R_i$. Thus

$$u_j(S_n) = \sum_{i=1}^n \mu_j(R_i)$$

and

$$\lim_{n \to \infty} \sum_{i=1}^n \mu_1(R_i) = \lim_{n \to \infty} \sum_{i=1}^n \mu_2(R_i).$$

Hence by using, in the process, the σ-additivity property of the measures we obtain

$$\mu_1\left(\bigcup_{i=1}^\infty S_i\right) = \mu_2\left(\bigcup_{i=1}^\infty S_i\right), \tag{1.2.4}$$

i.e., $\bigcup_{i=1}^\infty S_i \in \mathscr{M}_c$. For the case $S_m \supset S_{m+1}$, we write

$$S_1 = \left(\bigcap_{k=1}^\infty S_k\right) \bigcup_{j=1}^\infty (S_j - S_{j+1}),$$

where note that the sets $(\bigcap_{k=1}^\infty S_k), (S_1 - S_2), (S_2 - S_3), \ldots$ are pairwise disjoint. Thus we may write

$$\mu_j(S_1) = \mu_j\left(\bigcap_{k=1}^\infty S_k\right) + \lim_{n \to \infty} \sum_{i=1}^n \mu_j(S_i - S_{i+1}). \tag{1.2.5}$$

Now $S_i = S_{i+1} \cup (S_i - S_{i+1})$ and is written as the union of two disjoint sets; hence

$$\mu_j(S_i) = \mu_j(S_{i+1}) + \mu_j(S_i - S_{i+1})$$

or

$$\mu_j(S_i - S_{i+1}) = \mu_j(S_i) - \mu_j(S_{i+1}), \tag{1.2.6}$$

where all these measures are finite. Going back to (1.2.5), we obtain

$$\mu_j(S_1) = \mu_j\left(\bigcap_{k=1}^{\infty} S_k\right) + \mu_j(S_1) - \lim_{n \to \infty} \mu_j(S_n).$$

Since $\mu_j(S_1) < \infty$ and $\mu_1(S_n) = \mu_2(S_n)$, we finally obtain

$$\mu_1\left(\bigcap_{k=1}^{\infty} S_k\right) = \mu_2\left(\bigcap_{k=1}^{\infty} S_k\right), \qquad (1.2.7)$$

and $\bigcap_{k=1}^{\infty} S_k \in \mathcal{M}_c$. Therefore $\mathcal{M}_c$ is a monotone class of subsets of B^n containing $\mathscr{E} \cap B^n$. From Lemma 1.1.2 we know that

$$\mathscr{C}_\sigma(\mathscr{E}) \cap B^n = \hat{\mathscr{C}}_\sigma(\mathscr{E} \cap B^n) = \hat{\mathcal{M}}_c(\mathscr{E} \cap B^n).$$

Therefore $\mathcal{M}_c \supset \mathscr{C}_\sigma(\mathscr{E}) \cap B^n$ and for any set $S \in \mathscr{C}_\sigma(\mathscr{E}) \cap B^n$, $\mu_1(S) = \mu_2(S)$. Hence for any $A \in \mathscr{C}_\sigma(\mathscr{E})$, with $A^n = A \cap B^n \in \mathscr{C}_\sigma(\mathscr{E}) \cap B^n$, we may write

$$\mu_1(A) = \sum_{n=1}^{\infty} \mu_1(A^n) = \sum_{n=1}^{\infty} \mu_2(A^n)$$

because $\mu_1(A^n) = \mu_2(A^n)$, or

$$\mu_1(A) = \mu_2(A), \qquad (1.2.8)$$

which establishes the result of the lemma.

It is worth recalling that a box is a k-cell, and hence is in $\mathscr{E}$, and that any open set $S \subset \mathbb{R}^k$ may be written as the union of boxes (see Section 1.1). Therefore any open set S is in $\mathscr{C}_\sigma(\mathscr{E})$, and every closed set S^c is in $\mathscr{C}_\sigma(\mathscr{E})$, by definition of the latter. It is important to know the relation between the family M of all μ^*-measurable sets and $\mathscr{C}_\sigma(\mathscr{E})$. We already know that $M \supset \mathscr{C}_\sigma(\mathscr{E})$ and that the set function μ^* in (1.2.2) when restricted to M is a measure. We also know that the measure defined on $\mathscr{C}_\sigma(\mathscr{E})$ is unique. For simplicity of notation, we denote the latter measure by μ as well and the former (i.e., μ^* restricted to M) by $\bar{\mu}$. We define the concept of a measurable cover of a set in M. For any set $E \in M$, $S \in \mathscr{C}_\sigma(\mathscr{E})$ is a measurable cover of E if $S \supset E$, and for any $G \in \mathscr{C}_\sigma(\mathscr{E})$ such that $G \subset S - E$, we have $\mu(G) = 0$. The following two lemmas embody important results relating M and $\mathscr{C}_\sigma(\mathscr{E})$, $\bar{\mu}$ and μ.

Lemma 1.2.2:   *For any set $E \in M$, there exists a set $S \in \mathscr{C}_\sigma(\mathscr{E})$ such that $\bar{\mu}(E) = \mu(S)$ and such that S is a measurable cover.*

Lemma 1.2.3:   *Any subset E in M may be written as the disjoint union of a set B in $\mathscr{C}_\sigma(\mathscr{E})$ and a subset E_0 of a set in $\mathscr{C}_\sigma(\mathscr{E})$ with $\mu^*(E_0) = \bar{\mu}(E_0) = 0$, i.e., with $E_0 \in M$. That is, $E = B \cup E_0$, $B \in \mathscr{C}_\sigma(\mathscr{E})$, and $\bar{\mu}(E) = \mu(B)$. Conversely any set $B \cup E_0$, with $B \in \mathscr{C}_\sigma(\mathscr{E})$ and E_0 some subset of a set in $\mathscr{C}_\sigma(\mathscr{E})$ with $\mu^*(E_0) = 0$, belongs to M.*

To distinguish between the σ-algebras $\mathscr{C}_\sigma(\mathscr{E})$ and M, we call the sets in $\mathscr{C}_\sigma(\mathscr{E})$ as Borel sets and the sets in M as the Lebesgue measurable sets. Lemma 1.2.3 states, in particular, that every Lebesgue measurable set may be written as the disjoint union of a Borel set and a Lebesgue measurable set with the latter of Lebesgue measure $\bar{\mu}$ zero. We also note that if we define a measure $\hat{\mu}$ on a σ-algebra $\mathscr{C}_\sigma$ as complete by the condition that if $S \in \mathscr{C}_\sigma$ with $\hat{\mu}(S) = 0$ and that $R \subset S$ imply $R \in \mathscr{C}_\sigma$, then we see that the Lebesgue measure $\bar{\mu}$ is complete. Because if $S \in M$ and $\bar{\mu}(S) = 0$, then for any $R \subset S$ we have $\mu^*(R) \leq \bar{\mu}(S) = 0$, i.e., $R \in M$ [see Lemma 1.2.1(iv)]. Thus we may conclude that the Lebesgue measure $\bar{\mu}$ is the completion of the measure μ which is restricted to the Borel sets in $\mathscr{C}_\sigma(\mathscr{E})$. From now on we denote the Lebesgue measure defined for the Lebesgue measurable sets, which include the Borel sets, by μ also. We also call the Lebesgue measurable set, simply, measurable. The triplet $(\mathbb{R}^k, M, \mu)$ is called a measure space. Finally we say that a certain property holds almost everywhere (a.e.) on a set E if there is a set $S \subset E$ of measure zero, $\mu(S) = 0$, such that the property holds for all $x \in E - S$. For example, $f(x) = 0$ a.e. on E if there is a set $S \subset E$ with $f(x) = 0$ for all $x \in E - S$ and $\mu(S) = 0$.

The following lemma states the translational invariance property of μ.

Lemma 1.2.4:   *Let $y \in X \equiv \mathbb{R}^k$ and $S \in M$; then $S + y \in M$ and $\mu(S + y) = \mu(S)$, where $S + y = \{x + y : x \in S\}$.*

1.2.2   Integration

A real-valued function f defined in $\mathbb{R}^k$ is called a measurable function (or just measurable) if for each real number α, the set $\{x : f(x) \leq \alpha\}$ is measurable. A complex-valued function f defined in $\mathbb{R}^k$ is measurable if its real and imaginary parts are measurable.

Lemma 1.2.5:   *If a real-valued function f defined on a measurable set D is continuous, then f is measurable.*

The above result is easy to see. Consider the set $\{x : f(x) \leq \alpha\}$ for any real number α. Suppose x_p is a limit point of this set. Then, by definition, any neighborhood of x_p contains a point of $\{x : f(x) \leq \alpha\}$, i.e., contains a point y such that $f(y) \leq \alpha$. Since $f(x)$ is continuous, it follows that $f(x_p) \leq \alpha$, i.e., $x_p \in \{x : f(x) \leq \alpha\}$. That is, $\{x : f(x) \leq \alpha\}$ contains all of its limit points, and therefore it is closed. By using the fact that the measurable sets contain all the open sets and their complements, the closed sets, we conclude that $\{x : f(x) \leq \alpha\}$ is measurable. If a real-valued function $f(x)$ may take on the values $\pm\infty$, we must also require that the sets $\{x : f(x) = +\infty\}$ and $\{x : f(x) = -\infty\}$ be measurable for the measurability of f. If f, f_1, and f_2 are measurable functions on some measurable sets D, then $f_1 f_2$, $f_1 + cf_2$ for all c, $|f|$, f^2, $f^+ = \max(f, 0)$, $f^- = -\min(f, 0)$ are also measurable on D.

We define a simple function as any function of the form

$$g(x) = \sum_{i=1}^{n} a_i \chi_{S_i}(x),$$

for any finite n, for any finite real numbers a_i, and for any pairwise disjoint measurable sets $S_1, \ldots, S_n$. $\chi_{S_i}(x)$ is called the characteristic function of the set S_i and is defined by $\chi_{S_i}(x) = 1$ if $x \in S_i$ and $\chi_{S_i}(x) = 0$ if $x \notin S_i$. χ_{S_i} is obviously measurable. Let $S = \bigcup_{i=1}^{n} S_i$; then for $x \notin S$, $g(x) = 0$. If we want, we may write $g(x) = \sum_{i=1}^{n} a_i \chi_{S_i}(x) + 0 \cdot \chi_{S^c}(x)$, where we recall that S^c is also a measurable set. The simple function $g(x)$ is said to be integrable on a measurable set E if $\mu(E \cap \{x : g(x) \neq 0\}) < \infty$. In this case the integral of $g(x)$ on E is defined by

$$\int_E g \, d\mu = \sum_{i=1}^{n} a_i \mu(E \cap S_i). \tag{1.2.9}$$

If $x \in \{x : g(x) \neq 0\}$, then we may find an $i \in [1, \ldots, n]$ such that $x \in S_i$. The condition $\mu(E \cap \{x : g(x) \neq 0\}) < \infty$ is equivalent to stating that g is integrable on E if $\mu(E \cap S_i) < \infty$, for the corresponding $a_i \neq 0$, by using the non-negative property of μ and the fact that $\bigcup_{i=1}^{n}(E \cap S_i) = E \cap \{x : g(x) \neq 0\}$. For those sets S_i in (1.2.9) for which $\mu(S_i) = 0$ we may effectively set the corresponding $a_i = 0$.

Now we consider the definition of the integral of any nonnegative measurable function f. Consider any sequence $\{R_n\}$ of increasing measurable sets[6] $R_1 \subset R_2 \subset \cdots$, of finite measures $\mu(R_n) < \infty$ for $n < \infty$, and such that $\lim_{n \to \infty} R_n = \mathbb{R}^k$. For a fixed n, we define the pairwise disjoint sets

$$S_n = R_n \cap E \cap \{x : n \le f(x)\},$$

$$S_{n, i} = R_n \cap E \cap \left\{x : \frac{i - 1}{2^n} \le f(x) < \frac{i}{2^n}\right\}, \qquad i = 1, 2, \ldots, n2^n,$$

and define the simple functions

$$g_n(x) = \sum_{i=1}^{n2^n} \frac{(i - 1)}{2^n} \chi_{S_{n, i}}(x) + n \chi_{S_n}(x),$$

where E is some measurable set. We note that the sequence $\{g_n\}$ is non-decreasing: $0 \le g_1 \le g_2 \le \cdots \le f$ such that $\lim_{n \to \infty} g_n(x) = f(x)$ for all

---

[6] One particular example of this family of open (and hence measurable) sets is $\{R_n\}$, with $R_n = \{x : |x| < n\}$, $n = 1, 2, \ldots$. We shall not, however, make use of any particular realization of such sets.

$x \in E$. It is easy to see that the $g_n(x)$ are integrable on E. To this end note that $\{x : g_n(x) > 0\} \subset \bigcup_i S_{n,i} \cup S_n$ and that $\mu(S_{n,i}) \le \mu(R_n) < \infty$, $\mu(S_n) < \mu(R_n) < \infty$. We then define the integral of f on E by

$$\int_E f \, d\mu = \lim_{n \to \infty} \int_E g_n \, d\mu, \tag{1.2.10}$$

and if the expression on the right-hand side is finite, then we say that f is integrable on E. An important consequence of the definition of the integral of f is that its value, if it exists, does not depend on the particular sequence of simple functions. More precisely, if $\{h_n\}$ is a nondecreasing sequence of simple functions: $0 \le h_1 \le h_2 \le \cdots \le f$ such that $\lim_{n \to \infty} h_n(x) = f(x)$ for all $x \in E$, $h_n(x)$ integrable on E, and that $\lim_{n \to \infty} \int h_n \, d\mu < \infty$, then

$$\lim_{n \to \infty} \int g_n \, d\mu = \lim_{n \to \infty} \int h_n \, d\mu. \tag{1.2.11}$$

We say that a real-valued measurable function is integrable on a measurable set E if its positive f^+ and negative parts f^- are integrable on E, and we define

$$\int_E f \, d\mu = \int_E f^+ \, d\mu - \int_E f^- \, d\mu. \tag{1.2.12}$$

Similarly we say that complex-valued function h is integrable on a measurable set E if its real part $\operatorname{Re} h$ and imaginary part $\operatorname{Im} h$ are integrable on E, and we define

$$\int_E h \, d\mu = \int_E (\operatorname{Re} h) \, d\mu + i \int_E (\operatorname{Im} h) \, d\mu. \tag{1.2.13}$$

The following results are of importance.

Lemma 1.2.6: *If f is a (Lebesgue) measurable function, then there exists a Borel measurable function g such that $f = g$ a.e.*

Lemma 1.2.7: *Let E be a measurable set. Let f, f_1, and f_2 below be complex-valued measurable functions.*

(i) *If $E = \bigcup_{j=1}^n E_j$, where $E_1, \ldots, E_n$ are pairwise disjoint measurable sets, and f is integrable on each of the E_i, then f is integrable on E and*

$$\int_E f \, d\mu = \sum_{j=1}^n \int_{E_j} f \, d\mu.$$

(ii) *For any complex numbers a_1 and a_2, $a_1 f_1 + a_2 f_2$ is integrable on E if f_1 and f_2 are, and*

$$\int_E (a_1 f_1 + a_2 f_2) \, d\mu = a_1 \int_E f_1 \, d\mu + a_2 \int_E f_2 \, d\mu.$$

(iii)   *Let S be any subset of E such that $\mu(S) = 0$ (therefore S is also measurable); then if f is integrable on E,*

$$\int_E f \, d\mu = \int_{E \cap S^c} f \, d\mu.$$

This, in particular, means, according to Lemma 1.2.3, that if we write $E = B \cup E_0$ as a disjoint union, where B is a Borel set and E_0 ($\subset E$) is a subset of some Borel set with $\mu(E_0) = 0$, then

$$\int_E f \, d\mu = \int_B f \, d\mu.$$

(iv)   *If $f_1 = f_2$ a.e. on E and f_2 is integrable on E, then f_1 is integrable on E and*

$$\int_E f_1 \, d\mu = \int_E f_2 \, d\mu.$$

(v)   *Let $|f| \leq g$ a.e. on E, where g is integrable on E; then f and $|f|$ are integrable on E and*

$$\left| \int_E f \, d\mu \right| \leq \int_E |f| \, d\mu \leq \int_E g \, d\mu.$$

(vi)   *Let $E = \bigcup_{j=1}^n D_j$, where the D_j are measurable but not necessarily pairwise disjoint; then if $|f|$ is integrable on each D_j, $|f|$ is integrable on E and*

$$\int_E |f| \, d\mu \leq \sum_{j=1}^n \int_{D_j} |f| \, d\mu.$$

Theorem 1.2.2:   (i)   *Let $\{f_n\}$ be a nondecreasing sequence of nonnegative functions: $0 \leq f_1 \leq f_2 \leq \cdots$ such that $\lim_{n \to \infty} f_n(x) = f(x)$ a.e. on a measurable set E. If the f_n are integrable on E and $\lim_{n \to \infty} \int_E f_n \, d\mu < \infty$, then f is integrable on E and*

$$\int_E f \, d\mu = \lim_{n \to \infty} \int_E f_n \, d\mu.$$

(ii)   *Suppose $\{g_n\}$ is a sequence of complex measurable functions such that $\lim_{n \to \infty} g_n(x) = g(x)$ a.e. on a measurable set E. If there is a nonnegative function $f(x)$ integrable on a measurable set E such that $|g_n(x)| \leq f(x)$ a.e. on E for all n, then g is integrable on E and*

$$\lim_{n \to \infty} \int_E g_n \, d\mu = \int_E g \, d\mu.$$

(iii)   *Let $\{f_n\}$ be a sequence of nonnegative functions each integrable on a measurable set E. If $f(x) = \lim_{N \to \infty} \sum_{n=1}^{N} f_n(x)$ a.e. on E, and $\sum_{n=1}^{\infty} \int_E f_n \, d\mu < \infty$, then*

$$\int_E f \, d\mu = \sum_{n=1}^{\infty} \int_E f_n \, d\mu.$$

Part (i) of the above theorem is known as the "Lebesgue monotone convergence theorem," and part (ii) is known as the "Lebesgue dominated convergence theorem."

Now we are ready to develop and prove a basic theorem of *multiple integrals*.

1.2.3   Fubini's Theorem

We are interested in studying the relation between the measure spaces $(\mathbb{R}^{k_1}, M_1, \mu_1)$, $(\mathbb{R}^{k_2}, M_2, \mu_2)$, and $(\mathbb{R}^{k_1+k_2}, M, \mu)$, where $\mathbb{R}^{k_1+k_2} = \mathbb{R}^{k_1} \times \mathbb{R}^{k_2}$. Let $\mathscr{C}_\sigma(\mathscr{E}_1), \mathscr{C}_\sigma(\mathscr{E}_2)$, and $\mathscr{C}_\sigma(\mathscr{E})$ be the Borel sets in M_1, M_2, and M, respectively. $\mathscr{E}_1, \mathscr{E}_2$, and $\mathscr{E}$ are constructed out of k_1-, k_2-, and $(k_1 + k_2)$-cells as defined in the beginning of this section. From Lemma 1.1.3(ii) we know that $\mathscr{C}_\sigma(\mathscr{E}) = \mathscr{C}_\sigma(\mathscr{F})$, where $\mathscr{F}$ is an algebra, defined in Section 1.1, and the latter consists of all finite disjoint unions of the cartesian products of sets from $\mathscr{C}_\sigma(\mathscr{E}_1)$ and $\mathscr{C}_\sigma(\mathscr{E}_2)$. That is, if $F \in \mathscr{F}$, then $F = \bigcup_{i=1}^{n} F_i$, $F_i \cap F_j \neq \varnothing$ for $i \neq j$, and $F_i = A_i \times B_i$, $A_i \in \mathscr{C}_\sigma(\mathscr{E}_1)$ and $B_i \in \mathscr{C}_\sigma(\mathscr{E}_2)$. Lemma 1.1.3(ii) also leads to the fact that if f is $\mathscr{C}_\sigma(\mathscr{F})$-measurable, then f_x is $\mathscr{C}_\sigma(\mathscr{E}_2)$-measurable and f_y is $\mathscr{C}_\sigma(\mathscr{E}_1)$-measurable, where $f_x(y) = f(x, y)$ for x fixed in $\mathbb{R}^{k_1}$ and $f_y(x) = f(x, y)$ for y fixed in $\mathbb{R}^{k_2}$.

In the following lemma we establish the connection between $\mu(A)$ for any $A \in \mathscr{C}_\sigma(\mathscr{F})$ and the measures μ_1 and μ_2. To this end we receall that $A_x \in \mathscr{C}_\sigma(\mathscr{E}_2)$ and $A_y \in \mathscr{C}_\sigma(\mathscr{E}_1)$ [see Lemma 1.1.3(ii)].

Lemma 1.2.8:   *Let $A \in \mathscr{C}_\sigma(\mathscr{F})$. We write $\mathbb{R}^{k_1} = \bigcup_{n=1}^{\infty} B_1^n$ and $\mathbb{R}^{k_2} = \bigcup_{n=1}^{\infty} B_2^n$, where the B_1^n (and similarly the B_2^n) are disjoint boxes of finite μ_1 (μ_2) measure. For every $x \in \mathbb{R}^{k_1}$ let*

$$\phi_1^m(x) = \mu_2(A_x \cap B_2^m), \tag{1.2.14}$$

and for every $y \in \mathbb{R}^{k_2}$ let

$$\phi_2^n(y) = \mu_1(A_y \cap B_1^n); \tag{1.2.15}$$

then ϕ_1^m is μ_1-measurable and ϕ_2^n is μ_2-measurable and

$$\int_{B_1^m} \phi_1^m \, d\mu_1 = \int_{B_2^m} \phi_2^n \, d\mu_2. \tag{1.2.16}$$

Consider any set of the form $Q \cap B_1^n \times B_2^m$, where $Q = C \times D$ with $C \in \mathscr{C}_\sigma(\mathscr{E}_1)$ and $D \in \mathscr{C}_\sigma(\mathscr{E}_2)$. Then for $x \in B_1^n$

$$\mu_2(Q_x \cap B_2^m) = \int \chi_{Q \cap B_1^n \times B_2^m}(x, y) \, d\mu_2(y)$$

$$= \chi_{C \cap B_1^n}(x) \int \chi_{D \cap B_2^m}(y) \, d\mu_2(y)$$

$$= \chi_{C \cap B_1^n}(x) \mu_2(D \cap B_2^m).$$

Similarly for $y \in B_2^m$,

$$\mu_1(Q_y \cap B_1^n) = \chi_{D \cap B_2^m}(y) \mu_1(C \cap B_1^n).$$

Therefore

$$\int_{B_1^n} \mu_2(Q_x \cap B_2^m) \, d\mu_1 = \mu_1(C \cap B_1^n) \mu_2(D \cap B_2^m)$$

and

$$\int_{B_2^m} \mu_1(Q_y \cap B_1^n) \, d\mu_2 = \mu_1(C \cap B_1^n) \mu_2(D \cap B_2^m),$$

and the statements of the lemma are true for all sets Q of the form $Q = C \times D$ with $C \in \mathscr{C}_\sigma(\mathscr{E}_1)$ and $D \in \mathscr{C}_\sigma(\mathscr{E}_2)$. The latter implies that the lemma is true for all $Q \in \mathscr{F}$ by the application of the definition of $\mathscr{F}$ (see Section 1.1).

We now introduce the following family of sets D^{nm} consisting of all sets in $\mathscr{C}_\sigma(\mathscr{F}) \cap B_1^n \times B_2^m$ such that the statements of the lemma are true for such sets. The family D^{nm} is obviously nonempty as we have just shown that $\mathscr{F} \cap B_1^n \times B_2^m \subset D^{nm}$ and the lemma is true for all the sets in $\mathscr{F} \cap B_1^n \times B_2^m$. We first show that D^{nm} is a monotone class in $B_1^n \times B_2^m$.

Let $S_1 \subset S_2 \subset \cdots$, with the $S_i \in D^{nm}$. Then by definition of the family D^{nm},

$$\int_{B_1^n} \phi_{1i} \, d\mu_1 = \int_{B_2^m} \phi_{2i} \, d\mu_2, \tag{1.2.17}$$

where $\phi_{1i}(x) = \mu_2((S_i)_x)$ and $\phi_{2i}(y) = \mu_1((S_i)_y)$. Also $(S_i)_x \subset (S_{i+1})_x$. Then from the analysis leading to (1.2.4),

$$\lim_{i \to \infty} \mu_2((S_i)_x) = \mu_2 \left(\bigcup_{i=1}^{\infty} (S_i)_x \right). \tag{1.2.18}$$

Finally, the sequence $\{\phi_{1i}\}$ is monotonically nondecreasing; therefore by the application of the Lebesgue monotone convergence theorem [Theorem 1.2.2(i)], we conclude that

$$\lim_{i \to \infty} \int \phi_{1i} \, d\mu_1 = \int \lim_{i \to \infty} \phi_{1i} \, d\mu_1. \tag{1.2.19}$$

By using (1.2.18), repeating the analysis for the sequence $\{\phi_{2i}\}$, and finally, using (1.2.17), we arrive at

$$\int_{B_1^n} \mu_2\left(\bigcup_{i=1}^{\infty}(S_i)_x\right) d\mu_1 = \int_{B_2^m} \mu_1\left(\bigcup_{i=1}^{\infty}(S_i)_y\right) d\mu_2, \qquad (1.2.20)$$

i.e., $\bigcup_{i=1}^{\infty} S_i \in D^{nm}$. For $S_1 \supset S_2 \supset \cdots$, with the $S_i \in D^{nm}$, we have $(S_i)_x \supset (S_{i+1})_x$ and the analysis leading to (1.2.7) implies that

$$\lim_{i\to\infty} \mu_2((S_i)_x) = \mu_2\left(\bigcap_{i=1}^{\infty}(S_i)_x\right), \qquad (1.2.21)$$

where we recall that $\mu_2((S_i)_x) < \infty$ since $\mu_2((S_i)_x) \le \mu_2(B_2^m) < \infty$. Also, $\mu_2((S_i)_x) \ge \mu_2((S_{i+1})_x)$ and, in particular, $\mu_2((S_i)_x) \ge \mu_2(\bigcap_{i=1}^{\infty}(S_i)_x)$. Thus by the application of the Lebesgue dominated convergence theorem [Theorem 1.2.2(ii)], we have that (1.2.19) is again true, and hence finally arrive at the conclusion that $\bigcap_{i=1}^{\infty} S_i \in D^{nm}$ by repeating the analysis for $\mu_1((S_i)_y)$ and upon using (1.2.17). We have thus shown that D^{nm} is a monotone class in $B_1^n \times B_2^m$ containing the family $\mathscr{F} \cap B_1^n \times B_2^m$, i.e.,

$$D^{nm} \supset \hat{\mathscr{C}}_\sigma(\mathscr{F} \cap B_1^n \times B_2^m) \quad \text{or} \quad D^{nm} \supset \mathscr{C}_\sigma(\mathscr{F}) \cap B_1^n \times B_2^m$$

[see (i) and (ii) of Lemma 1.1.2]. Hence for any set $Q \in \mathscr{C}_\sigma(\mathscr{F})$ the statements of the lemma are true, i.e., for any $Q \in \mathscr{C}_\sigma(\mathscr{F})$,

$$\int_{B_1^n} \mu_2(Q_x \cap B_2^m) d\mu_1 = \int_{B_2^m} \mu_1(Q_y \cap B_1^n) d\mu_2, \qquad (1.2.22)$$

where $Q^{nm} = Q \cap B_1^n \times B_2^m$.

In particular, we note that if $Q_1, Q_2, \ldots$ are pairwise disjoint sets in D^{nm}, then if we let $P_r = \bigcup_{i=1}^r Q_i$, we have $P_1 \subset P_2 \subset \cdots$ and hence $\bigcup_{i=1}^{\infty} P_i = \bigcup_{i=1}^{\infty} Q_i \in D^{nm}$.

Now we are ready to define a measure for any $Q \in \mathscr{C}_\sigma(\mathscr{E})$ by the expression

$$\mu(Q) = \sum_{n,m=1}^{\infty} \mu(Q \cap B_1^n \times B_2^m), \qquad Q^{nm} = Q \cap B_1^n \times B_2^m, \quad (1.2.23)$$

where

$$\mu(Q^{nm}) = \int_{B_1^n} \mu_2(Q_x \cap B_2^m) d\mu_1 = \int_{B_2^m} \mu_1(Q_y \cap B_1^n) d\mu_2. \qquad (1.2.24)$$

It is readily verified that (1.2.23) defines a measure, satisfying, in particular, the σ-additivity property of a measure. We note that $\mu(Q)$ is an extended real-valued set function (i.e., the value $+\infty$ for it is permissible). From the uniqueness theorem (Theorem 1.2.1), then, the measure in (1.2.23) *coincides* with the measure μ in $(\mathbb{R}^{k_1+k_2}, M, \mu)$ when restricted to the Borel sets, and the measure μ in $(\mathbb{R}^{k_1 + k_2}, M, \mu)$, more generally, is the completion of the measure defined in (1.2.23).

Now we prove the following lemma.

Lemma 1.2.9: *If f is M-measurable and $f = 0$ a.e. with respect to the measure μ in $(\mathbb{R}^{k_1 + k_2}, M, \mu)$, then for almost all $x \in \mathbb{R}^{k_1}$, $f_x = 0$ a.e. with respect to the measure μ_2. Similarly, for almost all $y \in \mathbb{R}^{k_2}$, $f_y = 0$ a.e. with respect to the measure μ_1.*

Let $E = \{(x, y) : f(x, y) \neq 0\}$. Since $E \in M$, it follows from Lemma 1.2.2 that we may find a cover $S \supset E$, $S \in \mathscr{C}_\sigma(\mathscr{F})$, with $\bar{\mu}(E) = \mu(S) = 0$.[7] Lemma 1.2.8 implies that

$$\mu(S) = \sum_{n, m = 1}^{\infty} \int_{B_1^n} \mu_2(S_x^m) \, d\mu_1 = 0,$$

or

$$\mu(S^{nm}) = \int_{B_1^n} \mu_2(S_x^m) \, d\mu_1 = 0,$$

written in terms of the boxes B_1^n, B_2^m, i.e., $A^n = A \cap B_1^n$, if $A \in \mathscr{C}_\sigma(\mathscr{E}_1)$ and $C^{nm} = C \cap B_1^n \times B_2^m$ if $C \in \mathscr{C}_\sigma(\mathscr{F})$, etc. Let N^n be the set of all points $x \in B_1^n$ for which $\mu_2(S_x^m) \neq 0$.

Then

$$\mu(S^{nm}) = \int_{N^n} \mu_2(S_x^m) \, d\mu_1 = 0,$$

i.e., $\mu_1(N^n) = 0$. For $x \in B_1^n - N^n$, $\mu_2(S_x^m) = 0$, by definition. Now $S_x^m \supset E_x^m$, and thus for $x \in B_1^n - N^n$, $\bar{\mu}_2(E_x^m) = 0$ and $E_x^m \in M_2$ since the latter is complete. Therefore, for $x \in B_1^n - N^n$, $f_x(y) = 0$ for $y \in B_2^m - E_x^m$. Let $N = \bigcup_{n=1}^{\infty} N^n$ and $E_x = \bigcup_{m=1}^{\infty} E_x^m$; then for $x \notin N$, $f_x(y) = 0$ for $y \notin E_x$, which is the statement of the lemma since $\mu_1(N) = \sum_{n=1}^{\infty} \mu_1(N^n) = 0$, and for $x \notin N$, $\mu_2(E_x) = \sum_{m=1}^{\infty} \mu_2(E_x^m) = 0$. A similar statement holds for $f_y(x)$.

Theorem 1.2.3   (Fubini's theorem):   *If f is integrable on $\mathbb{R}^{k_1 + k_2}$, then*

$$\psi_1(x) = \int_{\mathbb{R}^{k_2}} f_x \, d\mu_2, \qquad \psi_2(y) = \int_{\mathbb{R}^{k_1}} f_y \, d\mu_1 \qquad (1.2.25)$$

are defined a.e. with respect to the measures μ_1 and μ_2, respectively, and

$$\int_{\mathbb{R}^{k_1 + k_2}} f \, d\mu = \int_{\mathbb{R}^{k_1}} \psi_1(x) \, d\mu_1 = \int_{\mathbb{R}^{k_2}} \psi_2(y) \, d\mu_2. \qquad (1.2.26)$$

---

[7] Here we write $\bar{\mu}(E)$ in order not to confuse the Lebesgue measure with the Borel measure.

Let f be M-measurable; then according to Lemma 1.2.6 there exists a $\mathscr{C}_\sigma(\mathscr{F})$-measurable function g such that $f = g$ a.e. Then by an application of Lemma 1.2.9, $f_x = g_x$ a.e. with respect to the μ_2-measure for almost all x and $f_y = g_y$ a.e. with respect to the μ_1-measure for almost all y. Suppose $Q \in M$ is of finite $\bar{\mu}$-measure. Let $f = \chi_Q$; then according to Lemma 1.2.2 there exists a Borel set $S \supset Q$ such that

$$\bar{\mu}(Q) = \mu(S) = \sum_{n,m=1}^\infty \int_{\mathbb{R}^{k_1+k_2}} \chi_{S^{nm}} \, d\mu$$

$$= \sum_{n,m=1}^\infty \int_{B_1^n} \mu_2(S_x \cap B_2^m) \, d\mu_1$$

$$= \sum_{n,m=1}^\infty \int_{B_2^m} \mu_1(S_y \cap B_1^n) \, d\mu_2 < \infty. \qquad (1.2.27)$$

Applying Theorem 1.2.2(iii) to the series $\sum_{n,m=1}^\infty \chi_{S^{nm}} = \chi_S$ gives

$$\bar{\mu}(Q) = \mu(S) = \int_{\mathbb{R}^{k_1+k_2}} \chi_S \, d\mu. \qquad (1.2.28)$$

Again applying Theorem 1.2.2(iii) to the (absolutely) convergent series $\sum_{n,m=1}^\infty \chi_{B_1^n} \mu_2(S_x \cap B_2^m) = \mu_2(S_x) < \infty$ a.e. with respect to μ_1,

$$\bar{\mu}(Q) = \mu(S) = \int_{\mathbb{R}^{k_1}} \mu_2(S_x) \, d\mu_1. \qquad (1.2.29)$$

Similarly,

$$\bar{\mu}(Q) = \mu(S) = \int_{\mathbb{R}^{k_2}} \mu_1(S_y) \, d\mu_2. \qquad (1.2.30)$$

From (1.2.28)–(1.2.30) we see that the theorem is true for all $f = \chi_Q$, where $f = g$ a.e. with respect to μ and $g = \chi_S$ and $\bar{\mu}(Q) = \mu(S) < \infty$. This in turn implies that the theorem is true for all simple functions.

Quite generally, suppose f is nonnegative and let $\{s^n\}$ be a nondecreasing sequence of nonnegative simple functions such that $\lim_{n\to\infty} s^n = f$ a.e. with respect to μ. Then we have just shown that

$$\int_{\mathbb{R}^{k_1+k_2}} s^n \, d\mu = \int_{\mathbb{R}^{k_1}} \psi_1^n(x) \, d\mu_1 = \int_{\mathbb{R}^{k_2}} \psi_2^n(y) \, d\mu_2, \qquad (1.2.31)$$

where

$$\psi_1^n(x) = \int_{\mathbb{R}^{k_2}} s_x^n \, d\mu_2, \qquad (1.2.32)$$

$$\psi_2^n(y) = \int_{\mathbb{R}^{k_1}} s_y^n \, d\mu_1. \qquad (1.2.33)$$

By applying the monotone convergence theorem [Theorem 1.2.2(i)] to $\{s_x^n\}$, $\{\psi_1^n\}$, $\{s_y^n\}$, $\{\psi_2^n\}$, $\{s^n\}$, we conclude that the statement of the theorem is true for all nonnegative functions. The general result then follows by applying the above to the positive and negative parts of the real and imaginary parts of an integrable function f.

Theorem 1.2.4   (Fubini–Tonelli's theorem):   *Let f be a complex M-measurable function; if*

$$\int_{\mathbb{R}^{k_1}} \left(\int_{\mathbb{R}^{k_2}} |f|_x \, d\mu_2 \right) d\mu_1 < \infty \qquad \left(\text{or} \int_{\mathbb{R}^{k_2}} \left(\int_{\mathbb{R}^{k_1}} |f|_y \, d\mu_1 \right) d\mu_2 < \infty \right),$$

(1.2.34)

then

$$\int_{\mathbb{R}^{k_1+k_2}} f \, d\mu = \int_{\mathbb{R}^{k_2}} \left(\int_{\mathbb{R}^{k_1}} f_y \, d\mu_1 \right) d\mu_2 = \int_{\mathbb{R}^{k_1}} \left(\int_{\mathbb{R}^{k_2}} f_x \, d\mu_2 \right) d\mu_1. \quad (1.2.35)$$

As in the proof of Theorem 1.2.3, the above theorem is true if $f = \chi_Q$ with $\mu(Q) < \infty$, or in general if f is a simple function. If f is nonnegative, then let $\{f^n\}$ be a nondecreasing sequence of nonnegative simple functions such that $\lim_{n \to \infty} f^n = f$ a.e. with respect to μ. Then

$$\int_{\mathbb{R}^{k_1+k_2}} f^n \, d\mu = \int_{\mathbb{R}^{k_2}} \left(\int_{\mathbb{R}^{k_1}} f_y^n \, d\mu_1 \right) d\mu_2 = \int_{\mathbb{R}^{k_1}} \left(\int_{\mathbb{R}^{k_2}} f_x^n \, d\mu_2 \right) d\mu_1. \quad (1.2.36)$$

By applying the monotone convergence theorem [Theorem 1.2.2(i)] to the sequences $\{f^n\}$, $\{\int_{\mathbb{R}^{k_1}} f_y^n \, d\mu_1\}$, $\{f_y^n\}$, $\{\int_{\mathbb{R}^{k_2}} f_x^n \, d\mu_2\}$, $\{f_x^n\}$, we obtain the stated result in (1.2.35) if $\lim_{n \to \infty} \int_{\mathbb{R}^{k_1+k_2}} f^n \, d\mu \, (= \int_{\mathbb{R}^{k_1+k_2}} f \, d\mu) < \infty$.

Now consider, in general, a complex M-measurable function f such that (1.2.34) is true. Since $|f|$ is an M-measurable real-valued nonnegative function, we have just shown that

$$\int_{\mathbb{R}^{k_1+k_2}} |f| \, d\mu = \int_{\mathbb{R}^{k_2}} \left(\int_{\mathbb{R}^{k_1}} |f|_y \, d\mu_1 \right) d\mu_2 = \int_{\mathbb{R}^{k_1}} \left(\int_{\mathbb{R}^{k_2}} |f|_x \, d\mu_2 \right) d\mu_1 \quad (<\infty).$$

(1.2.37)

From the estimate on the left-hand side of Lemma 1.2.7(v), we have that f is integrable. The stated result in the theorem then follows from Fubini's theorem (Theorem 1.2.3).

The importance of Theorem 1.2.4 cannot be overemphasized.[8]

---

[8] An elementary example where the interchange of the orders of integration give different results(!) is the following:

$$\int_1^\infty dx \int_1^\infty dy \, f(x, y) = +\frac{\pi}{4}, \qquad \int_1^\infty dy \int_1^\infty dx \, f(x, y) = -\frac{\pi}{4},$$

where $f(x, y) = (y^2 - x^2)/(x^2 + y^2)^2$. Many other examples may be also constructed.

1.3 GEOMETRY IN $\mathbb{R}^k$

We consider subspaces of $\mathbb{R}^k$. We recall that a subspace S of $\mathbb{R}^k$ is a nonempty subset of $\mathbb{R}^k$ such that if x and x' are in S, then $\alpha x + x'$ is also in S for all real α. If we choose any vectors $x_1, \ldots, x_s$ in $\mathbb{R}^k$, then the set of all linear combinations of these vectors is, by definition, a subspace of $\mathbb{R}^k$ called the subspace generated by $x_1, \ldots, x_s$. The dimension of a subspace S, written dim S, is the maximum number of linearly independent vectors that can be found in S. Suppose the dimension of a subspace S is r. Let $\{x_1, \ldots, x_r\}$ be any basis of S. Each vector $x \in S$ may be then written as $x = \sum_{i=1}^{r} a_i x_i$, where the reals a_i are uniquely determined. With respect to the basis $\{x_i, \ldots, x_r\}$, we may put each vector $x \in S$ in one-to-one correspondence with r-tuple $(a_1, \ldots, a_r)$ of real numbers, and we denote this correspondence by $x \leftrightarrows (a_1, \ldots, a_r)$. This correspondence preserves all linear relations if we define for all reals a and b: $a(a_1, \ldots, a_r) + b(b_1, \ldots, b_r) = (aa_1 + bb_1, \ldots, aa_r + bb_r)$ since $ax + by = \sum_{i=1}^{r}(aa_i + bb_i)x_i$, where $y = \sum_{i=1}^{r} b_i x_i$. We say that S is isomorphic to $\mathbb{R}^r$, with the latter consisting of r-tuples of real numbers, as there is a correspondence $\leftrightarrows$ between S and the set of all r-tuples of real numbers that preserves all linear relations.

We say that a space I is the direct sum of its subspaces I_1 and I_2, written $I = I_1 \oplus I_2$, if every vector $x \in I$ may be written *uniquely* as $x = x_1 + x_2$, with $x_1 \in I_1$ and $x_2 \in I_2$. A direct consequence of this definition is that the subspaces I_1 and I_2 are disjoint, i.e., they have only the zero vector in common. Conversely, if we have three subspaces I, I_1, I_2 such that every vector $x \in I$ may be written as $x = x_1 + x_2$, with $x_1 \in I_1, x_2 \in I_2$ and the subspaces I_1, I_2 disjoint, then $I = I_1 \oplus I_2$. It is readily checked that if $I = I_1 \oplus I_2$, then dim $I = $ dim $I_1 + $ dim I_2. Also, since with each vector $x \in I$ we may assign a unique vector x_1 in I_1 (and similarly a vector x_2 in I_2), we may introduce a linear mapping $\Lambda(I_2)$ that takes every vector $x \in I$ into its unique counterpart in I_1, i.e., $\Lambda(I_2)x = x_1$. The latter is obviously a linear mapping; it is called the *projection on I_1 along I_2*.

Given a space I and a subspace S of I, there is no unique way of choosing a complement of S in I. That is, one may write $I = S \oplus I' = S \oplus I''$, where I' and I'' may be different. A particular example of a complement of a subspace S of I is the so-called orthogonal complement. Let the standard components of a vector $x \in \mathbb{R}^k$ be denoted $(x_1, \ldots, x_k)$ (i.e., the components of x with respect to the standard basis of $\mathbb{R}^k$). For any vectors $x, x' \in \mathbb{R}^k$, we introduce the quantity (inner product) $\langle x | x' \rangle \equiv \sum_{i=1}^{k} x_i x_i'$. Then the orthogonal complement of $S \subset I$ in I is defined by

$$S^{\perp} = \{x \in I : \langle x | x' \rangle = 0 \text{ for all } x' \in S\},$$

and we may write $I = S \oplus S^{\perp}$. For example, suppose that $I \subset \mathbb{R}^4$ and that I is spanned by the vectors $(1, 1, 0, 1)$, $(1, 1, 1, 1)$, and $(0, 1, 0, 1)$. Let $S \subset I$ be a subspace spanned by the vector $(0, 1, 1, 1)$; then the subspace $S^{\perp}$ spanned by vectors $(0, 1, -2, 1)$ and $(1, -1, 2, -1)$ is the orthogonal complement of S in I.

Let $\mathbb{R}^k = I \oplus E$, where I is a subspace of $\mathbb{R}^k$ and E a complement of I in $\mathbb{R}^k$. We sometimes write $I = I_1 \oplus I_2$ and $\mathbb{R}^k = I_1 \oplus E_2$. Then if $S' \subset \mathbb{R}^k$, we note that $\Lambda(I_1)S' \subset E_2$ and $\Lambda(I)S' \subset E$.

Lemma 1.3.1:   *Suppose that $S \subset E$ and hence, in particular, that S and I are disjoint. If $\Lambda(I)S' = S$, then $S' \subset I \oplus S$, and $\dim S \leq \dim S' \leq \dim I + \dim S$.*

$\Lambda(I)S' = S$ means that for any vector $x' \in S'$, $\Lambda(I)x' \in S$. Any vector $x' \in S'$ may be written $x' = x_1 + x_2$, with $x_1 \in I$ and $x_2 \in S \subset E$, i.e., $x' \in I \oplus S$, which implies that $S' \subset I \oplus S$. Let $\{x^1, \ldots, x^r\}$ be a basis of S. For each $x^i \in \{x^1, \ldots, x^r\}$ we may introduce a vector $x'_i \in S'$ such that $x'_i = x^i + y_i$, with $y_i \in I$. Consider the expression $\sum_{i=1}^r a_i x'_i = 0$, or $\Lambda(I) \sum_{i=1}^r a_i x'_i = \sum_{i=1}^r a_i x^i = 0$, which implies that $a_1 = \cdots = a_r = 0$. Accordingly the vectors $x'_1, \ldots, x'_r$ are linearly independent vectors in S'. Hence $\dim S \leq \dim S'$. Also, $S' \subset I \oplus S$; therefore $\dim S \leq \dim S' \leq \dim S + \dim I$.

Lemma 1.3.2:   *Suppose $S \subset E$. Then $S'' \subset I \oplus S$ if and only if there is some subspace $S' \subset S$ such that $\Lambda(I)S'' = S'$.*

Suppose there is some subspace $S' \subset S$ such that $\Lambda(I)S'' = S'$; then we show that $S'' \subset I \oplus S$. If $y \in S''$, then y is of the form $y = y_1 + y_2$, where $y_1 \in S'$ and $y_2 \in I$ with $\Lambda(I)y = y_1$, and hence $y \in I \oplus S$, i.e., $S'' \subset I \oplus S$. Conversely, suppose that $S'' \subset I \oplus S$; i.e., we may write $y = y_1 + y_2$ with $y_1 \in S$ and $y_2 \in I$. Let $\{x_1, \ldots, x_s\}$ be a basis of S''; then $x_i = x_i^1 + x_i^2$, where $x_i^1 \in S$ and $x_i^2 \in I$. Any $x'' \in S''$ may be written $x'' = \sum_{i=1}^s \alpha_i x_i$, for some reals α_i, with $\Lambda(I) \sum_{i=1}^s \alpha_i x_i = \sum_{i=1}^s \alpha_i x_i^1$. We may then choose S' to be a subspace generated by the vectors $\{x_1^1, \ldots, x_s^1\}$. Since any vector $x' \in S'$ is of the form $x' = \sum_{i=1}^s a_i x_i^1$, and $x' \in S$, it follows that $S' \subset S$.

Lemma 1.3.3:   *Let x be a vector in I. Let J be a subspace generated by the vector x; i.e., any vector in J is of the form αx for some real α. Let $S \subset E$. Then $\Lambda(J)S' = S$ if and only if S' is spanned by the vectors in $\{x_1, \ldots, x_r, x\}$ or by the vectors in $\{x_1 + \alpha_1 x, \ldots, x_r + \alpha_r x\}$, where $\{x_1, \ldots, x_r\}$ is a basis of S and $\alpha_1, \ldots, \alpha_r$ are some real numbers.*

If S' is spanned either by the vectors in $\{x_1, \ldots, x_r, x\}$ or by the vectors in $\{x_1 + \alpha_1 x, \ldots, x_r + \alpha_r x\}$, for some reals α_i, then clearly $\Lambda(J)S' = S$. On the other hand, if $\Lambda(J)S' = S$, then from Lemma 1.3.1, $\dim S' = \dim S + 1$, or $\dim S' = \dim S$, since $\dim J = 1$. Any vector $x' \in S'$ is of the form $x' = \sum_{i=1}^r \alpha'_i x_i + \beta x$, where $\{x_1, \ldots, x_r\}$ is a basis of S and $\alpha'_1, \ldots, \alpha'_r, \beta$ are reals.

That is, $x' \in S'$ is either a linear combination of the vectors $x_1, \ldots, x_r, x$ or of the vectors $x_1 + \alpha_1 x, \ldots, x_r + \alpha_r x$, for some reals $\alpha_1, \ldots, \alpha_r$, corresponding to the two alternatives dim $S' = $ dim $S + 1$ or dim $S' = $ dim S, respectively.

Lemma 1.3.4:   *Let $S \subset E$ and $I = I_1 \oplus I_2$. Then the condition on S'' through $\Lambda(I_2)S'' = S'$ with S' satisfying the condition $\Lambda(I_1)S' = S$ is equivalent to the condition on S'' through $\Lambda(I_1 \oplus I_2)S'' = S$.*

Any vector $x'' \in S''$ is of the form $x'' = x' + x_2$, where $x' \in S'$ and $x_2 \in I_2$. On the other hand, x' must be of the form $x' = x + x_1$, where $x \in S$ and $x_1 \in I_1$. That is, any vector $x'' \in S''$ is of the form $x'' = x + x_1 + x_2$ with $x \in S$, $x_1 \in I_1$, and $x_2 \in I_2$, or $x'' = \hat{x} + x$, with $\hat{x} \in I_1 \oplus I_2$, and $\hat{x} = x_1 + x_2$. Thus if $x'' \in S''$, then $\Lambda(I_1 \oplus I_2)x'' = x \in S$, i.e., $\Lambda(I_1 \oplus I_2)S'' \subset S$. On the other hand, let $x \in S$; then we may introduce a vector $x' \in S'$ such that $x' = x + x_1$, with $x_1 \in I_1$. For such an $x' \in S'$ we may introduce a vector $x'' \in S''$ such that $x'' = x' + x_2$, with $x_2 \in I_2$, i.e., for any $x \in S$ we may introduce a vector $x'' = x + x_1 + x_2$ with $x_1 \in I_1$ and $x_2 \in I_2$. That is, if $x \in S$, then

$$x \in \Lambda(I_1)\Lambda(I_2)S'' = \Lambda(I_1 \oplus I_2)S'',$$

and hence $S \subset \Lambda(I_1 \oplus I_2)S''$, which with the relation $\Lambda(I_1 \oplus I_2)S'' \subset S$ established above implies that $\Lambda(I_1 \oplus I_2)S'' = S$.

We shall usually use the notation $\mathbf{x}$ for a vector $x \in \mathbb{R}^k$. If $\mathbf{L}_1, \ldots, \mathbf{L}_r$ are some vectors in $\mathbb{R}^k$ spanning a subspace $S \subset \mathbb{R}^k$, we will often use the notation $S \equiv \{\mathbf{L}_1, \ldots, \mathbf{L}_r\}$.

NOTES

Excellent references for Sections 1.1 and 1.2 are Rudin (1964), Royden (1963), Hoffman (1975), Munroe (1959), Berberian (1965), Halmos (1974a), and Rudin (1966); the latter two references are particularly recommended for the topics covered. For Section 1.3 we refer the reader to the classic paper by Weinberg (1960) and the work of Hoffman (1975) and Halmos (1974b).

Chapter 2 / CLASS B_n FUNCTIONS AND FEYNMAN INTEGRALS

Feynman integrands and Feynman integrals will be shown to belong to a special class of functions called the class B_n of functions. This class of functions is defined in Section 2.1. In Section 2.2 we study the structure of a Feynman integrand and show that it belongs to this class. This latter property will be important when studying the convergence of Feynman integrals in the next chapter, where it is shown, in particular, that the absolutely integrable integrals as well belong to such a class.

Consider two functions f and g of k real variables $x_1, \ldots, x_k$, If we can find real positive constants $b_1, \ldots, b_k : b_1 > 1, \ldots, b_k > 1$, and we can find a strictly positive constant C, independent of $x_1, \ldots, x_k$, such that for

$$|x_1| \geq b_1, \qquad \ldots, \qquad |x_k| \geq b_k, \tag{2.1}$$

we have

$$|f(x_1, \ldots, x_k)| \leq C|g(x_1, \ldots, x_k)|, \tag{2.2}$$

then we denote the relation (2.2) symbolically by

$$f(x_1, \ldots, x_k) = 0(g(x_1, \ldots, x_k)), \qquad |x_1| \to \infty, \qquad \ldots, \qquad |x_k| \to \infty. \tag{2.3}$$

If $\mathbf{L}_1, \ldots, \mathbf{L}_r$ ($r \leq k$) are independent vectors spanning a subspace $S \subset \mathbb{R}^k$, we use the notation

$$S \equiv \{\mathbf{L}_1, \ldots, \mathbf{L}_r\}. \tag{2.4}$$

29

2.1 DEFINITION OF CLASS B_n FUNCTIONS

A function $f(\mathbf{P})$, with $\mathbf{P} \in \mathbb{R}^n$, is said to belong to a class $B_n(I)$ of functions if for all choices of a nonzero subspace $S \subset \mathbb{R}^n$, $m \leq n$ independent vectors $\mathbf{L}_1, \ldots, \mathbf{L}_m$ in $\mathbb{R}^n$ and a bounded region $W \subset \mathbb{R}^n$, such that $|f(\mathbf{P})| \neq \infty$, there exist a pair of coefficients (real numbers) $\alpha(S)$ and $\beta(S)$ with the latter [i.e., the coefficients $\beta(S)$] nonnegative integers and

$$f(\mathbf{L}_1 \eta_1 \cdots \eta_m + \mathbf{L}_2 \eta_2 \cdots \eta_m + \cdots + \mathbf{L}_m \eta_m + \mathbf{C})$$

$$= 0\left\{\eta_1^{\alpha(\{\mathbf{L}_1\})} \cdots \eta_m^{\alpha(\{\mathbf{L}_1, \ldots, \mathbf{L}_m\})} \sum_{\gamma_1, \ldots, \gamma_m} (\ln \eta_{\pi_1})^{\gamma_1} \cdots (\ln \eta_{\pi_m})^{\gamma_m}\right\}, \quad (2.1.1)$$

where $\eta_1, \ldots, \eta_m$ are real and positive such that $\eta_1, \ldots, \eta_m \to \infty$ independently and $\mathbf{C} \in W$. The sums in (2.1.1) are over all nonnegative integers $\gamma_1, \ldots, \gamma_m$ such that

$$\sum_{i=1}^{k} \gamma_i \leq \beta(\{\mathbf{L}_1, \ldots, \mathbf{L}_{\pi_k}\}) \qquad (2.1.2)$$

for all $1 \leq k \leq m$, and the coefficients β have been arranged in increasing order

$$\beta(\{\mathbf{L}_1, \ldots, \mathbf{L}_{\pi_1}\}) \leq \cdots \leq \beta(\{\mathbf{L}_1, \ldots, \mathbf{L}_{\pi_m}\}), \qquad (2.1.3)$$

where $\{\pi_1, \ldots, \pi_m\}$ is a permutation of the integers in $\{1, \ldots, m\}$. As part of the definition of $B_n(I)$ we also associate with it some subspace $I \subset \mathbb{R}^n$ such that the condition $|f(\mathbf{P})| \neq \infty$, for some $\mathbf{P} \in \mathbb{R}^n$, implies that $|f(\mathbf{P} + \mathbf{P}')| \neq \infty$ for all finite $\mathbf{P}' \in I$. For the zero subspace we shall use simply the notation B_n for $B_n(\{\mathbf{0}\})$ in the sequel. As discussed above, the conditions $\eta_1, \ldots, \eta_m \to \infty$ mean that there exist real numbers $b_1 > 1, \ldots, b_m > 1$ such that $\eta_1 \geq b_1, \ldots, \eta_m \geq b_m$. The constants $b_1, \ldots, b_m$, and $C > 0$ as in (2.2), are independent of $\eta_1, \ldots, \eta_m$, but may depend on $\mathbf{L}_1, \ldots, \mathbf{L}_m$ and W.

The coefficients $\alpha(S)$ and $\beta(S)$ are, respectively, called *power and logarithmic asymptotic coefficients*. In the language of Feynman integrands, some fixed subspace I is chosen and is associated with the integration variables. We shall then conveniently introduce the orthogonal complement E of I in $\mathbb{R}^k$ and associate it with the so-called external momenta and the masses of the Feynman integrals.

2.2 STRUCTURE OF FEYNMAN INTEGRALS

We are interested in Feynman integrals, which are integrals of the following form and are absolutely convergent:

$$\mathscr{F}_\varepsilon(P, \mu) = \int_{\mathbb{R}^{4n}} dK \, \mathscr{I}(P, K, \mu, \varepsilon), \qquad (2.2.1)$$

where

$$K = (k_1^0, \ldots, k_n^3), \qquad P = (p_1^0, \ldots, p_m^3), \qquad \mu = (\mu^1, \ldots, \mu^\rho). \quad (2.2.2)$$

P denotes the so-called set of external momentum components of a graph, K the set of its internal momentum components (i.e., the integration variables), and μ the set of its masses.[1]

The integrands $\mathscr{I}(P, K, \mu, \varepsilon)$ are of the form

$$\mathscr{I}(P, K, \mu, \varepsilon) = \mathscr{P}(P, K, \mu, \varepsilon) \bigg/ \prod_{l=1}^{L} [Q_l^2 + \mu_l^2 - i\varepsilon(\mathbf{Q}_l^2 + \mu_l^2)], \qquad \varepsilon > 0,$$

$$(2.2.3)$$

where $\mathscr{P}(P, K, \mu, \varepsilon)$ are polynomials in the variables in (2.2.2), in ε, and, in general, may even be polynomials in the $(\mu^i)^{-1}$ as well. The role of the $i\varepsilon$ factor will be discussed shortly. The Q_l are of the form

$$Q_l = \sum_{j=1}^{n} a_{lj} k_j + \sum_{j=1}^{m} b_{lj} p_j. \qquad (2.2.4)$$

Each μ_l coincides with one of the masses μ^i in (2.2.2). It is assumed that $\mu^i > 0$ for all $i = 1, \ldots, \rho$. In our metric $Q_l^2 = \mathbf{Q}_l^2 - Q_0^2$. Because of the $i\varepsilon$ factor with $\varepsilon > 0$ (and $\mu_l > 0$), the denominators $[Q_l^2 + \mu_l^2 - i\varepsilon(\mathbf{Q}_l^2 + \mu_l^2)]$ in (2.2.3) never vanish for all $(Q_l^0, Q_l^1, Q_l^2, Q_l^3) \in \mathbb{R}^4$.

Using the relation

$$[Q_l^2 + \mu_l^2 - i\varepsilon(\mathbf{Q}_l^2 + \mu_l^2)]^{-1} = [Q_l^2 + \mu_l^2 + i\varepsilon(\mathbf{Q}_l^2 + \mu_l^2)]$$
$$\times [(Q_l^2 + \mu_l^2)^2 + \varepsilon^2(\mathbf{Q}_l^2 + \mu_l^2)^2]^{-1}, \quad (2.2.5)$$

we may rewrite (2.2.3) in the form

$$\mathscr{I}(P, K, \mu, \varepsilon) = \mathscr{I}_1(P, K, \mu, \varepsilon) + i\mathscr{I}_2(P, K, \mu, \varepsilon), \qquad (2.2.6)$$

where $\mathscr{I}_1$ and $\mathscr{I}_2$ are real and they are given as the ratio of two polynomials having the same denominator (a polynomial), which never vanishes with the $\mu^i > 0$ ($\varepsilon > 0$). In particular, we note that since a polynomial is a continuous function, this implies that $\mathscr{I}_1$, $\mathscr{I}_2$, and hence also $\mathscr{I}$, are continuous and thus measurable as functions of the variables in K with the $\mu^i > 0$ ($\varepsilon > 0$).

The role of the $i\varepsilon$ factor will become clear. To this end we consider the following expression: $|x(1 - i\varepsilon) - 1|$ with $x \geq 0$ and $\varepsilon > 0$. As a function of x, $|x(1 - i\varepsilon) - 1|$ attains its minimum when $x = (1 + \varepsilon^2)^{-1}$, yielding to $(1 - x)^2 + \varepsilon^2 x^2 \geq \varepsilon^2(1 + \varepsilon^2)^{-1}$, i.e.,

$$|x(1 - i\varepsilon) - 1|^{-1} \leq \sqrt{1 + 1/\varepsilon^2}. \qquad (2.2.7)$$

---

[1] Details on subgraphs and external and internal variables will be given in Chapter 5. These details are not needed here. The significance of (2.2.1) as a multiple integral and its absolute convergence will be discussed in Chapter 3.

We also have

$$x|x(1 - i\varepsilon) - 1|^{-1} \le (\varepsilon)^{-1}, \qquad (2.2.8)$$

and hence from (2.2.7) and (2.2.8) we may write

$$|x(1 - i\varepsilon) - 1|^{-1} \le (x + 1)^{-1}(1/\varepsilon + \sqrt{1 + 1/\varepsilon^2}). \qquad (2.2.9)$$

On the other hand,

$$|x(1 - i\varepsilon) - 1| = \sqrt{1 - 2x + x^2(1 + \varepsilon^2)} \le \sqrt{1 + 2x + x^2(1 + \varepsilon^2)}$$
$$\le \sqrt{1 + \varepsilon^2}(x + 1). \qquad (2.2.10)$$

From (2.2.9) and (2.2.10) we may then write

$$(x + 1)^{-1}(\sqrt{1 + \varepsilon^2})^{-1} \le |x(1 - i\varepsilon) - 1|^{-1}$$
$$\le (x + 1)^{-1}(1/\varepsilon + \sqrt{1 + 1/\varepsilon^2}). \qquad (2.2.11)$$

In particular, by choosing $x = (\mathbf{Q}^2 + \mu^2)/Q_0^2$, we obtain from (2.2.11) the following useful lemma ($\varepsilon > 0$):

Lemma 2.2.1

$$(Q_E^2 + \mu^2)^{-1}(\sqrt{1 + \varepsilon^2})^{-1} \le |Q^2 + \mu^2 - i\varepsilon(\mathbf{Q}^2 + \mu^2)|^{-1}$$
$$\le (Q_E^2 + \mu^2)^{-1}(1/\varepsilon + \sqrt{1 + 1/\varepsilon^2}), \qquad (2.2.12)$$

where

$$Q_E^2 = \mathbf{Q}^2 + Q_0^2, \qquad (2.2.13)$$

and denotes the Euclidean counterpart of Q^2.

For convenience, the denominators $[Q_{iE}^2 + \mu_i^2]$ will be called the Euclidean counterparts of the denominators $[Q_i^2 + \mu_i^2 - i\varepsilon(\mathbf{Q}_i^2 + \mu_i^2)]$.

We define

$$\hat{\mathscr{I}}(P, K, \mu, \varepsilon) = \mathscr{P}(P, K, \mu, \varepsilon) \Big/ \prod_{l=1}^{L} (Q_{iE}^2 + \mu_i^2), \qquad (2.2.14)$$

as obtained from (2.2.3) by replacing the denominators in $\mathscr{I}$ by their Euclidean counterparts $(Q_{iE}^2 + \mu_i^2)$, as given in (2.2.13). From the inequalities in (2.2.12), we may then find constants $\bar{G}_\varepsilon > 0$, $\underline{G}_\varepsilon > 0$, depending only on ε, with $\varepsilon > 0$, such that

$$\underline{G}_\varepsilon |\hat{\mathscr{I}}(P, K, \mu, \varepsilon)| \le |\mathscr{I}(P, K, \mu, \varepsilon)| \le \bar{G}_\varepsilon |\hat{\mathscr{I}}(P, K, \mu, \varepsilon)|, \qquad \varepsilon > 0. \qquad (2.2.15)$$

If we define

$$\hat{\mathscr{F}}_\varepsilon(P, \mu) = \int_{\mathbb{R}^{4n}} dK \, \hat{\mathscr{I}}(P, K, \mu, \varepsilon), \qquad (2.2.16)$$

then we conclude from (2.2.15), with $\varepsilon > 0$, that the absolute convergence of an integral $\hat{\mathscr{F}}_\varepsilon$ in (2.2.16) implies the absolute convergence of the integral $\mathscr{F}_\varepsilon$ in (2.2.1) and vice versa.

The polynomials $\mathscr{P}(P, K, \mu, 0)$ are Lorentz covariant. On the other hand, the presence of the denominators, for example, in (2.2.3), clearly destroys the Lorentz covariant of $\mathscr{F}_\varepsilon(P, \mu)$; therefore we have finally to take the limit $\varepsilon \to +0$ to establish the Lorentz covariance of the Feynman integrals in (2.2.1).

Finally, we wish to establish that, with $\varepsilon > 0$, the Feynman integrands $\mathscr{I}(P, K, \mu, \varepsilon)$ belong to class $B_{4n+4m+\rho}(I)$. From the right-hand side of the inequality (2.2.15) we conclude that it is sufficient to show that $\hat{\mathscr{I}}(P, K, \mu, \varepsilon)$ belong to class $B_{4n+4m+\rho}(I)$ to arrive at this conclusion.

To this end we introduce a $(4n + 4m + \rho)$-vector $\mathbf{P}$ in a Euclidean space $\mathbb{R}^{4n+4m+\rho}$ such that the $4n + 4m + \rho$ independent variables in (2.2.2) may be written as some linear combinations of the components (the standard coordinates) of $\mathbf{P}$. We rewrite the polynomial $\mathscr{P}(P, K, \mu, \varepsilon)$ in (2.2.14) [and (2.2.3)]

$$\mathscr{P}(P, K, \mu, \varepsilon) = \prod_{i=1}^{\rho} (\mu^i)^{-\sigma_i}\tilde{\mathscr{P}}(P, K, \mu, \varepsilon), \tag{2.2.17}$$

where $\tilde{\mathscr{P}}(P, K, \mu, \varepsilon)$ is a polynomial in its arguments and the σ_i are some positive integers. The polynomial $\tilde{\mathscr{P}}(P, K, \mu, \varepsilon)$ is of the general form

$$\tilde{\mathscr{P}}(P, K, \mu, \varepsilon) = \sum_i A^i_{s_i t_i u_i} k^{s_i} p^{t_i} \mu^{u_i}, \tag{2.2.18}$$

$$s_i = (s^i_{01}, \ldots, s^i_{3n}), \qquad s^i_{jk} \geq 0,$$
$$t_i = (t^i_{01}, \ldots, t^i_{3m}), \qquad t^i_{jk} \geq 0, \tag{2.2.19}$$
$$u_i = (u^i_1, \ldots, u^i_\rho), \qquad u^i_j \geq 0,$$

and

$$k^{s_i} \equiv (k^0_1)^{s^i_{01}} \cdots (k^3_n)^{s^i_{3n}},$$
$$p^{t_i} \equiv (p^0_1)^{t^i_{01}} \cdots (p^3_m)^{t^i_{3m}}, \tag{2.2.20}$$
$$\mu^{u_i} \equiv (\mu^1)^{u^i_1} \cdots (\mu^\rho)^{u^i_\rho},$$

with $A^i_{s_i t_i u_i}$ some suitable coefficients that may depend on ε. The sum over i in (2.2.18) goes over a finite number of terms.

Let $\mathbf{L}_1, \ldots, \mathbf{L}_k$ be any $k \leq 4n + 4m + \rho$ independent vectors in $\mathbb{R}^{4n+4m+\rho}$, and consider the vector $\mathbf{P}$ having the form

$$\mathbf{P} = \mathbf{L}_1 \eta_1 \eta_2 \cdots \eta_k + \mathbf{L}_2 \eta_2 \cdots \eta_k + \cdots + \mathbf{L}_k \eta_k + \mathbf{C}, \tag{2.2.21}$$

where $\eta_1, \ldots, \eta_k$ are real and positive and $\mathbf{C}$ is confined to a bounded region W in $\mathbb{R}^{4n+4m+p}$ such that $\mu^i \neq 0$ for all $i = 1, \ldots, p$. We first prove the following important lemma.

Lemma 2.2.2:   *Consider an expression of the form*

$$a_0 + a_1 x_1^{-1} + a_2 (x_1 x_2)^{-1} + \cdots + a_n (x_1 x_2 \cdots x_n)^{-1}, \qquad (2.2.22)$$

where $x_1 > 0, \ldots, x_n > 0$, the a_i are real (noninfinite) with arbitrary signs, and $a_0 \neq 0$. Then we may find constants $b_1, \ldots, b_n, m_0, M_0$:

$$b_1 > 1, \qquad \ldots, \qquad b_n > 1, \qquad (2.2.23)$$

$$M_0 \geq m_0 > 0, \qquad (2.2.24)$$

such that for

$$x_1 \geq b_1, \qquad \ldots, \qquad x_n \geq b_n, \qquad (2.2.25)$$

we have

$$m_0 \leq \xi[a_0 + a_1 x_1^{-1} + \cdots + a_n (x_1 \cdots x_n)^{-1}] \leq M_0, \qquad (2.2.26)$$

where $\xi = \operatorname{sgn} a_0$, i.e., $\xi a_0 = |a_0|$.

The proof is partly by induction. Consider the expression

$$f(x_1) = a_0 + a_1 x_1^{-1}, \qquad (2.2.27)$$

where $a_0 \neq 0$. Let A_1 and c_1 be any two numbers such that $A_1 \geq |a_1|$, $0 < c_1 < |a_0|$, and $A_1/(|a_0| - c_1) > 1$. Let $b_1 \equiv A_1/(|a_0| - c_1)$. Then

$$\xi f(x_1) = |a_0| + \xi a_1 x_1^{-1} \geq |a_0| - |a_1| x_1^{-1} \geq |a_0| - A_1 x_1^{-1} \geq c_1 > 0$$
$$(2.2.28)$$

for $x_1 \geq b_1$.

Now suppose as an induction hypothesis that for an expression

$$\xi[a_0 + a_1 x_1^{-1} + \cdots + a_{n-1}(x_1 \cdots x_{n-1})^{-1}],$$

we can find constants $b_1 > 1, \ldots, b_{n-1} > 1$, and $c_{n-1} > 0$, such that for $x_1 \geq b_1, \ldots, x_{n-1} \geq b_{n-1}$,

$$\xi[a_0 + a_1 x_1^{-1} + \cdots + a_{n-1}(x_1 \cdots x_{n-1})^{-1}] \geq c_{n-1} > 0. \quad (2.2.29)$$

Accordingly, we have the following chain of inequalities:

$$\begin{aligned}
\xi[a_0 &+ a_1 x_1^{-1} + \cdots + a_{n-1}(x_1 \cdots x_{n-1})^{-1} + a_n(x_1 \cdots x_n)^{-1}] \\
&\geq c_{n-1} + \xi a_n (x_1 \cdots x_n)^{-1} \\
&\geq c_{n-1} - |a_n|(x_1 \cdots x_{n-1})^{-1} x_n^{-1} \\
&\geq c_{n-1} - |a_n|(b_1 \cdots b_{n-1})^{-1} x_n^{-1} \\
&> c_{n-1} - |a_n| x_n^{-1}.
\end{aligned} \qquad (2.2.30)$$

Therefore we may use the chain of inequalities in (2.2.28), find two numbers A_n and c_n and that $A_n \geq |a_n|$, and $0 < c_n < c_{n-1}$, and $A_n/(c_{n-1} - c_n) > 1$, and choose $b_n \equiv A_n/(c_{n-1} - c_n)$, where we recall by the induction hypothesis that $c_{n-1} > 0$, to conclude from (2.2.30) quite generally that for $x_1 \geq b_1 > 1$, $\ldots, x_n \geq b_n > 1$,

$$\xi[a_0 + a_1 x_1^{-1} + \cdots + a_n(x_1 \cdots x_n)^{-1}] \geq c_n, \qquad (2.2.31)$$

with $c_n > 0$. This establishes the left-hand side of the inequality (2.2.26) by choosing $m_0 = c_n$.

On the other hand, for $x_i \geq b_i > 1$, $i = 1, \ldots, n$, where $b_1, \ldots, b_n$ have been introduced above, we may choose M_0 any finite constant such that

$$M_0 \geq |a_0| + |a_1|(b_1)^{-1} + \cdots + |a_n|(b_1 \cdots b_n)^{-1}, \qquad (2.2.32)$$

which establishes the right-hand-side inequality in (2.2.26). This completes the proof of the lemma. The important thing to note in this lemma is that *no matter what* the values of the real finite constants $a_1, \ldots, a_n$ are, we may always find constants $b_1 > 1, \ldots, b_n > 1$, with the conditions in (2.2.25) on $x_1, \ldots, x_n$, such that the expression in (2.2.22) may be bounded as in (2.2.26).

The $4n + 4m + \rho$ independent variables in (2.2.2) may be written as some linear combinations z_i, $i = 1, 2, \ldots, (4n + 4m + \rho)$ of the components of $\mathbf{P}$ and we note from (2.2.21), in particular, that these z_i may, in general, depend on the parameters $\eta_1, \ldots, \eta_k$, i.e., $z_i = z_i(\eta_1, \ldots, \eta_k)$. We write

$$k_1^0 = z_1, \qquad \ldots, \qquad k_n^3 = z_{4n}, \qquad p_1^0 = z_{4n+1}, \qquad \ldots, \qquad p_m^3 = z_{4n+4m},$$

$$\qquad (2.2.33)$$

$$\mu^1 = z_{4n+4m+1}, \qquad \ldots, \qquad \mu^\rho = z_{4n+4m+\rho},$$

and hence we may write for the polynomial $\widetilde{\mathscr{P}}(P, K, \mu, \varepsilon)$ in a compact and suitable notation,

$$\widetilde{\mathscr{P}}(P, K, \mu, \varepsilon) = \sum_i A_{m_i}^i \prod_{j=1}^{4n+4m+\rho} (z_j)^{m_{ij}}, \qquad (2.2.34)$$

where the m_{ij} are some nonnegative integers and the i are restricted over those for which $A_{m_i}^i \neq 0$. Finally, we introduce vectors $\mathbf{V}_1, \ldots, \mathbf{V}_{4n+4m+\rho}$ in $\mathbb{R}^{4n+4m+\rho}$ such that

$$\mathbf{V}_j \cdot \mathbf{P} = z_j, \qquad j = 1, \ldots, 4n + 4m + \rho. \qquad (2.2.35)$$

Now suppose that for j fixed,

$$\mathbf{V}_j \cdot \mathbf{L}_1 = 0, \qquad \ldots, \qquad \mathbf{V}_j \cdot \mathbf{L}_{r(j)-1} = 0, \qquad \mathbf{V}_j \cdot \mathbf{L}_{r(j)} = c_{jr} \neq 0. \qquad (2.2.36)$$

Then we have

$$\mathbf{V}_j \cdot \mathbf{P} = c_{jr}\eta_{r(j)} \cdots \eta_k + \cdots + c_{jk}\eta_k + c_j, \qquad c_{jr} \neq 0, \qquad (2.2.37)$$

where $\mathbf{V}_j \cdot \mathbf{C} = c_j$. The expression (2.2.37) may be rewritten

$$\mathbf{V}_j \cdot \mathbf{P} = \eta_{r(j)} \cdots \eta_k(c_{jr} + \cdots + c_{jk}(\eta_{r(j)} \cdots \eta_{k-1})^{-1} + c_j(\eta_{r(j)} \cdots \eta_k)^{-1}),$$

$$(2.2.38)$$

with $c_{jr} \neq 0$. Therefore we may apply the inequalities (2.2.26) of Lemma 2.2.2 to conclude that we may find constants $b_{r(j)} > 1, \ldots, b_{k(j)} > 1$, $M_0^{(j)} > 0$, $m_0^{(j)} > 0$, such that for $\eta_{r(j)} \geq b_{r(j)}, \ldots, \eta_k \geq b_{k(j)}$,

$$m_0^{(j)}\eta_{r(j)} \cdots \eta_k \leq \xi\mathbf{V}_j \cdot \mathbf{P} = |\mathbf{V}_j \cdot \mathbf{P}| \leq M_0^{(j)}\eta_{r(j)} \cdots \eta_k, \qquad (2.2.39)$$

where $\xi = \operatorname{sgn} c_{jr}$. For $1 \leq s(j) < r(j)$ we introduce any finite fixed constants $b_{s(j)} > 1$. With such a notation, we may apply the right-hand side of the inequality (2.2.39) to (2.2.34) with (2.2.35) to conclude, after summing over i in (2.2.34), that we may find a strictly positive constant C such that

$$|\tilde{\mathscr{P}}(P, K, \mu, \varepsilon)| \leq C\eta_1^{\delta_1} \cdots \eta_k^{\delta_k}, \qquad (2.2.40)$$

for

$$\eta_r \geq b'_r \equiv \max_j b_{r(j)}, \qquad r = 1, \ldots, k, \qquad (2.2.41)$$

where $j \in [1, \ldots, 4n + 4m + \rho]$. The exponent δ_r is of the form

$$\delta_r = \max_i \left(\sum_j^{(r)} m_{ij}\right), \qquad (2.2.42)$$

where the index i was introduced in (2.2.34), and the summation in (2.2.42) is over all $j \in [1, \ldots, 4n + 4m + \rho]$ for which

$$\mathbf{V}_j \cdot \mathbf{L}_1 \neq 0 \qquad \text{and/or} \qquad \mathbf{V}_j \cdot \mathbf{L}_2 \neq 0 \qquad \text{and/or} \qquad \cdots$$

$$\text{and/or} \qquad \mathbf{V}_j \cdot \mathbf{L}_r \neq 0. \quad (2.2.43)$$

The condition in (2.2.43) may be equivalently stated as requiring that $\mathbf{V}_j$ not be orthogonal to the subspace $\{\mathbf{L}_1, \ldots, \mathbf{L}_r\}$. Therefore we may write

$$\delta_r = \delta(\{\mathbf{L}_1, \ldots, \mathbf{L}_r\}). \qquad (2.2.44)$$

In obtaining the estimate (2.2.40) we have not paid attention to the signs of the coefficients $A_{m_i}^i$ in (2.2.24). Such an estimate for the *numerator* of a Feynman integrand will be sufficient in all our applications in this book, including the convergence proof of renormalization in Chapter 5. We shall denote the estimated degree δ_r of the numerator $\tilde{\mathscr{P}}$ with respect to η_r simply by $\operatorname{degr}_{\eta_r} \tilde{\mathscr{P}}$.

Now we consider the denominators in (2.2.14). To this end we may use the labeling in (2.2.33) and we may rewrite (2.2.4) in the form

$$Q_l^\sigma = \sum_{j=1}^n a_{lj} z_{4(j-1)+\sigma+1} + \sum_{j=1}^m b_{lj} z_{4n+4(j-1)+\sigma+1}. \qquad (2.2.45)$$

By writing $a_{lj} = a(l, j)$, $b_{lj} = b(l, j)$ and redefining the indices of the summations, we may rewrite (2.2.45):

$$Q_l^\sigma = \sum_{i=\sigma+1}^{4n-3+\sigma} a\left(l, \frac{i-\sigma-1}{4}+1\right) z_i + \sum_{i=4n+\sigma+1}^{4n+4m-3+\sigma} b\left(l, \frac{i-4n-\sigma-1}{4}+1\right) z_i.$$

$$(2.2.46)$$

Define

$$c(l, i, \sigma) = \begin{cases} a\left(l, \dfrac{i-\sigma-1}{4}+1\right), & \sigma+1 \le i \le 4n-3+\sigma \\[2ex] b\left(l, \dfrac{i-4n-\sigma-1}{4}+1\right), & 4n+\sigma+1 \le i \\[1ex] & \qquad \le 4n+4m-3+\sigma \end{cases}$$

$$(2.2.47)$$

and let $c(l, i, \sigma) \equiv 0$ for $i = 4n - 2 + \sigma$, $4n - 1 + \sigma$, $4n + \sigma$. With these notations we may rewrite (2.2.46) as

$$Q_l^\sigma = \sum_{i=\sigma+1}^{4n+4m-3+\sigma} c(l, i, \sigma) z_i. \qquad (2.2.48)$$

For any $l \in [1, \ldots, L]$ there exists from (2.2.33) an integer $t(l) \in [1, \ldots, \rho]$ such that

$$\mu_l = z_{4n+4m+t(l)} = \sum_{i=4n+4m+1}^{4n+4m+\rho} \delta(i - 4n - 4m, t(l)) z_i, \qquad (2.2.49)$$

where $\delta(i, j)$ is the Kronecker delta: $\delta(i, i) = 1$ and $\delta(i, j) = 0$ for $i \ne j$. Let

$$\Delta(a) = \begin{cases} 1 & \text{if} \quad a \in [0, 1, 2, 3] \\ 0 & \text{if} \quad a = 4, \end{cases} \qquad (2.2.50)$$

$$\Delta^c(a) = 1 - \Delta(a), \qquad (2.2.51)$$

and define $Q_l^4 = \mu_l$. We may then combine (2.2.48) and (2.2.49) to write

$$Q_l^a = \sum_{i=a+1}^{4n+4m-3+a} \Delta(a) c(l, i, a) z_i + \sum_{i=4n+4m-3+a}^{4n+4m+\rho-4+a} \Delta^c(a) \delta(i - 4n - 4m, t(l)) z_i$$

$$(2.2.52)$$

for $a = 0, 1, 2, 3, 4$. Finally, upon writing $d(l, i, \sigma) = c(l, i, \sigma)$, $d(l, i, 4) = \delta(i - 4n - 4m, t(l))$, we obtain the convenient expression

$$Q_l^a = \sum_{i=a+1}^{4n+4m+\rho-4+a} d(l, i, a)z_i, \tag{2.2.53}$$

or by the understanding that we restrict in (2.2.53) only those i for which $d(l, i, a) \neq 0$, we write

$$Q_l^a = \sum_i d(l, i, a)z_i = \sum_i d(l, i, a)\mathbf{V}_i \cdot \mathbf{P} \equiv \mathbf{V}(al) \cdot \mathbf{P}, \tag{2.2.54}$$

where

$$\mathbf{V}(al) = \sum_i d(l, i, a)\mathbf{V}_i. \tag{2.2.55}$$

Suppose from (2.2.36) that

$$\mathbf{V}(al) \cdot \mathbf{L}_1 = 0, \quad \ldots, \quad \mathbf{V}(al) \cdot \mathbf{L}_{s(al)-1} = 0,$$

$$\mathbf{V}(al) \cdot \mathbf{L}_{s(al)} = h(s(al), a, l) \neq 0, \tag{2.2.56}$$

where

$$h(s(al), a, l) = {\sum_i}' d(l, i, a)c_{is}, \tag{2.2.57}$$

with the sum, in general, over some subset of the set of the i contributing in (2.2.54) and

$$\mathbf{V}(al) \cdot \mathbf{C} = h(al). \tag{2.2.58}$$

Equations (2.2.54)–(2.2.58) then imply

$$Q_l^a = \eta_{s(al)} \cdots \eta_k[h(s(al), a, l) + \cdots + (\eta_{s(al)} \cdots \eta_k)^{-1}h(al)], \tag{2.2.59}$$

with $h(s(al), a, l) \neq 0$. Let

$$\xi(s(al), a, l) = \operatorname{sgn} h(s(al), a, l); \tag{2.2.60}$$

we may then apply the left-hand side of (2.2.26) of Lemma 2.2.2 to the coefficient of $\eta_{s(al)} \cdots \eta_k$ within the square brackets in (2.2.59) to conclude that we may find constants $b(s(al), a, l) > 1, \ldots, b(k, a, l) > 1, m_0(al) > 0$, such that for $\eta_{s(al)} \geq b(s(al), a, l), \ldots, \eta_k \geq b(k, a, l)$,

$$\xi(s(al), a, l)Q_l^a \geq \eta_{s(al)} \cdots \eta_k m_0(al), \tag{2.2.61}$$

or

$$(Q_l^a)^2 \geq \eta_{s(al)}^2 \cdots \eta_k^2 m_0^2(al), \tag{2.2.62}$$

with $m_0^2(al) > 0$.

Let $J(l)$ be that subset of elements in the set $J = \{0, 1, 2, 3, 4\}$ such that for any $j \in J(l)$,

$$s(jl) = \min_{a \in J} s(al) \equiv s(l). \tag{2.2.63}$$

For future reference, we write

$$J(l) = \{j_1(l), j_2(l), \ldots\}. \tag{2.2.64}$$

For $a \in J - J(l) \equiv J_c(l)$, we introduce any finite fixed constants $b(s(al) - 1, a, l) > 1, \ldots, b(s(l), a, l) > 1$, and we define

$$\hat{b}(i, l) = \max_{a \in J} b(i, a, l), \tag{2.2.65}$$

with $s(l) \le i \le k$. Then for

$$\eta_i \ge \hat{b}(i, l), \tag{2.2.66}$$

with $i = s(l), s(l) + 1, \ldots, k$, we have from (2.2.62)

$$Q_{\mathrm{IE}}^2 + \mu_l^2 = \sum_{a \in J}(Q_l^a)^2 \ge \eta_{s(l)}^2 \cdots \eta_k^2 \left[\hat{m}_0^2(l) + \sum_{a \in J_c(l)} \frac{\eta_{s(al)}^2 \cdots \eta_k^2}{\eta_{s(l)}^2 \cdots \eta_k^2} m_0^2(al) \right], \tag{2.2.67}$$

where

$$\hat{m}_0^2(l) = \sum_{a \in J(l)} m_0^2(al) > 0. \tag{2.2.68}$$

We may again apply the left-hand side of (2.2.26) now to the coefficient of $\eta_{s(l)}^2 \cdots \eta_k^2$ within the square brackets in (2.2.67) to conclude that we may find constants $b^2(s(l), l) > 1, \ldots, b^2(k, l) > 1$, which may be so chosen that $|b(i, l)| \ge \hat{b}(i, l), s(l) \le i \le k$, and $m_0^2(l) > 0$, such that for

$$\eta_i^2 \ge b^2(i, l), \qquad i = s(l), s(l) + 1, \ldots, k, \tag{2.2.69}$$

we have from (2.2.67) that

$$Q_{\mathrm{IE}}^2 + \mu_l^2 \ge \eta_{s(l)}^2 \cdots \eta_k^2 m_0^2(l). \tag{2.2.70}$$

We introduce any finite fixed constants $|b(i, l)| > 1$ for $1 \le i < s(l)$. With this notation let

$$b_r'' = \max_l |b(r, l)|, \qquad r = 1, \ldots, k, \quad l = 1, \ldots, L. \tag{2.2.71}$$

Then from (2.2.70) and (2.2.71) we have for $\eta_r \ge b_r''$, $r = 1, \ldots, k$,

$$\prod_{l=1}^{L} (Q_{\mathrm{IE}}^2 + \mu_l^2)^{-1} \le C' \eta_1^{-\gamma_1} \cdots \eta_k^{-\gamma_k}, \tag{2.2.72}$$

where $C' = \prod_{l=1}^{L} (m_0^2(l))^{-1}$. The exponents γ_r are of the form

$$\gamma_r = 2 \sum_{l=1}^{L} \Delta_l^{(r)}, \tag{2.2.73}$$

where $\Delta_l^{(r)} = 1$, if and only if, for the fixed l in question,

$$\mathbf{V}(j_1(l)l) \cdot \mathbf{L}_1 \neq 0 \qquad \text{and/or} \qquad \cdots \qquad \text{and/or} \qquad \mathbf{V}(j_1(l)l) \cdot \mathbf{L}_r \neq 0$$
$$\text{and/or} \qquad \mathbf{V}(j_2(l)l) \cdot \mathbf{L}_2 \neq 0 \qquad \text{and/or} \qquad \cdots \qquad \text{and/or}$$
$$\mathbf{V}(j_2(l)l) \cdot \mathbf{L}_r \neq 0 \qquad \text{and/or} \qquad \cdots, \tag{2.2.74}$$

with $j_1(l), j_2(l), \ldots,$ the elements in the set $J(l)$, defined in (2.2.64), and depend on l, and $\Delta_l^{(r)} \equiv 0$ otherwise. The condition in (2.2.74) may be equivalently stated as having at least one of the vectors $\mathbf{V}(j_1(l)l), \mathbf{V}(j_2(l)l), \ldots$ not orthogonal to the subspace $\{\mathbf{L}_1, \ldots, \mathbf{L}_r\}$. These vectors are defined in terms of the vectors $\mathbf{V}_i$ in (2.2.55). If for a particular l, the corresponding expression $(Q_{lE}^2 + \mu_l^2)$ is independent of the parameters $\eta_1, \ldots, \eta_k$, then $\Delta_l^{(i)} \equiv 0$ for $i = 1, \ldots, k$. In such a case we may write the bound $Q_{lE}^2 + \mu_l^2 \geq \mu_l^2$ and take $m_0^2(l) = \mu_l^2$ for the corresponding l in C'. From (2.2.74) we may also write

$$\gamma_r = \gamma(\{\mathbf{L}_1, \ldots, \mathbf{L}_r\}). \tag{2.2.75}$$

γ_r coincides with the degree of $\prod_l (Q_l^2 + \mu_l^2)$, with respect to η_r, and will be denoted by $\deg r_{\eta_r} \prod_l (Q_l^2 + \mu_l^2)$.

We may repeat an analysis for the factor

$$\prod_{i=1}^{\rho} (\mu^i)^{-\sigma_i} \equiv \prod_{j=4n+4m+1}^{4n+4m+\rho} (z_j)^{-\sigma_{j-4n-4m}}$$

in (2.2.17) in a similar way to the one leading to (2.2.70) and (2.2.72). We may then find constants b_r''' such that for $\omega_r \neq 0$ [see (2.2.78)] $b_r''' > 1$, and for $\omega_r \equiv 0$ [see (2.2.78)] we introduce any finite fixed constants $b_r''' > 1$, $r = 1, \ldots k$, such that for

$$\eta_r \geq b_r''', \tag{2.2.76}$$

$$\prod_{j=4n+4m+1}^{4n+4m+\rho} (z_j)^{-\sigma_{j-4n-4m}} \leq C'' \eta_1^{-\omega_1} \cdots \eta_k^{-\omega_k}, \tag{2.2.77}$$

with $C'' > 0$,

$$\omega_r = \sum_j^{(r)} \sigma_{j-4n-4m}, \tag{2.2.78}$$

and ω_r coincides with the degree of $\prod_{i=1}^{\rho} (\mu^i)^{+\sigma_i}$ with respect to the parameter η_r. The sum in (2.2.78) is over all those $j \in [4n + 4m + 1, \ldots, 4n + 4m + \rho]$ for which (2.2.43) is true. If (2.2.43) is not true, then $\omega_r \equiv 0$. Again we may write

$$\omega_r = \omega(\{\mathbf{L}_1, \ldots, \mathbf{L}_r\}). \tag{2.2.79}$$

Finally, from (2.2.14), (2.2.17), (2.2.40), (2.2.44), (2.2.72), (2.2.75), (2.2.77), and (2.2.79),

$$|\hat{\mathscr{I}}(P, K, \mu, \varepsilon)| \le C_0 \eta_1^{\alpha(\{\mathbf{L}_1\})} \cdots \eta_k^{\alpha(\{\mathbf{L}_1, \dots, \mathbf{L}_k\})} \tag{2.2.80}$$

for

$$\eta_r \ge b_r \equiv \max[b_r', b_r'', b_r''']. \tag{2.2.81}$$

The parameters b_r', b_r'', b_r''' are defined, respectively, through (2.2.41), (2.2.71), and (2.2.76), and $C_0 = CC'C'' > 0$,

$$\alpha(\{\mathbf{L}_1, \dots, \mathbf{L}_r\}) = \delta(\{\mathbf{L}_1, \dots, \mathbf{L}_r\}) - \gamma(\{\mathbf{L}_1, \dots, \mathbf{L}_r\}) - \omega(\{\mathbf{L}_1, \dots, \mathbf{L}_r\}).^2 \tag{2.2.82}$$

The latter is identified with $\text{degr}_{\eta_r} \hat{\mathscr{I}}(P, K, \mu, \varepsilon)$ and $\text{degr}_{\eta_r} \mathscr{I}(P, K, \mu, \varepsilon)$.[3]

With the constraints $\mu^i > 0$, $i = 1, \dots, \rho$, one cannot consistently find vectors $\mathbf{L}_1, \dots, \mathbf{L}_k$ and $\mathbf{C}$ in (2.2.21) to make any of the denominators in the expression for $\hat{\mathscr{I}}$ in (2.2.14) [and any of the ones in (2.2.3) for $\varepsilon > 0$] vanish identically and thus possibly obtain that $|\hat{\mathscr{I}}| = \infty$ (and $|\mathscr{I}| = \infty$). Also, if I is a $4n$-dimensional subspace of $\mathbb{R}^{4n+4m+\rho}$ associated with the integration variables in K, then $|\hat{\mathscr{I}}(\mathbf{P})| \ne \infty$, with $\mathbf{P} \in \mathbb{R}^{4n+4m+\rho}$, implies that for any finite vector $\mathbf{P}' \in I$, $|\hat{\mathscr{I}}(\mathbf{P} + \mathbf{P}')| \ne \infty$ [and $|\mathscr{I}(\mathbf{P} + \mathbf{P}')| \ne \infty$, $\varepsilon > 0$] for $\mu^i > 0, i = 1, \dots, \rho$.

Thus we have established that $\hat{\mathscr{I}}(P, K, \mu, \varepsilon)$, in (2.2.14), and hence from (2.2.15) also $\mathscr{I}(P, K, \mu, \varepsilon)$, in (2.2.3) for $\varepsilon > 0$, belong to the class $B_{4n+3m+\rho}(I)$ with zero logarithmic asymptotic coefficient ($\beta \equiv 0$). Application of this useful result will be made later in the book. In particular, we shall learn in the next chapter that the integrals $\mathscr{F}_\varepsilon$, if absolutely convergent, also belong to such a class with, in general, nonzero logarithmic asymptotic coefficients as well.

Finally, by replacing the Minkowski metric $g_{\mu\nu}$ by the Euclidean metric $\eta_{\mu\nu}$, diag $\eta_{\mu\nu} = [1, 1, 1, 1]$, in the expression for $\mathscr{I}$ and setting $\varepsilon = 0$ in the latter we obtain what we call the Euclidean version of a Feynman integrand $\mathscr{I}_{\mathrm{E}}$:

$$\mathscr{I}_{\mathrm{E}}(P, K, \mu) = \mathscr{P}_{\mathrm{E}}(P, K, \mu) \bigg/ \prod_{l=1}^{L} [Q_{l\mathrm{E}}^2 + \mu_l^2]. \tag{2.2.83}$$

Again the same analysis as for $\hat{\mathscr{I}}$ establishes the fact that $\mathscr{I}_{\mathrm{E}}$ also belongs to class $B_{4n+4m+\rho}(I)$ with power asymptotic coefficients, with the $\eta_r \ge b_r$ for some constants $b_r > 1$, that may be identified with $\text{degr}_{\eta_r} \mathscr{I}_{\mathrm{E}}$, and having zero

---

[2] For any two polynomials $P_1(x)$ and $P_2(x)$, we define $\text{degr}_x[P_1(x)/P_2(x)] = \text{degr}_x P_1(x) - \text{degr}_x P_2(x)$.

[3] Note that $\text{degr}_{\eta_r}[Q_l^2 + \mu_l^2 - i\varepsilon(\mathbf{Q}_l^2 + \mu_l^2)] = \text{degr}_{\eta_r}[Q_{l\mathrm{E}}^2 + \mu_l^2]$, with $\varepsilon > 0$.

logarithmic coefficients. The Euclidean version of a Feynman integral is (up to a multiplicative constant) given by

$$\mathscr{F}_E(P, \mu) = \int_{\mathbb{R}^{4n}} dK \, \mathscr{I}_E(P, K, \mu). \tag{2.2.84}$$

What should be particularly noted in this chapter is that no matter what the values of the coefficients a_{lj}, b_{lj} in (2.2.4), we may always find constants $b_r > 1, r = 1, \ldots, k$, and introduce parameters $\eta_r, r = 1, \ldots, k$, as in (2.2.21), with $\eta_r \geq b_r$, such that we may bound the absolute value of $\hat{\mathscr{I}}$ (and similarly for $\mathscr{I}$, $\varepsilon > 0$, and $\mathscr{I}_E$) as in (2.2.80), with the coefficients $\alpha(\{L_1, \ldots, L_r\})$ denoting estimated degrees of $\hat{\mathscr{I}}$ with respect to the parameters η_r. This basic result follows essentially from the key lemma, Lemma 2.2.2. The greater generality of the definition of class $B_{4n+4m+\rho}(I)$ involving logarithmic coefficients will be important when studying Feynman integrals.

NOTES

The class B_n of functions was introduced by Fink (1967, 1968), as a subclass of functions introduced by Weinberg (1960), to deal not only with the power growth of Feynman integrals but also with their logarithmic growth. The $i\varepsilon$ factor in the denominators of Feynman integrands [see (2.2.3)] was first introduced by Zimmermann (1968), and the basic inequalities in Eq. (2.2.12) are due to him as well. The analysis in this chapter, and in particular Lemma 2.2.2, is based on Manoukian (1982b), and we have slightly modified the original definition of class B_n. The unification of the masses with the momenta as components of a vector **P** was also considered by Slavnov (1974).

Chapter 3 / THE POWER-COUNTING
THEOREM AND MORE

In this chapter we discuss and prove a powerful theorem in quantum field theory that will be used frequently throughout the book. The theorem gives a convergence criterion for Feynman integrals and establishes that the integrals themselves belong to class B_n-functions. It also yields the expression for the power and the logarithmic asymptotic coefficients of the integrals in terms of the corresponding ones of the Feynman integrands.

3.1 STATEMENT OF THE THEOREM

Let $f(\mathbf{P})$ be a class $B_n(I)$-function with the latter as defined in Section 2.1, with $\mathbf{P} \in \mathbb{R}^n$. Let I be a k-dimensional subspace of $\mathbb{R}^n$. Suppose $\mathbf{L}_1, \mathbf{L}_2, \ldots, \mathbf{L}_k$ constitute a set of orthonormal vectors spanning the subspace I. We consider iterated integrals of the form

$$f_{\mathbf{L}_1 \ldots \mathbf{L}_k}(\mathbf{P}) = \int_{-\infty}^{\infty} dy_1 \int_{-\infty}^{\infty} dy_2 \cdots \int_{-\infty}^{\infty} dy_k \, f(\mathbf{P} + \mathbf{L}_1 y_1 + \cdots + \mathbf{L}_k y_k).$$

$$(3.1.1)$$

The Fubini–Tonelli theorem (Theorem 1.2.4) then states, in particular, that the absolute integrability of a measurable function $f(\mathbf{P} + \mathbf{L}_1 y_1 + \cdots + \mathbf{L}_k y_k)$ in *any* order, in particular, in the order corresponding to the integrations over $y_k, \ldots, y_1$ as shown in (3.1.1), means that the value of the integral is independent of the particular vectors $\mathbf{L}_1, \ldots, \mathbf{L}_k$ and depends only on the subspace I. That is, in an obvious notation, we have (uniquely)

$$f_{\mathbf{L}_1 \ldots \mathbf{L}_k}(\mathbf{P}) = \int_{-\infty}^{\infty} dy_1 \int_{-\infty}^{\infty} dy_2 \cdots \int_{-\infty}^{\infty} dy_k \, f(\mathbf{P} + \mathbf{L}_1 y_1 + \cdots + \mathbf{L}_k y_k)$$

$$= \int_I d^k \mathbf{P}' \, f(\mathbf{P} + \mathbf{P}') = f_I(\mathbf{P}). \tag{3.1.2}$$

Furthermore, by the application of Lemma 1.2.4, we conclude that if we translate the vector $\mathbf{P}$ in $f_I(\mathbf{P})$ by a vector $\mathbf{P}'' \in I$, then the value of the integral $f_I(\mathbf{P})$ does not change: $f_I(\mathbf{P} + \mathbf{P}'') = f_I(\mathbf{P})$. The latter means that $f_I(\mathbf{P})$ depends only on the projection of $\mathbf{P}$ along the subspace I. For concreteness and in view of applications, we introduce a complement of I in $\mathbb{R}^n$, which we may conveniently take to be the orthogonal complement of I in $\mathbb{R}^n$, and we denote it by E: $\mathbb{R}^n = I \oplus E$ and restrict the vector $\mathbf{P} \in E$. With f taken to be a renormalized Feynman interand (Section 5.2) and with the absolute convergence of f_I (Theorem 3.1.1 and Section 5.3) established, we may then rigorously justify the usual computations of renormalized Feynman amplitudes through iterated integrals of the form (3.1.1). We may identify the components of the independent external momenta and the masses associated with a (proper and connected) graph G,[1] in general, as some linear combinations of the components of the vector $\mathbf{P} \in E$. I will be the subspace associated with the class $B_n(I)$. We suppose that f in (3.1.1) is locally (i.e., over any bounded region in I) absolutely integrable. We denote the power and logarithmic asymptotic coefficients of f by $\alpha(S)$ and $\beta(S)$, respectively, with $S \subset \mathbb{R}^n$.

Theorem 3.1.1: *For all finite* $\mathbf{P}$ *such that* $\Lambda(I)\mathbf{P} = \mathbf{P}$ *and* $|f(\mathbf{P})| \neq \infty$, *if for* $f(\mathbf{P} + \mathbf{P}')$, $\mathbf{P}' \in I$,[2]
[A]

$$D_I = \max_{S' \subset I} [\alpha(S') + \dim S'] < 0, \tag{3.1.3}$$

for all nonzero subspaces $S' \subset I$, *then* $f_I(\mathbf{P})$ *is absolutely convergent.*

If [A] *is true, then*

[B] $f_I(\mathbf{P}) \in B_{n-k}$ *with power asymptotic coefficients* $\alpha_I(S)$, $S \subset E$, *given by*

$$\alpha_I(S) = \max_{\Lambda(I)S' = S} [\alpha(S') + \dim S' - \dim S]. \tag{3.1.4}$$

---

[1] Definitions of internal and external momenta and proper and connected graphs will be given in Chapter 5. These definitions will not be needed in this chapter.

[2] Recall by definition of class $B_n(I)$ that $|f(\mathbf{P} + \mathbf{P}')| \neq \infty$ for all finite $\mathbf{P}' \in I$. Also note that for finite $\mathbf{P}''$, $|f(\mathbf{P}'')| \neq \infty$ if and only if $|f(\Lambda(I)\mathbf{P}'')| \neq \infty$.

let $\mathcal{M}$ be the set of all the maximizing subspaces S' in (3.1.4), i.e., for any $S' \in \mathcal{M}$,

$$\alpha_I(S) = \alpha(S') + \dim S' - \dim S, \tag{3.1.5}$$

then $\mathcal{M}$ is a **finite** set. A subspace $S' \in \mathcal{M}$ will be called **a maximizing subspace for the I integration relative to the subspace S**.

   [C]   *The logarithmic asymptotic coefficients of f_I are given by*

$$\beta_I(S) = \max_{S' \in \mathcal{M}} \beta(S') + \sum_{i=1}^{k} p_i, \tag{3.1.6}$$

where $\mathcal{M}$ is the set of all maximizing subspace as defined in [**B**]. *The parameters p_i, for $i = 1, \ldots, k$, are called the dimension numbers and may take on the values of 0 or 1. They are defined inductively as follows.*

   Let $I_1, I_2, \ldots, I_k$ be one-dimensional subspaces of I such that $I = I_1 \oplus I_2 \oplus \cdots \oplus I_k$. Then if **all** the maximizing subspaces for the I_1 integration relative to $S \subset E$, after performing the $I_2 \oplus \cdots \oplus I_k$ integration, have the **same** dimension, then $p_1 = 0$, otherwise $p_1 = 1$. A maximizing subspace S' for the I_1 integration relative to S, after performing the $I_2 \oplus \cdots \oplus I_k$ integration is defined by

$$\alpha_I(S) = \alpha_{I_2 \oplus \cdots \oplus I_k}(S') + \dim S' - \dim S, \tag{3.1.7}$$

where $\alpha_{I_2 \oplus \cdots \oplus I_k}(S')$ is defined similarly to $\alpha_I(S)$ in (3.1.4) with $\Lambda(I)$ replaced by $\Lambda(I_2 \oplus \cdots \oplus I_k)$. Similarly, if **all** the maximizing subspaces for the I_j integration, after performing the $I_{j+1} \oplus \cdots \oplus I_k$ integration, relative to any one of the maximizing subspaces for the I_{j-1} integration, after performing the $I_j \oplus \cdots \oplus I_k$ integration, have the **same** dimension, then $p_j = 0$; otherwise $p_j = 1$.

   The dimension numbers p_j, $j = 1, \ldots, k$, may be determined with respect to any decomposition of I into one-dimensional subspaces. That is, if $I = I_1 \oplus \cdots \oplus I_k$ and $I = I'_1 \oplus \cdots \oplus I'_k$, $\dim I_i = \dim I'_i = 1$, with the I_i and I'_i, in general, different, then

$$\sum_{i=1}^{k} p_i = \sum_{i=1}^{k} p'_i, \tag{3.1.8}$$

where the p'_i correspond to the dimension numbers of the decomposition: $I = I'_1 \oplus \cdots \oplus I'_k$. In the light of Feynman integrands [i.e., $\mathscr{I}_{\mathrm{E}}$ in (2.2.83), $\hat{\mathscr{I}}$ in (2.2.14), and with $\varepsilon > 0$ for $\mathscr{I}$ in (2.2.3)], the condition $|f(\mathbf{P})| \neq \infty$ is automatically satisfied as long as $\mathbf{P}$ is restricted so that $\mu^i \neq 0$ for all $i = 1, \ldots, p$, and the theorem is then immediately applicable.

   This completes the statement of the theorem. We first prove the theorem in one dimension, i.e., when $\dim I = 1$; we then generalize the proof to an arbitrary finite number of dimensions by induction.

3.2   PROOF OF THE THEOREM IN ONE DIMENSION

In this section we prove Theorem 3.1.1 in one dimension for I, i.e., for the integral

$$f_I(\mathbf{P}) = \int_{-\infty}^{\infty} dy\, f(\mathbf{P} + \mathbf{L}y), \tag{3.2.1}$$

when $I = \{\mathbf{L}\}$, dim $I = 1$, where $|f(\mathbf{P})| \neq \infty$.

The absolute convergence criterion [A] of the theorem is easy to prove. We recall that f is assumed to be locally absolutely integrable. Let $\mathbf{P}$ be confined to a finite region W in $\mathbb{R}^n$. By definition of class $B_n(I)$ (Section 2.1), we may find a constant $b > 1$ such that for $|y| \geq b$

$$|f(\mathbf{P} + \mathbf{L}y)| \leq M(|y|)^{\alpha(\{\mathbf{L}\})} \sum_{\gamma=0}^{\beta(\{\mathbf{L}\})} (\ln|y|)^{\gamma}, \tag{3.2.2}$$

for some finite constant $M > 0$. Obviously, if

$$\alpha(\{\mathbf{L}\}) + 1 < 0, \tag{3.2.3}$$

then the integral in (3.2.1) is absolutely convergent. As the only nonnull subspace of I is I itself, we may write for (3.2.3)

$$D_I = \alpha(\{\mathbf{L}\}) + 1 < 0, \tag{3.2.4}$$

which is the statement of the criterion [A] in the theorem.

Now consider the vector $\mathbf{P} + \mathbf{L}y$ in the form

$$\mathbf{P} + \mathbf{L}y = \mathbf{L}_1\eta_1\eta_2 \cdots \eta_m + \mathbf{L}_2\eta_2 \cdots \eta_m + \cdots + \mathbf{L}_m\eta_m + \mathbf{L}y + \mathbf{C}, \tag{3.2.5}$$

where $\mathbf{C}$ is confined to a finite region in $\mathbb{R}^n$, and η_i, $i = 1, \ldots, m$, $0 \leq m \leq n - 1$, are real and positive. By definition of class $B_n(I)$, given the subspace I associated with $B_n(I)$, the condition $|f(\mathbf{P})| \neq \infty$ implies $|f(\mathbf{P} + \mathbf{L}y)| \neq \infty$.

The basic idea of the proof of the remaining part of the theorem is as follows. We replace the interval $(-\infty, \infty)$ of integration over y by a finite number of subintervals over each of which the class $B_n(I)$-property of f may be applied and the integrations over the bound of $|f|$ on each subinterval may be then explicitly carried out. Let

$$y = Z\eta_1\eta_2 \cdots \eta_m; \tag{3.2.6}$$

then we may find constants $b_0 > 1$, $b_1 > 1$, $\ldots$, $b_m > 1$, such that with

$$|Z| \geq b_0, \qquad \eta_1 \geq b_1, \qquad \ldots, \qquad \eta_m \geq b_m, \tag{3.2.7}$$

f satisfies an inequality as in (2.1.1) with asymptotic coefficients

$$\alpha(\{L\}), \qquad \alpha(\{L, L_1\}), \qquad \cdots, \qquad \alpha(\{L, L_1, \ldots, L_m\}),$$
$$\beta(\{L\}), \qquad \beta(\{L, L_1\}), \qquad \ldots, \qquad \beta(\{L, L_1, \ldots, L_m\}), \qquad (3.2.8)$$

and where

$$\mathbf{P} + \mathbf{L}y = \pm\mathbf{L}|Z|\eta_1\eta_2 \cdots \eta_m + \mathbf{L}_1\eta_1\eta_2 \cdots \eta_m + \cdots + \mathbf{L}_m\eta_m + \mathbf{C}. \quad (3.2.9)$$

Accordingly, we introduce the following pair of intervals to which y, as defined in (3.2.6), and with the restrictions in (3.2.7), may belong:

$$J^\pm = \{y : y = Z\eta_1 \cdots \eta_m, |Z| = \pm Z \geq b_0\}. \qquad (3.2.10)$$

Now we consider the region with $|y| \leq b_0\eta_1\eta_2 \cdots \eta_m$. To this end we may use the Heine–Borel theorem (Section 1.1) to cover the closed and bounded set $[-b_0, b_0]$ by a *finite* number of open intervals $(U_{i_1} + \lambda_{i_1}, U_{i_1} - \lambda_{i_1})$, where $|U_{i_1}| \leq b_0$ and we determine the λ_{i_1} consistently. The index i_1 runs over a finite number of set of integers (Section 1.1).

We write

$$y = U_{i_1}\eta_1\eta_2 \cdots \eta_m + Z\eta_2 \cdots \eta_m, \qquad (3.2.11)$$

and hence $\mathbf{P} + \mathbf{L}y$ may be written in the form

$$\mathbf{P} + \mathbf{L}y = (\mathbf{L}_1 + U_{i_1}\mathbf{L})(\eta_1/|Z|)|Z|\eta_2 \cdots \eta_m \pm \mathbf{L}|Z|\eta_2 \cdots \eta_m$$
$$+ \mathbf{L}_2\eta_2 \cdots \eta_m + \cdots + \mathbf{L}_m\eta_m + \mathbf{C}. \qquad (3.2.12)$$

Thus we may find constants $b_0(i_1) > 1, b_1(i_1) > 1, \ldots, b_m(i_1) > 1$ such that for

$$\eta_1/|Z| \geq b_1(i_1), \qquad |Z| \geq b_0(i_1),$$
$$\eta_j \geq b_j(i_1), \qquad j = 2, 3, \ldots, m, \qquad (3.2.13)$$

$f(\mathbf{P} + \mathbf{L}\eta)$ satisfies an inequality as in (2.1.1) with asymptotic coefficients

$$\alpha(i_1), \qquad \alpha(\{L, L_1, \ldots, L_j\}),$$
$$\beta(i_1), \qquad \beta(\{L, L_1, \ldots, L_j\}), \qquad (3.2.14)$$

for $j = 1, \ldots, m$, where we have used the notation $\alpha(i_1) \equiv \alpha(\{L_1 + U_{i_1}L\})$ and $\beta(i_1) \equiv \beta(\{L_1 + U_{i_1}L\})$. From the conditions (3.2.13) we may take λ_{i_1} such that

$$0 < \lambda_{i_1} < b_1^{-1}(i_1). \qquad (3.2.15)$$

Hence we may introduce another pair of intervals:

$$J_{i_1}^\pm = \{y : y = U_{i_1}\eta_1\eta_2 \cdots \eta_m + Z\eta_2 \cdots \eta_m, b_0(i_1) \leq |Z| = \pm Z \leq \eta_1\lambda_{i_1}\}, \qquad (3.2.16)$$

where λ_{i_1} satisfies (3.2.15) and $|U_{i_1}| \leq b_0$.

We may again consider the closed and bounded set $[-b_0(i_1), b_0(i_1)]$ and cover it by a finite number of open intervals, as done above, and thus continuing in this manner we may generate the following pairs of intervals:

$$J^{\pm}_{i_1,\dots,i_r} = \{y : y = U_{i_1}\eta_1 \cdots \eta_m + \cdots + U_{i_1,\dots,i_r}\eta_r \cdots \eta_m + Z\eta_{r+1} \cdots \eta_m,$$
$$b_0(i_1,\dots,i_r) \le |Z| = \pm Z \le \eta_r \lambda_{i_1,\dots,i_r}\}, \tag{3.2.17}$$

for $1 \le r \le m$, where

$$|U_{i_1,\dots,i_r}| \le b_0(i_1,\dots,i_r), \qquad 0 < \lambda_{i_1,\dots,i_r} < b_r^{-1}(i_1,\dots,i_r), \tag{3.2.18}$$

with the $b_0(i_1,\dots,i_r) > 1$ and $b_r(i_1,\dots,i_r) > 1$ some suitable constants.

Finally, we consider the interval $[-b_0(i_1,\dots,i_m), b_0(i_1,\dots,i_m)]$ and introduce the following pair of intervals:

$$J^{0\pm}_{i_1,\dots,i_m} = \{y : y = U_{i_1}\eta_1 \cdots \eta_m + \cdots + U_{i_1,\dots,i_m}\eta_m + Z,$$
$$0 \le |Z| = \pm Z \le b_0(i_1,\dots,i_m)\}. \tag{3.2.19}$$

Thus we have shown that $y \in \mathbb{R}^1$ may fall in at least one of the following intervals:

$$J^{\pm} = \{y : y = Z\eta_1 \cdots \eta_m, |Z| = \pm Z \ge b_0\},$$

$$J^{\pm}_{i_1,\dots,i_r} = \{y : y = U_{i_1}\eta_1 \cdots \eta_m + \cdots + U_{i_1,\dots,i_r}\eta_r \cdots \eta_m + Z\eta_{r+1} \cdots \eta_m,$$

$$\times \, b_0(i_1,\dots,i_r) \le |Z| = \pm Z \le \eta_r \lambda_{i_1,\dots,i_r}\}, \qquad r = 1,\dots,m,$$

$$J^{0\pm}_{i_1,\dots,i_m} = \{y : y = U_{i_1}\eta_1 \cdots \eta_m + \cdots + U_{i_1,\dots,i_m}\eta_m + Z,$$

$$\times \, 0 \le |Z| = \pm Z \le b_0(i_1,\dots,i_m)\}, \tag{3.2.20}$$

where the indices $i_1,\dots,i_m$ run over a finite number of integers.

We consider the integration of $|f|$ over y in each of the intervals in (3.2.20). We may then use the bound

$$|f_I(\mathbf{P})| \le \sum_{\pm} \int_{J^{\pm}} dy \, |f(\mathbf{P} + \mathbf{L}y)| + \sum_{\pm} \sum_{r=1}^{m} \sum_{i_1,\dots,i_r} \int_{J^{\pm}_{i_1,\dots,i_r}} dy \, |f(\mathbf{P} + \mathbf{L}y)|$$

$$+ \sum_{\pm} \sum_{i_1,\dots,i_m} \int_{J^{0\pm}_{i_1,\dots,i_m}} dy \, |f(\mathbf{P} + \mathbf{L}y)|, \tag{3.2.21}$$

where the indices $i_1,\dots,i_m$ run over a *finite* number of integers, to complete the proof of the remaining part of the theorem in one dimension. To this end note that the argument $\mathbf{P} + \mathbf{L}y$ of f corresponding to the terms in (3.2.21) is of the general form

$$(\mathbf{L}_1 + u_1\mathbf{L})\eta_1 \cdots \eta_r \eta_0 \eta_{r+1} \cdots \eta_m + \cdots + (\mathbf{L}_r + u_r\mathbf{L})\eta_r \eta_0 \eta_{r+1} \cdots \eta_m$$
$$\pm \mathbf{L}\eta_0\eta_{r+1} \cdots \eta_m + \mathbf{L}_{r+1}\eta_{r+1} \cdots \eta_m + \cdots + \mathbf{L}_m\eta_m + \mathbf{C}, \tag{3.2.22}$$

where $u_1, u_2, \ldots, u_r$ are real numbers and take on the finite set of values $U_{i_1}, \ldots, U_{i_1, \ldots, i_r}$, where $1 \leq r \leq m$. Also note that

$$L_1 + u_1 L, \qquad \ldots, \qquad L_r + u_r L, L, L_{r+1}, \qquad \ldots, \qquad L_m \quad (3.2.23)$$

are $(m + 1)$ independent vectors.

Without loss of generality and for simplicity of notation we assume that we have the following ordering of the logarithmic asymptotic coefficients corresponding to a typical term in (3.2.21):

$$\beta(\{L_1 + u_1 L\}) \leq \cdots \leq \beta(\{L_1 + u_1 L, \ldots, L_r + u_r L\}) \leq \beta(\{L, L_1, \ldots, L_r\})$$
$$\leq \cdots \leq \beta(\{L, L_1, \ldots, L_m\}). \quad (3.2.24)$$

Now we consider the contribution of each interval on the right-hand side of (3.2.21).

[I] Let $y \in J^{\pm}$. The vector $P + Ly$ is then of the form given in (3.2.9). Thus we may find constants $b_r > 1$, $r = 1, \ldots, m$, $M > 0$, such that for $\eta_r \geq b_r$,

$$\int_{J\pm} dy \, |f(P + Ly)| = \int_{b_0}^{\infty} d|Z| \, |f(\pm L|Z|\eta_1 \cdots \eta_m + \cdots$$
$$+ L_m \eta_m + C)|\eta_1 \cdots \eta_m$$
$$\leq M \eta_1^{\alpha(\{L, L_1\})} \cdots \eta_m^{\alpha(\{L, L_1, \ldots, L_m\})}$$
$$\times \sum_{\gamma_1, \ldots, \gamma_m} (\ln \eta_1)^{\gamma_1} \cdots (\ln \eta_m)^{\gamma_m}, \quad (3.2.25)$$

where the sum is over all nonnegative integers $\gamma_1, \ldots, \gamma_m$ such that

$$\sum_{i=1}^{k} \gamma_i \leq \beta(\{L, L_1, \ldots, L_k\}), \qquad 1 \leq k \leq m. \quad (3.2.26)$$

In writing the expression on the extreme right-hand side of (3.2.25) we have used the fact that the integral

$$\int_{b_0}^{\infty} |Z|^{\alpha(\{L\})} (\ln |Z|)^{\gamma_0} \, d|Z| \quad (3.2.27)$$

is, with the hypothesis $\alpha(\{L\}) + 1 < 0$, a finite constant which we have absorbed in the constant M.

[II] Let $y \in J_{i_1, \ldots, i_r}^{\pm}$, $1 \leq r \leq m$. The corresponding $P + Ly$ may be written in the form

$$P + Ly = (L_1 + U_{i_1} L)\eta_1 \cdots \eta_{r-1}(\eta_r/|Z|)|Z|\eta_{r+1} \cdots \eta_m$$
$$+ \cdots + (L_r + U_{i_1, \ldots, i_r} L)(\eta_r/|Z|)|Z|\eta_{r+1} \cdots \eta_m$$
$$\pm L|Z|\eta_{r+1} \cdots \eta_m + L_{r+1}\eta_{r+1} \cdots \eta_m + \cdots + L_m \eta_m + C.$$
$$(3.2.28)$$

Accordingly, with

$$\eta_l \geq b_l(i_1, \ldots, i_r) > 1 \qquad (l \neq r),$$

$$\eta_r/|Z| \geq b_r(i_1, \ldots, i_r) > 1,$$

$$|Z| \geq b_0(i_1, \ldots, i_r) > 1, \qquad (3.2.29)$$

$$M(i_1, \ldots, i_r) > 0,$$

$$0 < \lambda_{i_1, \ldots, i_r} < b_r^{-1}(i_1, \ldots, i_r)$$

[see definition of $J_{i_1, \ldots, i_r}^{\pm}$ in (3.2.17)], we may write

$$\int_{J_{i_1, \ldots, i_r}^{\pm}} dy \, |f(\mathbf{P} + \mathbf{L}y)|$$

$$= \int_{b_0(i_1, \ldots, i_r)}^{\eta_r \lambda_{i_1, \ldots, i_r}} d|Z| \, |f(\mathbf{P} + \mathbf{L}y)| \eta_{r+1} \cdots \eta_m$$

$$\leq M(i_1, \ldots, i_r) \eta_1^{\alpha(i_1)} \cdots \eta_{r-1}^{\alpha(i_1, \ldots, i_{r-1})} \eta_{r+1}^{\alpha(\{\mathbf{L}, \mathbf{L}_1, \ldots, \mathbf{L}_{r+1}\})} \cdots \eta_m^{\alpha(\{\mathbf{L}, \mathbf{L}_1, \ldots, \mathbf{L}_m\})}$$

$$\times \sum_{\gamma} (\ln \eta_1)^{\gamma_1} \cdots (\ln \eta_{r-1})^{\gamma_{r-1}} (\ln \eta_{r+1})^{\gamma_{r+1}} \cdots (\ln \eta_m)^{\gamma_m}$$

$$\times \int_{b_0(i_1, \ldots, i_r)}^{\eta_r \lambda_{i_1, \ldots, i_r}} d|Z| \left(\frac{\eta_r}{|Z|}\right)^{\alpha(i_1, \ldots, i_r)}$$

$$\times \left(\ln \frac{\eta_r}{|Z|}\right)^{\gamma_r} |Z|^{\alpha(\{\mathbf{L}, \mathbf{L}_1, \ldots, \mathbf{L}_r\})} (\ln |Z|)^{\gamma_0}. \qquad (3.2.30)$$

The sum in (3.2.30) is over all nonnegative integers $\gamma_1, \ldots, \gamma_r, \gamma_0, \gamma_{r+1}, \ldots, \gamma_m$ such that

$$\sum_{i=1}^{k} \gamma_i \leq \beta(i_1, \ldots, i_k), \qquad 1 \leq k \leq r,$$

$$\sum_{i=1}^{r} \gamma_i + \gamma_0 \leq \beta(\{\mathbf{L}, \mathbf{L}_1, \ldots, \mathbf{L}_r\}), \qquad (3.2.31)$$

$$\sum_{i=1}^{k} \gamma_i + \gamma_0 + \sum_{i=r+1}^{k} \gamma_i \leq \beta(\{\mathbf{L}, \mathbf{L}_1, \ldots, \mathbf{L}_k\}), \qquad r+1 \leq k \leq m.$$

In evaluating the integral on the extreme right-hand side of the inequality (3.2.30), three possibilities arise. To see this let $1 > \lambda > 0$, $b > 1$, and β, β' nonnegative integers; then for $\eta \to \infty$ we have

$$\int_b^{\eta \lambda} d|Z| \left(\frac{\eta}{|Z|}\right)^{\alpha} \left(\ln \frac{\eta}{|Z|}\right)^{\beta} |Z|^{\alpha'} (\ln |Z|)^{\beta'}$$

$$= \begin{cases} 0(\eta^{\alpha}(\ln \eta)^{\beta + \beta' + 1}) & \text{if} \quad \alpha' + 1 = \alpha \\ 0(\eta^{\alpha}(\ln \eta)^{\beta}) & \text{if} \quad \alpha' + 1 < \alpha \\ 0(\eta^{\alpha' + 1}(\ln \eta)^{\beta'}) & \text{if} \quad \alpha' + 1 > \alpha. \end{cases} \qquad (3.2.32)$$

Therefore we may find constants $c(i_1, \ldots, i_n) > 1$ and $N(i_1, \ldots, i_r) > 0$ such that

$$\int_{J^{\pm}_{i_1, \ldots, i_r}} dy \, |f(\mathbf{P} + \mathbf{L}y)| \leq M(i_1, \ldots, i_r) N(i_1, \ldots, i_r) \eta_1^{\alpha(i_1)} \cdots \eta_{r-1}^{\alpha(i_1, \ldots, i_{r-1})}$$

$$\times \, \eta_r^{\alpha_r} \eta_{r+1}^{\alpha(\{\mathbf{L}, \mathbf{L}_1, \ldots, \mathbf{L}_{r+1}\})} \cdots \eta_m^{\alpha(\{\mathbf{L}, \mathbf{L}_1, \ldots, \mathbf{L}_m\})}$$

$$\times \sum_{\gamma_1, \ldots, \gamma_m} (\ln \eta_1)^{\gamma_1} \cdots (\ln \eta_r)^{\gamma_r} \cdots (\ln \eta_m)^{\gamma_m},$$

$$(3.2.33)$$

for $\eta_l \geq b_l(i_1, \ldots, i_r)$, $l \neq r$, and $\eta_r \geq c(i_1, \ldots, i_r) > 1$. The coefficient α_r and the sums over the relabelled nonnegative integers $\gamma_1, \ldots, \gamma_m$ will now depend in which of the three categories in (3.2.32) we are. We spell out each case separately.

(a) If $\alpha(\{\mathbf{L}, \mathbf{L}_1, \ldots, \mathbf{L}_r\}) + 1 = \alpha(i_1, \ldots, i_r)$, then we have from (3.2.32) and (3.2.33),

$$\alpha_r = \alpha(i_1, \ldots, i_r) = \alpha(\{\mathbf{L}, \mathbf{L}_1, \ldots, \mathbf{L}_r\}) + 1. \qquad (3.2.34)$$

The sum in (3.2.33) is, from (3.2.31) and (3.2.32), over all nonnegative integers $\gamma_1, \ldots, \gamma_m$ such that

$$\sum_{i=1}^{k} \gamma_i \leq \beta(i_1, \ldots, i_k), \qquad 1 \leq k \leq r - 1,$$

$$\sum_{i=1}^{r} \gamma_i \leq \max[\beta(i_1, \ldots, i_r), \beta(\{\mathbf{L}, \mathbf{L}_1, \ldots, \mathbf{L}_r\})] + 1,^3 \qquad (3.2.35)$$

$$\sum_{i=1}^{k} \gamma_i \leq \beta(\{\mathbf{L}, \mathbf{L}_1, \ldots, \mathbf{L}_k\}) + 1, \qquad r + 1 \leq k \leq m.$$

(b) If $\alpha(\{\mathbf{L}, \mathbf{L}_1, \ldots, \mathbf{L}_r\}) + 1 < \alpha(i_1, \ldots, i_r)$, then from (3.2.32) we have

$$\alpha_r = \alpha(i_1, \ldots, i_r), \qquad (3.2.36)$$

and from (3.2.31) and (3.2.32), the sum in (3.2.33) is over all nonnegative integers $\gamma_1, \ldots, \gamma_m$ such that

$$\sum_{i=1}^{k} \gamma_i \leq \beta(i_1, \ldots, i_k), \qquad 1 \leq k \leq r,$$

$$\sum_{i=1}^{k} \gamma_i \leq \beta(\{\mathbf{L}, \mathbf{L}_1, \ldots, \mathbf{L}_k\}), \qquad r + 1 \leq k \leq m. \qquad (3.2.37)$$

---

[3] Note that because of the assumption on the ordering in (3.2.24) we could have written $\sum_{i=1}^{r} \gamma_i \leq \beta(\{\mathbf{L}, \mathbf{L}, \ldots, \mathbf{L}_r\}) + 1$ in (3.2.35) since

$$\max[\beta(i_1, \ldots, i_r), \beta(\{\mathbf{L}, \mathbf{L}_1, \ldots, \mathbf{L}_r\})] = \beta(\{\mathbf{L}, \mathbf{L}_1, \ldots, \mathbf{L}_r\}).$$

The form as given in (3.2.35), for $\sum_{i=1}^{r} \gamma_i$, is, however, more useful for generalization when the ordering condition in (3.2.24) is relaxed.

(c)   If $\alpha(\{\mathbf{L}, \mathbf{L}_1, \ldots, \mathbf{L}_r\}) + 1 > \alpha(i_1, \ldots, i_r)$, then from (3.2.32)

$$\alpha_r = \alpha(\{\mathbf{L}, \mathbf{L}_1, \ldots, \mathbf{L}_r\}) + 1. \tag{3.2.38}$$

From (3.2.31) and (3.2.32), the sum in (3.2.33) is now over all nonnegative integers $\gamma_1, \ldots, \gamma_m$ such that

$$\sum_{i=1}^{k} \gamma_i \le \beta(i_1, \ldots, i_k), \qquad 1 \le k \le r - 1,$$

$$\sum_{i=1}^{k} \gamma_i \le \beta(\{\mathbf{L}, \mathbf{L}_1, \ldots, \mathbf{L}_k\}), \qquad r \le k \le m. \tag{3.2.39}$$

[III]   Let $y \in J^{0\pm}_{i_1, \ldots, i_m}$. Then the vector $\mathbf{P} + \mathbf{L}y$ may be written

$$\mathbf{P} + \mathbf{L}y = (\mathbf{L}_1 + U_{i_1}\mathbf{L})\eta_1 \cdots \eta_m + \cdots + (\mathbf{L}_m + U_{i_1, \ldots, i_m}\mathbf{L})\eta_m + \mathbf{C}', \tag{3.2.40}$$

where

$$\mathbf{C}' = \mathbf{C} \pm \mathbf{L}|Z|, \qquad 0 \le |Z| \le b_0(i_1, \ldots, i_m), \tag{3.2.41}$$

and thus $\mathbf{C}'$ is confined to a finite region in $\mathbb{R}^n$. Accordingly, we may find constants $M'(i_1, \ldots, i_m) > 0$, $b_l'(i_1, \ldots, i_m) > 1$, $l = 1, \ldots, m$, such that for $\eta_l \ge b_l'(i_1, \ldots, i_m)$,

$$|f(\mathbf{P} + \mathbf{L}y)| \le M'(i_1, \ldots, i_m)\eta_1^{\alpha(i_1)} \cdots \eta_m^{\alpha(i_1, \ldots, i_m)} \sum_{\gamma} (\ln \eta_1)^{\gamma_1} \cdots (\ln \eta_m)^{\gamma_m}, \tag{3.2.42}$$

where the sum is over all nonnegative integers $\gamma_1, \ldots, \gamma_m$ such that

$$\sum_{i=1}^{k} \gamma_i \le \beta(i_1, \ldots, i_k), \qquad 1 \le k \le m. \tag{3.2.43}$$

Hence

$$\sum_{\pm} \int_{J^{0\pm}_{i_1, \ldots, i_m}} dy \, |f(\mathbf{P} + \mathbf{L}y)| \le 2b_0(i_1, \ldots, i_m)M'(i_1, \ldots, i_m)\eta_1^{\alpha(i_1)} \cdots \eta_m^{\alpha(i_1, \ldots, i_m)}$$

$$\times \sum_{\gamma} (\ln \eta_1)^{\gamma_1} \cdots (\ln \eta_m)^{\gamma_m}, \tag{3.2.44}$$

where the sum is over all nonnegative integers $\gamma_1, \ldots, \gamma_m$ satisfying (3.2.43).

Now we use the inequality (3.2.21), and the estimates in [I]–[III] to sum over the *finite* number of terms in (3.2.21) as $i_1, \ldots, i_m$ are made to vary over their corresponding sets of finite integers. Thus we may find some new constants $M > 0$, $b_1 > 1, \ldots, b_m > 1$, and take $\eta_1 \ge b_1, \ldots, \eta_m \ge b_m$. The power asymptotic coefficients $\alpha_I(\{\mathbf{L}_1, \ldots, \mathbf{L}_r\})$ of f_I are then given from (3.2.21), (3.2.25), (3.2.33)–(3.2.44):

$$\alpha_I(\{\mathbf{L}_1, \ldots, \mathbf{L}_r\}) = \max\{\alpha(\{\mathbf{L}_1 + u_1\mathbf{L}, \ldots, \mathbf{L}_r + u_r\mathbf{L}\}), \alpha(\{\mathbf{L}, \mathbf{L}_1, \ldots, \mathbf{L}_r\}) + 1\}, \tag{3.2.45}$$

where $u_1, \ldots, u_r$ take on the finite set of values $U_{i_1}, \ldots, U_{i_1, \ldots, i_r}$, and the $i_1, \ldots, i_r$ vary over a finite set of integers. Accordingly, the set of maximizing subspaces for $\alpha_I(\{\mathbf{L}_1, \ldots, \mathbf{L}_r\})$ in (3.2.45) is a finite set. According to Lemma 1.3.3, (3.2.45) may be rewritten

$$\alpha_I(S_r) = \max_{\Lambda(I)S' = S_r} [\alpha(S') + \dim S' - \dim S_r], \tag{3.2.46}$$

where $S_r \equiv \{\mathbf{L}_1, \ldots, \mathbf{L}_r\}$ and we note, in particular, $\dim S' = \dim S_r$ or $\dim S' = \dim S_r + 1$ in (3.2.46), as $\{\mathbf{L}_1 + u_1\mathbf{L}, \ldots, \mathbf{L}_r + u_r\mathbf{L}\}$ has dimension equal to $\dim S_r$ and $\{\mathbf{L}, \mathbf{L}_1, \ldots, \mathbf{L}_r\}$ has dimension equal to $(\dim S_r + 1)$. Also the set $\mathcal{M}$ [see (3.1.5)] of the maximizing subspaces for the I integration, i.e., over y, relative to S_r is a *finite* set.

Now we turn to the logarithmic asymptotic coefficients $\beta(\{\mathbf{L}_1, \ldots, \mathbf{L}_r\})$, $1 \le r \le m$. From (3.2.26), (3.2.33)–(3.2.35), (3.2.37)–(3.2.39), and (3.2.43) and the inequality (3.2.21), we obtain

(i) $$\beta_I(S_r) = \max\{\max[\beta(i_1, \ldots, i_r), \beta(\{\mathbf{L}, \mathbf{L}_1, \ldots, \mathbf{L}_r\})]\} + 1 \tag{3.2.47}$$

if

$$\alpha_I(\{\mathbf{L}_1, \ldots, \mathbf{L}_r\}) = \alpha(i_1, \ldots, i_r) = \alpha(\{\mathbf{L}, \mathbf{L}_1, \ldots, \mathbf{L}_r\}) + 1,^4 \tag{3.2.48}$$

for some $u_i, \ldots, u_r$, i.e., if all the maximizing subspaces in $\mathcal{M}$ are of the form $\{\mathbf{L}_1 + u_1\mathbf{L}, \ldots, \mathbf{L}_r + u_r\mathbf{L}\}$ *and* $\{\mathbf{L}, \mathbf{L}_1, \ldots, \mathbf{L}_r\}$, for some $u_1, \ldots, u_r$. We recall with the definition of the dimension number p in Section 3.1; $p = 1$ in this case, as the dimension of $\{\mathbf{L}_1 + u_1\mathbf{L}, \ldots, \mathbf{L}_r + u_r\mathbf{L}\}$ and $\{\mathbf{L}, \mathbf{L}_1, \ldots, \mathbf{L}_r\}$ are different (they differ by one). The $u_1, \ldots, u_r$ take on values from the finite set of values $U_{i_1}, \ldots, U_{i_r}$ and $i_1, \ldots, i_r$ vary over a finite set of positive integers, all corresponding to the maximizing subspaces for the I integration relative to S_r, i.e., corresponding to the subspaces in $\mathcal{M}$.

(ii) $$\beta_I(S_r) = \max\{\beta(i_1, \ldots, i_r)\} \tag{3.2.49}$$

if

$$\alpha_I(\{\mathbf{L}_1, \ldots, \mathbf{L}_r\}) = \alpha(i_1, \ldots, i_r) \tag{3.2.50}$$

for some $u_1, \ldots, u_r$ taking on values from the set $U_{i_1}, \ldots, U_{i_1, \ldots, i_m}$, and $\alpha(\{\mathbf{L}, \mathbf{L}_1, \ldots, \mathbf{L}_r\}) + 1 < \alpha(i_1, \ldots, i_r)$ for the corresponding $u_1, \ldots, u_r$. That

---

[4] Recall that $\alpha(i_1, \ldots, i_r)$ is of the form $\alpha(\{\mathbf{L}_1 + u_1\mathbf{L}, \ldots, \mathbf{L}_r + u_r\mathbf{L}\})$.

is, all the maximizing subspaces in $\mathcal{M}$ are of the form $\{\mathbf{L}_1 + u_1\mathbf{L}, \ldots, \mathbf{L}_r + u_r\mathbf{L}\}$ for some $u_1, \ldots, u_r$. They all have the same dimension and hence $p = 0$, by definition, in this case.

$\quad$(iii)$\quad \beta_I(S_r) = \beta(\{\mathbf{L}, \mathbf{L}_1, \ldots, \mathbf{L}_r\})$ $\qquad\qquad\qquad$ (3.2.51)

if

$$\alpha_I(\{\mathbf{L}_1, \ldots, \mathbf{L}_r\}) = \alpha(\{\mathbf{L}, \mathbf{L}_1, \ldots, \mathbf{L}_r\}) + 1 \qquad (3.2.52)$$

and $\alpha(\{\mathbf{L}_1 + u_1\mathbf{L}, \ldots, \mathbf{L}_r + u_r\mathbf{L}\}) < \alpha(\{\mathbf{L}, \mathbf{L}_1, \ldots, \mathbf{L}_r\}) + 1$ for all $u_1, \ldots, u_r$ taking values from the finite set of values $U_{i_1}, \ldots, U_{i_1, \ldots, i_r}$ with $i_1, \ldots, i_r$ varying over a finite set of positive integers. That is, $\{\mathbf{L}, \mathbf{L}_1, \ldots, \mathbf{L}_r\}$ is the only maximizing subspace in $\mathcal{M}$, and $p = 0$, by definition, in this case.

$\quad$From (i)–(iii), with the corresponding expressions (3.2.47), (3.2.49), (3.2.51), and the dimension numbers worked out above for each case, we may then write

$$\beta_I(S_r) = \max_{S' \in \mathcal{M}} \beta(S') + p, \qquad (3.2.53)$$

which coincides with (3.1.6) for $k = 1$. This completes the proof of Theorem 3.1.1 in one dimension, i.e., when $k = 1$.

3.3 PROOF OF THE THEOREM IN AN ARBITRARY FINITE NUMBER OF DIMENSIONS

$\quad$In this section we prove Theorem 3.1.1 in an arbitrary finite number of dimensions for I by induction.

3.3.1 The Absolute Convergence Criterion and the Power Asymptotic Coefficients

$\quad$We first define the following class of B_n^0-functions.

Definition of class B_n^0:$\quad$A function $f(\mathbf{P})$, with $\mathbf{P} \in \mathbb{R}^n$, is said to *belong to a class $B_n^0(I)$* if for all choices of a nonzero subspace $S \subset \mathbb{R}^n$, $m \leq n$, independent vectors $\mathbf{L}_1, \ldots, \mathbf{L}_m$ in $\mathbb{R}^n$ and a bounded region $W \subset \mathbb{R}^n$, such that $|f(\mathbf{P})| \neq \infty$, there exist coefficients $\alpha(S)$ and

$$f(\mathbf{L}_1\eta_1\eta_2\cdots\eta_m + \mathbf{L}_2\eta_2\cdots\eta_m + \mathbf{L}_m\eta_m + \mathbf{C})$$
$$= 0\left\{\eta_1^{\alpha(\{\mathbf{L}_1\})}\cdots\eta_m^{\alpha(\{\mathbf{L}_1,\ldots,\mathbf{L}_m\})} \sum_{N_1,\ldots,N_m} \sum_{\substack{\gamma_1,\ldots,\gamma_m \\ 0\leq\gamma_i\leq N_i}} (\ln\eta_1)^{\gamma_1}\cdots(\ln\eta_m)^{\gamma_m}\right\},$$

$$(3.3.1)$$

where $\eta_1, \ldots, \eta_m$ are real and positive such that $\eta_1, \ldots, \eta_m \to \infty$, independently, $C \in W$, and each of $N_1, \ldots, N_m$ is made to vary over any arbitrary finite number of nonnegative integers. We also associate with $B_n^0(I)$ a subspace $I \subset \mathbb{R}^n$ such that the condition $|f(\mathbf{P})| \neq \infty$, for some $\mathbf{P} \in \mathbb{R}^n$, implies that $|f(\mathbf{P} + \mathbf{P}')| \neq \infty$ for all finite $\mathbf{P}' \in I$. As before $\alpha(S)$ will be called a power asymptotic coefficient. Obviously we have $B_n(I) \subset B_n^0(I)$. To simplify the notation we shall denote $B_n^0(\{0\})$ simply by B_n^0.

Suppose that f is locally integrable in I. We first prove the following theorem.

Theorem 3.3.1:   *Let $f(\mathbf{P}) \in B_n(I)$, and hence also $f(\mathbf{P}) \in B_n^0(I)$, with asymptotic coefficients $\alpha(S)$ and $\beta(S)$, $S \subset \mathbb{R}^n$. For all finite $\mathbf{P}$ such that $\Lambda(I)\mathbf{P} = \mathbf{P}$ and $|f(\mathbf{P})| \neq \infty$, if for $f(\mathbf{P} + \mathbf{P}')$, $\mathbf{P}' \in I$,*

$$[A] \quad D_I \equiv \max_{S' \subset I}[\alpha(S') + \dim S'] < 0, \qquad (3.3.2)$$

then $f_I(\mathbf{P})$ is absolutely convergent.
If [A] *is true, then*

$$[B] \quad f_I(\mathbf{P}) \in B_{n-k}^0, \dim I = k, \text{ with power asymptotic coefficients } \alpha_I(S),$$
$S \subset E$:

$$\alpha_I(S) = \max_{\Lambda(I)S' = S}[\alpha(S') + \dim S' - \dim S]. \qquad (3.3.3)$$

The proof of the theorem is by induction. Suppose first $\dim I = 1$. Since $f \in B_n(I) \subset B_n^0(I)$, we may use the definition of the former class and apply the proof of Theorem 3.1.1 for $\dim I = 1$ as given in Section 3.2, word for word, corresponding to each of the terms in the summand in (3.3.1), i.e., to

$$\eta_1^{\alpha(\{\mathbf{L}_1\})} \cdots \eta_m^{\alpha(\{\mathbf{L}_1, \ldots, \mathbf{L}_m\})}(\ln \eta_1)^{\gamma_1} \cdots (\ln \eta_m)^{\gamma_m}$$

for fixed integers $\gamma_1, \ldots, \gamma_m$, to conclude after summing on $\gamma_1, \ldots, \gamma_m$ and $N_1, \ldots, N_m$ that f_I is absolutely convergent if $D_I < 0$, as given in (3.3.2), and that $f_I \in B_{n-1}^0$ with power asymptotic coefficients, as given in (3.3.3). Finally, we note that if $\dim I > 1$, then we may write $I = I' \oplus I''$, with $\dim I' = 1$. Let $f \in B_n^0(I)$. Upon writing

$$f_{I'}(\mathbf{P}) = \int_{I'} f(\mathbf{P} + \mathbf{P}')$$

symbolically, we note by definition of $f(\mathbf{P}) \in B_n^0(I)$ that $|f(\mathbf{P})| \neq \infty$ implies $|f(\mathbf{P} + \mathbf{P}')| \neq \infty$ with $\mathbf{P}' \in I' \subset I$ and $|f(\mathbf{P} + \mathbf{P}' + \mathbf{P}'')| \neq \infty$ with $\mathbf{P}'' \in I'' \subset I$. Accordingly, if (3.3.2) is true with I in the latter replaced by I', we have, by repeating the proof in Section 3.2, with $\dim I' = 1$, $|f(\mathbf{P})| \neq \infty$, that $|f_{I'}(\mathbf{P})| \neq \infty$ and $|f_{I'}(\mathbf{P} + \mathbf{P}'')| \neq \infty$ with $\mathbf{P}'' \in I'' (\subset I)$, and $f_{I'}(\mathbf{P}) \in B_{n-1}^0(I'')$.

Now we generalize the proof of Theorem 3.3.1 for dim $I > 1$. As an induction hypothesis, suppose that Theorem 3.3.1 is true whenever a subspace of integration of I has dimensionality $\leq k'$, and then we prove the theorem for dim $I = k' + 1$.

Let I be a $(k' + 1)$-dimensional subspace of $\mathbb{R}^n$. We decompose I as follows:

$$I = I_1 \oplus I_2, \tag{3.3.4}$$

where dim $I_1 = k_1$ and dim $I_2 = k_2$, with $k_1 = 1$, $k_2 = k'$, and hence $k_1 + k_2 = k' + 1$.

Let $f \in B_n(I) \subset B_n^0(I)$; write

$$f_I(\mathbf{P}) = \int_{I_1} d^{k_1}\mathbf{P}' \, f_{I_2}(\mathbf{P} + \mathbf{P}'), \tag{3.3.5}$$

and we now justify the validity of the definition of $f_I(\mathbf{P})$, given in (3.3.5), as an absolutely convergent integral by induction.

According to our *induction hypothesis*, f_{I_2} is absolutely convergent (from [A]) if

$$D_{I_2}(f) = \max_{S'' \subset I_2} [\alpha(S'') + \dim S''] < 0. \tag{3.3.6}$$

From [B], if $D_{I_2}(f) < 0$, then $f_{I_2} \in B_{n-k_2}^0(I_1)$ [with B_{n-k}^0 replaced by $B_{n-k_2}^0(I_1)$] and

$$\alpha_{I_2}(S') = \max_{\Lambda(I_2)S'' = S'} [\alpha(S'') + \dim S'' - \dim S']. \tag{3.3.7}$$

Since $f_{I_2} \in B_{n-k_2}^0(I_1)$ and dim $I_1 \leq k'$, we may again use [A], by our induction hypothesis, to conclude that f_I as given in (3.3.5) is absolutely convergent if

$$D_{I_1}(f_{I_2}) = \max_{S' \subset I_1} [\alpha_{I_2}(S') + \dim S'] < 0. \tag{3.3.8}$$

Accordingly, the integral

$$\int_{I_1} d^{k_1}\mathbf{P}' \int_{I_2} d^{k_2}\mathbf{P}'' \, f(\mathbf{P} + \mathbf{P}' + \mathbf{P}'') \tag{3.3.9}$$

is absolutely convergent if both (3.3.6) and (3.3.8) are true. Hence by the Fubini–Tonelli theorem (Theorem 1.2.4), the integral

$$f_I(\mathbf{P}) = \int_I d^{k'+1}\mathbf{P}' \, f(\mathbf{P} + \mathbf{P}') \tag{3.3.10}$$

is absolutely convergent, and its value coincides with the value of the iterated integral in (3.3.9) if

$$D_I = \max[D_{I_2}(f), D_{I_1}(f_{I_2})] < 0, \qquad (3.3.11)$$

i.e., when both (3.3.6) and (3.3.8) are true.

Now from (3.3.7) and (3.3.8) we may rewrite for the latter

$$D_{I_1}(f_{I_2}) = \max_{\Lambda(I_2)S'' \equiv S' \subset I_1} [\alpha(S'') + \dim S'']. \qquad (3.3.12)$$

We may apply Lemma 1.3.2 to (3.3.12) and (3.3.6) to rewrite the condition (3.3.11):

$$D_I = \max_{S'' \subset I}[\alpha(S'') + \dim S''] < 0, \qquad (3.3.13)$$

where we have used the fact that $I = I_1 \oplus I_2$. Equation (3.3.13) is nothing but the criterion [A] of the theorem for dim $I = k' + 1$.

Again, since dim $I_1 = 1 \le k'$ and $f_{I_2} \in B^0_{n-k_2}(I_1)$, our induction hypothesis implies

$$\alpha_I(S) = \max_{\Lambda(I_1)S' = S} [\alpha_{I_2}(S') + \dim S' - \dim S], \qquad (3.3.14)$$

where $\alpha_{I_2}(S')$ is given in (3.3.7). Accordingly, $\alpha_I(S)$ may be rewritten

$$\alpha_I(S) = \max_{\Lambda(I_1)S' = S} \left\{ \max_{\Lambda(I_2)S'' = S'} [\alpha(S'') + \dim S'' - \dim S] \right\}. \qquad (3.3.15)$$

By an immediate application of Lemma 1.3.4, we then have from (3.3.15)

$$\alpha_I(S) = \max_{\Lambda(I)S'' = S} [\alpha(S'') + \dim S'' - \dim S]. \qquad (3.3.16)$$

Therefore we have proved parts [A] and [B] of the theorem for dim $I = k' + 1$ as well, with $f_I \in B^0_{n-k'-1}$, where the logarithmic factors are treaded as in Section 3.2 since $k_1 = 1$ and $f_{I_2} \in B^0_{n-k'}(I_1)$, thus completing its proof for all finite $k = \dim I > 1$ by induction.

We complete this subsection by showing that the set $\mathcal{M}$ of all the maximizing subspaces for the I integration relative to S in (3.3.3), as defined through (3.1.5) is a finite set; i.e., there is a finite number of subspaces in $\mathcal{M}$.

To this end, as before, we decompose I into two arbitrary disjoint subspaces $I_1, I_2: I = I_1 \oplus I_2$. We reconsider Eqs. (3.3.7), (3.3.14), and (3.3.16):

$$\alpha_{I_2}(S') = \max_{\Lambda(I_2)S'' = S'} [\alpha(S'') + \dim S'' - \dim S'], \qquad (3.3.17)$$

$$\alpha_I(S) = \max_{\Lambda(I_1)S' = S} [\alpha_{I_2}(S') + \dim S' - \dim S]$$

$$= \max_{\Lambda(I)S'' = S} [\alpha(S'') + \dim S'' - \dim S], \qquad (3.3.18)$$

where

$$S \subset E, \qquad \mathbb{R}^n = I \oplus E,$$

$$S' \subset E_2, \qquad \mathbb{R}^n = I_2 \oplus E_2, \qquad (3.3.19)$$

$$S'' \subset \mathbb{R}^n.$$

Since we have already established the absolute convergence of f_I in (3.3.10), under the condition (3.3.2), we may in evaluating it use the iterated integral (3.3.9), and we may further change the orders of integration given in the latter at will. We have already established in Section 3.2, for dim $I = 1$, that $\mathscr{M}$ consists of only a finite number of (maximizing) subspaces. As an induction hypothesis suppose that the latter is also true whenever we have a subspace $I' \subset I$ with dim $I' < k$. In particular, we suppose there are only a finite number of maximizing subspaces for the I_1 integration relative to S after performing the I_2 integration. Denote the set of these maximizing subspaces by $\{S'_\mu\}_\mu$, the latter means that

$$\alpha_I(S) = \alpha_{I_2}(S'_\mu) + \dim S'_\mu - \dim S,$$

$$\Lambda(I_1)S'_\mu = S. \qquad (3.3.20)$$

For each S'_σ, suppose, again by the induction hypothesis, that there is only a finite number of maximizing subspaces for the I_2 integration relative to $S'_\sigma \in \{S'_\mu\}_\mu$. Denote the set of these maximizing subspaces by $\{S'_{\sigma\nu}\}_\nu$; then the latter means that

$$\alpha_{I_2}(S'_\sigma) = \alpha(S'_{\sigma\nu}) + \dim S'_{\sigma\nu} - \dim S'_\sigma,$$

$$\Lambda(I_2)S'_{\sigma\nu} = S'_\sigma. \qquad (3.3.21)$$

In the above, dim $I_1 < k$ and dim $I_2 < k$. We then prove the following lemma.

Lemma 3.3.1: *The set $\{S'_{\mu\nu}\}_{\mu\nu}$ is precisely the set $\mathscr{M}$ of all maximizing subspaces for the I integration relative to S, i.e.,*

$$\alpha_I(S) = \alpha(S'_{\sigma\rho}) + \dim S'_{\sigma\rho} - \dim S,$$

$$\Lambda(I)S'_{\sigma\rho} = S, \qquad (3.3.22)$$

with $S'_{\sigma\rho} \in \{S'_{\mu\nu}\}_{\mu\nu}$, and any maximizing subspace for the I integration relative to S, i.e., satisfying (3.3.22), necessarily belongs to $\{S'_{\mu\nu}\}_\nu$. Since μ and ν run over a finite set of integers, we establish, in particular, the finite property of $\mathscr{M}$ for the I integration as well by induction with dim $I = k$.

The proof of the above lemma is elementary. First we note that any $S'_{\sigma\rho} \in \{S'_{\mu\nu}\}_{\mu\nu}$ is a maximizing subspace for the I integration to S, as from (3.3.20) and (3.3.21) we have

$$\alpha_I(S) = (\alpha(S'_{\sigma\rho}) + \dim S'_{\sigma\rho} - \dim S'_\sigma) + \dim S'_\sigma - \dim S$$

$$= \alpha(S'_{\sigma\rho}) + \dim S'_{\sigma\rho} - \dim S, \tag{3.3.23}$$

and

$$\Lambda(I_1)(\Lambda(I_2)S'_{\sigma\rho}) = S$$

or

$$\Lambda(I_1 \oplus I_2)S'_{\sigma\rho} = S. \tag{3.3.24}$$

Now suppose S''_0 is a maximizing subspace for the I integration relative to S, i.e.,

$$\alpha_I(S) = \alpha(S''_0) + \dim S''_0 - \dim S, \tag{3.3.25}$$

and let

$$S'_0 = \Lambda(I_2)S''_0; \tag{3.3.26}$$

then

$$\Lambda(I_1)S'_0 = S = \Lambda(I)S''_0. \tag{3.3.27}$$

On the other hand, we note that

$$\alpha_I(S) = \max_{\Lambda(I_1)S'=S} [\alpha_{I_2}(S') + \dim S' - \dim S] \geq \alpha_{I_2}(S'_0) + \dim S'_0 - \dim S$$

$$= \max_{\Lambda(I_2)S''=S'_0} [\alpha(S'') + \dim S'' - \dim S'_0] + \dim S'_0 - \dim S$$

$$\geq \alpha(S''_0) + \dim S''_0 - \dim S = \alpha_I(S), \tag{3.3.28}$$

where in writing the last equality in (3.3.28) we have used (3.3.25). The chain of inequalities in (3.3.28) implies

$$\alpha_I(S) = \alpha_{I_2}(S'_0) + \dim S'_0 - \dim S, \tag{3.3.29}$$

$$\alpha_{I_2}(S'_0) = \alpha(S''_0) + \dim S''_0 - \dim S'_0, \tag{3.3.30}$$

as the first and the last terms in this chain are the same quantities, and hence all the inequalities in it are reduced to equalities. Equation (3.3.29) implies that S'_0 is a maximizing subspace for the I_1 integration relative to S after performing the I_2 integration. The latter then implies that $S'_0 \in \{S'_\mu\}_\mu$, by definition of $\{S'_\mu\}_\mu$ as the set of all the maximizing subspaces for the I_1 integration relative to S after performing the I_2 integration. On the other hand, for each $S'_\sigma \in \{S'_\mu\}_\mu$, we have defined $\{S'_{\mu\nu}\}_{\mu\nu}$ as the set of all the maximizing subspaces for the I_2 integration relative to S, and consequently from (3.3.20) S''_0 must be in $\{S'_{\mu\nu}\}_{\mu\nu}$, as S'_0 is in $\{S'_\mu\}_\mu$. This completes the proof of the lemma.

3.3.2   The Logarithmic Asymptotic Coefficients and the Class B_{n-k} Property of f_I

In Section 3.3.1 we have established that if $f \in B_n(I)$ ($\subset B_n^0(I)$), and $D_I < 0$ as defined in (3.3.2), then $f_I \in B_{n-k}^0$, with power asymptotic coefficients as given in (3.3.3). In this subsection we complete the proof of Theorem 3.1.1 by showing that f_I also belongs to B_{n-k} with logarithmic coefficients as given in (3.1.6).

Definition 3.3.1:   Let $I = I_1 \oplus I_2 \oplus \cdots \oplus I_k$, with dim $I_j = 1$, $j = 1$, $2, \ldots, k$, let $\{S_{\mu_1}\}_{\mu_1}$ be the set of all the maximizing subspaces for the I_1 integration relative to $S \subset E$, after performing the $I_2 \oplus \cdots \oplus I_k$ integration. Recursively, we define $\{S_{\mu_1, \ldots, \mu_j}\}_{\mu_j}$ to be the set of all the maximizing subspaces for the I_j integration, after performing the $I_{j+1} \oplus \cdots \oplus I_k$ integration, relative to $S_{\mu_1, \ldots, \mu_{j-1}}$ as one of the maximizing subspaces in $\{S_{\sigma_1, \ldots, \sigma_{j-1}}\}_{\sigma_1, \ldots, \sigma_{j-1}}$ for the I_{j-1} integration, after performing the $I_j \oplus \cdots \oplus I_k$ integration, with $j = 2, \ldots, k$. For each fixed j, $j = 2, \ldots, k$, choose any[5] subspace $S_{\mu_1, \ldots, \mu_{j-1}} \in \{S_{\sigma_1, \ldots, \sigma_{j-1}}\}_{\sigma_{j-1}}$. We may then define the dimension numbers computed relative to the chosen maximizing subspaces as follows:

$$p_j = \begin{cases} 0 & \text{if all the subspaces in } \{S_{\mu_1, \ldots, \mu_j}\}_{\mu_j} \text{ have the same dimension} \\ 1 & \text{otherwise} \end{cases}$$

(3.3.31)

for $j = 1, 2, \ldots, k$.

Obviously p_1 is computed relative to the subspace $S \subset E$.

We now prove the following important lemma.

---

[5] According to Fink (1967, 1968), the dimension numbers p_j in (3.3.31) are independent of the maximizing subspaces for the I_{j-1} integration relative to which they are computed. Accordingly, the definition of the parameters p_j as given here is consistent. Refer to Fink (1967, 1968) for a proof of this statement.

Lemma 3.3.2: *If* $I = I'_1 \oplus I'_2 \oplus \cdots \oplus I'_k$ *is any other decomposition of* I, *dim* $I'_j = 1$, $j = 1, \ldots, k$, *with corresponding dimensions numbers* $p'_1, \ldots, p'_k$, *then* $\sum_{i=1}^{k} p_i = \sum_{i=1}^{k} p'_i$, *and the logarithmic asymptotic coefficients are given as in* (3.1.6).

The proof is by induction. Let $\{S_\mu\}$ be the set of all maximizing subspaces for the I integration relative to S after performing the $I_2 \oplus \cdots \oplus I_k$ integration. Let $\{\tilde{S}_{\nu\mu}\}_\nu$ be the set of all maximizing subspaces for the $I_2 \oplus \cdots \oplus I_k$ integration relative to S_μ. According to Lemma 3.3.1, the set $\{\tilde{S}_{\nu\mu}\}_{\nu\mu}$ constitutes all of the maximizing subspaces for the I integration relative to S. As an induction hypothesis, suppose that $f_{I_2\oplus\cdots\oplus I_k} \in B_{n-k+1}(I_1)$ with logarithmic asymptotic coefficients:

$$\beta_{I_2\oplus\cdots\oplus I_k}(S_\mu) = \max_{\nu} \beta(\tilde{S}_{\nu\mu}) + \sum_{i=2}^{k} p_i. \tag{3.3.32}$$

Since dim $I_1 = 1$, we conclude from (3.2.53) that

$$\beta_I(S) = \max_{\mu} \beta_{I_2\oplus\cdots\oplus I_k}(S_\mu) + p_1, \tag{3.3.33}$$

or from (3.3.32)

$$\beta_I(S) = \max_{\nu,\mu} \beta(\tilde{S}_{\nu\mu}) + \sum_{i=1}^{k} p_i = \max_{S'' \in \mathcal{M}} \beta(S'') + \sum_{i=1}^{k} p_i, \tag{3.3.34}$$

and $f_I \in B_{n-k}$.

Finally, let $I = I'_1 \oplus \cdots \oplus I'_k$ be another decomposition of I into one-dimensional components. Let $\{T_\sigma\}_\sigma$ be the set of all maximizing subspaces for the I'_1 integration relative to S after performing the $I'_2 \oplus \cdots \oplus I'_k$ integration. Let $\{T_{\rho\sigma}\}_\rho$ be the set of all maximizing subspaces for the $I'_2 \oplus \cdots \oplus I'_k$ integration relative to T_σ. Then according to Lemma 3.3.1 we have again that $\{T_{\rho\sigma}\}_{\rho\sigma} = \mathcal{M} = \{\tilde{S}_{\nu\mu}\}_{\nu\mu}$. With $D_I < 0$, we have from Theorem 3.1.1 [A], by the application of the Fubini–Tonelli theorem (Theorem 1.2.4), that $\int_{I_1\oplus\cdots\oplus I_k} f = \int_{I'_1\oplus\cdots\oplus I'_k} f$, written in a symbolic notation. Hence we conclude (by induction as above) that

$$\beta_I(S) = \max_{\rho,\sigma} \beta(T_{\rho\sigma}) + \sum_{i=1}^{k} p'_i = \max_{\nu,\mu} \beta(\tilde{S}_{\nu\mu}) + \sum_{i=1}^{k} p'_i, \tag{3.3.35}$$

or from (3.3.34) that $\sum_{i=1}^{k} p_i = \sum_{i=1}^{k} p'_i$. This completes the proof of the lemma.

By completing the proof of the above lemma we have thus finally completed the proof of Theorem 3.1.1.

NOTES

The power-counting theorem is due to Weinberg (1960), where the convergence criterion and the expression for the power asymptotic coefficients are obtained. This work was then generalized by Fink (1967, 1968) to obtain the expression for the logarithmic asymptotic coefficients by defining in the process the class B_n of functions. We have followed their treatment quite closely, but have slightly modified the original definition of class B_n.

Chapter 4 / $\varepsilon \rightarrow +0$ LIMIT OF FEYNMAN INTEGRALS AND LORENTZ COVARIANCE

The purpose of this chapter is to establish the $\varepsilon \rightarrow +0$ limit of Feynman integrals of the type defined in Section 2.2, (2.2.1)–(2.2.4), which are absolutely convergent for $\varepsilon > 0$. This result is essential, as the presence of the factors $i\varepsilon(\mathbf{Q}_l^2 + \mu_l^2)$ in the denominators of, for example, the integrands in (2.2.3) clearly break the Lorentz covariance of the integrals. We show that the limit $\varepsilon \rightarrow +0$ of absolutely convergent Feynman integrals, (2.2.1), exists in the sense of distributions in Minkowski space. To this end we give a few definitions.

4.1 CLASS $\mathscr{S}(\mathbb{R}^k)$ FUNCTIONS (TEST FUNCTIONS)

A function $f(x)$, $x \in \mathbb{R}^k$, belongs to class (or space) $\mathscr{S}(\mathbb{R}^k)$ of functions, if $f(x)$ is infinitely continuously differentiable in x,[1] and $f(x)$ together with all its derivatives of arbitrary order vanish for $|x| \rightarrow \infty$ faster than any power of $1/|x|$, where $|x| = (\sum_{i=1}^{k} x_i^2)^{1/2}$. This class of functions is usually called the class of test functions and was introduced by Schwartz (1978). The property of an $\mathscr{S}(\mathbb{R}^k)$-function f may be precisely stated as follows. For *any* pair of k-tuples of positive integers $n = (n_1, n_2, \ldots, n_k)$, $m = (m_1, m_2, \ldots, m_k)$, we can find a real positive constant $C_{n,m}^f$ depending on n, m, and f such that

$$| x^n D^m f(x) | < C_{n,m}^f, \tag{4.1.1}$$

---

[1] Functions with the property given so far are called $\mathscr{C}^\infty$-functions.

where

$$x^n \equiv \prod_{i=1}^{k} (x_i)^{n_i}, \qquad D^m f(x) \equiv \frac{\partial^{|m|}}{\partial x_1^{m_1} \cdots \partial x_k^{m_k}} f(x), \qquad |m| \equiv \sum_{i=1}^{k} m_i. \qquad (4.1.2)$$

let $\mathscr{F}_\varepsilon(P, \mu)$ be an absolutely convergent Feynman integral of the general type defined in (2.2.1) with $\varepsilon > 0$. $P = (p_1^0, \ldots, p_m^3)$ denotes the components of the independent external momenta of a graph.

We show that for *any* $f(P) \in \mathscr{S}(\mathbb{R}^{4m})$, with

$$T_\varepsilon(f) = \int_{\mathbb{R}^{4m}} dP \, f(P) \mathscr{F}_\varepsilon(P, \mu), \qquad (4.1.3)$$

the limit $\varepsilon \rightarrow +0$ of $T_\varepsilon(f)$ exists. The existence of such a limit is called convergence of $\mathscr{F}_\varepsilon(P, \mu)$ for $\varepsilon \rightarrow +0$ as a tempered distribution.[2]

We first give a generalization of the so-called Lagrange interpolating formula.

Consider a polynomial of degree n in the real $x \in \mathbb{R}^1$,

$$P(x) = \sum_{j=0}^{n} a_j(x)^j. \qquad (4.1.4)$$

Let $x^0, x^1, \ldots, x^n$ be $n + 1$ distinct values for x. We may then introduce new coefficients $b_0, b_1, \ldots, b_n$ to rewrite (4.1.4) in the form

$$P(x) = \sum_{j=0}^{n} b_j \prod_{\substack{i=0 \\ i \neq j}}^{n} (x - x^i). \qquad (4.1.5)$$

We may solve for b_j as follows. Let $x = x^j$ in (4.1.5) for some j; then

$$b_j = P(x^j) \prod_{\substack{i=0 \\ i \neq j}}^{n} (x^j - x^i)^{-1}. \qquad (4.1.6)$$

---

[2] We shall not go into the details of tempered distributions T as the class of all continuous linear functionals on $\mathscr{S}(\mathbb{R}^k)$. By linearity it is meant that for any complex number a, $T(af) = aT(f)$, etc., and by continuity it is meant that $T(f_i) \rightarrow T(f)$ whenever $\|f - f_i\|_{n,m} \rightarrow 0$, $i \rightarrow \infty$, for all n, and m, where

$$\|f\|_{n,m} = \sup_{\substack{x \\ |n'| \leq n, |m'| \leq m}} |x^{n'} D^{m'} f(x)|, \qquad f, f_i \in \mathscr{S}(\mathbb{R}^k).$$

The class (or space) of tempered distributions is denoted $\mathscr{S}'(\mathbb{R}^k)$, to which the limit of T_ε for $\varepsilon \rightarrow +0$ in (4.1.3), if it exists, belongs with $k = 4m$. For such details see Schwartz (1978). Here it suffices to establish the existence of the limit $\varepsilon \rightarrow +0$ of $T_\varepsilon(f)$ for all $f(P) \in \mathscr{S}(\mathbb{R}^{4m})$.

Hence we may equivalently write for $P(x)$[3]

$$P(x) = \sum_{j=0}^{n} P(x^j) \prod_{\substack{i=0 \\ i \neq j}}^{n} \left(\frac{x - x^i}{x^j - x^i}\right)$$

$$\equiv \sum_{j=0}^{n} P(x^j) \sum_{i=0}^{n} (x)^i A_{ij}^{(x)}. \qquad (4.1.7)$$

The last equality defines the coefficients $A_{ij}^{(x)}$. The superscript (x) on $A_{ij}^{(x)}$ is just to remind us that these *constants* depend on the values $x^0, \ldots, x^n$ given to the variable x. Equation (4.1.7) also defines the coefficients a_j in (4.1.4) in terms of the $A_{ij}^{(x)}$ as follows:

$$a_i = \sum_{j=0}^{n} P(x^j) A_{ij}^{(x)}. \qquad (4.1.8)$$

Now consider a function $Q(Y; x_1, \ldots, x_m)$ that is a polynomial of degrees $d_1, \ldots, d_m$ in the real variables $x_1, \ldots, x_m$; $x_i \in \mathbb{R}^1$, respectively, and also depends on a variable $Y \in \mathbb{R}^k$:

$$Q(Y; x_1, \ldots, x_m) = \sum_{0 \leq l_1 \leq d_1} \cdots \sum_{0 \leq l_m \leq d_m} a_{l_1, \ldots, l_m}(Y)(x_1)^{l_1} \cdots (x_m)^{l_m}. \qquad (4.1.9)$$

Applying formula (4.1.7) to (4.1.9) for the variable x_m, we obtain

$$Q(Y; x_1, \ldots, x_m) = \sum_{j_m = 0}^{d_m} Q(Y; x_1, \ldots, x_{m-1}, x_m^{j_m}) \sum_{l_m = 0}^{d_m} (x_m)^{l_m} A_{l_m j_m}^{(x_m)}, \qquad (4.1.10)$$

where $x_m^0, \ldots, x_m^{d_m}$ denote $d_m + 1$ distinct values for x_m. Applying (4.1.7) $m - 1$ more times to (4.1.10) for the remaining variables $x_{m-1}, \ldots, x_1$, we obtain for the coefficients $a_{l_1 \ldots l_m}(Y)$ the expression

$$a_{l_1, \ldots, l_m}(Y) = \sum_{j_1 = 0}^{d_1} \cdots \sum_{j_m = 0}^{d_m} Q(Y; x_1^{j_1}, \ldots, x_m^{j_m}) A_{l_1 j_1}^{(x_1)} \cdots A_{l_m j_m}^{(x_m)}, \qquad (4.1.11)$$

where $(x_1^0, \ldots, x_m^0), \ldots, (x_1^{d_1}, \ldots, x_m^{d_m})$ denote $(d_1 + 1)(d_2 + 1) \cdots (d_m + 1)$ distinct points in $\mathbb{R}^m$. This gives a direct generalization of (4.1.8) in $\mathbb{R}^m$.

---

[3] The first equality in (4.1.7) is known as Lagrange's interpolating formula (cf. Isaacson and Keller, 1966).

Equation (4.1.11) leads to a very useful result. It says, in particular, that if $Q(Y; x_1, \ldots, x_m)$ is absolutely integrable in Y at $(d_1 + 1) \times \cdots \times (d_m + 1)$ distinct points for $(x_1, \ldots, x_m)$ in $\mathbb{R}^m$, then from

$$\int_{\mathbb{R}^k} dY |a_{l_1, \ldots, l_m}(Y)| \leq \sum_{j_1=0}^{d_1} \cdots \sum_{j_m=0}^{d_m} |A_{l_1 j_1}^{(x_1)} \cdots A_{l_m j_m}^{(x_m)}|$$

$$\times \int_{\mathbb{R}^k} dY |Q(Y; x_1^{j_1}, \ldots, x_m^{j_m})| < \infty, \qquad (4.1.12)$$

$a_{l_1, \ldots, l_m}(Y)$ is also absolutely integrable.

4.2   BASIC ESTIMATES, $\varepsilon \to +0$ LIMIT, AND LORENTZ COVARIANCE

We rewrite a Feynman integral as

$$\mathscr{F}_\varepsilon(P, \mu) = \int_{\mathbb{R}^{4m}} dK \, \mathscr{P}(P, K, \mu, \varepsilon) \bigg/ \prod_{l=1}^{L} D_l \qquad (4.2.1)$$

with

$$D_l = [Q_l^2 + \mu_l^2 - i\varepsilon(\mathbf{Q}_l^2 + \mu_l^2)]. \qquad (4.2.2)$$

The momenta Q_l are of the form

$$Q_l = k^l + q^l, \qquad (4.2.3)$$

with

$$k^l = \sum_{j=1}^{n} a_{lj} k_j, \qquad q^l = \sum_{j=1}^{m} b_{lj} p_j, \qquad (4.2.4)$$

where $K = (k_1^0, \ldots, k_n^3)$, $P = (p_1^0, \ldots, p_m^3)$.

We use the so-called Feynman parameter representation[4]:

$$\prod_{l=1}^{L} D_l^{-1} = (L - 1)! \int_{\mathscr{D}} d\alpha \left(\sum_{l=1}^{L} \alpha_l D_l \right)^{-L}, \qquad (4.2.5)$$

where

$$\mathscr{D} = \left\{ \alpha = (\alpha_1, \ldots, \alpha_L) : \alpha_l \geq 0, \sum_{l=1}^{L} \alpha_l = 1 \right\}, \qquad (4.2.6)$$

---

[4] Cf. Feynman (1949b).

which is easily proved by induction in L. Accordingly, we may rewrite $\mathcal{F}_\varepsilon(P, \mu)$:

$$\mathcal{F}_\varepsilon(P, \mu) = (L - 1)! \int_{\mathbb{R}^{4m}} dK \, \mathcal{P}(P, K, \mu, \varepsilon) \int_{\mathcal{D}} d\alpha \left(\sum_{l=1}^{L} \alpha_l D_l \right)^{-L}. \quad (4.2.7)$$

We note that

$$\sum_{l=1}^{L} \alpha_l Q_l^2 = \sum_{j, j'=1}^{n} k_j A_{jj'} k_{j'} + 2 \sum_{j=1}^{n} B_j k_j + C, \quad (4.2.8)$$

$$\sum_{l=1}^{L} \alpha_l \mathbf{Q}_l^2 = \sum_{j, j'=1}^{n} \mathbf{k}_j \cdot (A_{jj'} \mathbf{k}_{j'}) + 2 \sum_{j=1}^{n} \mathbf{B}_j \cdot \mathbf{k}_j + C_0, \quad (4.2.9)$$

where

$$B_j^\mu = \sum_{l=1}^{L} \alpha_l a_{lj} q^{l\mu}, \quad (4.2.10)$$

$$C = \sum_{l=1}^{L} \alpha_l (q^l)^2, \quad \text{ } (4.2.11)$$

$$C_0 = \sum_{l=1}^{L} \alpha_l (\mathbf{q}^l)^2, \quad (4.2.12)$$

$$A_{jj'} = \sum_{l=1}^{L} \alpha_l a_{lj} a_{lj'} = \sum_{l=1}^{L} (\sqrt{\alpha_l} a_{lj})(\sqrt{\alpha_l} a_{lj'}). \quad (4.2.13)$$

If the Feynman integral $\mathcal{F}_\varepsilon(P, \mu)$, $\varepsilon > 0$, in (4.2.7) is absolutely convergent, then we may interchange the order of the K and α integrations. To see that this may be done, consider the inequality (2.2.9),

$$|x(1 - i\varepsilon) - 1|^{-1} \le (x + 1)^{-1}(1/\varepsilon + \sqrt{1 + 1/\varepsilon^2}), \quad x \ge 0, \quad \varepsilon > 0,$$

and set

$$x = \left[\sum_{l=1}^{L} \alpha_l (\mathbf{Q}_l^2 + \mu_l^2) \middle/ \sum_{l=1}^{L} \alpha_l (Q_l^0)^2 \right], \quad (4.2.14)$$

to obtain

$$\left| \sum_{l=1}^{L} \alpha_l D_l \right|^{-1} \le \left(\sum_{l=1}^{L} \alpha_l D_{l\mathrm{E}} \right)^{-1} (1/\varepsilon + \sqrt{1 + 1/\varepsilon^2}), \quad (4.2.15)$$

where

$$D_{l\mathrm{E}} = Q_{l\mathrm{E}}^2 + \mu_l^2 \equiv \mathbf{Q}_l^2 + (Q_l^0)^2 + \mu_l^2. \quad (4.2.16)$$

Accordingly, we have

$$(L - 1)! \int_{\mathbb{R}^{4n}} dK \, |\mathscr{P}(P, K, \mu, \varepsilon)| \int_{\mathscr{D}} d\alpha \, \left| \sum_{l=1}^{L} \alpha_l D_l \right|^{-L}$$

$$\leq (L - 1)! G_\varepsilon \int_{\mathbb{R}^{4n}} dK \, |\mathscr{P}(P, K, \mu, \varepsilon)| \int_{\mathscr{D}} d\alpha \left(\sum_{l=1}^{L} \alpha_l D_{lE} \right)^{-L}$$

$$= G_\varepsilon \int_{\mathbb{R}^{4n}} dK \, |\hat{\mathscr{I}}(P, K, \mu, \varepsilon)|, \qquad (4.2.17)$$

where we have used (4.2.5), with D_l replaced by D_{lE}, to integrate over α, and the definition of the integrand $\hat{\mathscr{I}}(P, K, \mu, \varepsilon)$ given in (2.2.14), and where $G_\varepsilon > 0$ is a constant. Now we know from the right-hand side of inequality (2.2.15) that the absolute convergence of $\hat{\mathscr{F}}_\varepsilon(P, \mu)$, $\varepsilon > 0$, implies, in particular, the absolute convergence of $\mathscr{F}_\varepsilon(P, \mu)$. Hence the right-hand side of the inequality (4.2.17) is bounded, i.e.,

$$(L - 1)! \int_{\mathbb{R}^{4n}} dK \, |\mathscr{P}(P, K, \mu, \varepsilon)| \int_{\mathscr{D}} d\alpha \, \left| \sum_{l=1}^{L} \alpha_l D_l \right|^{-L} < \infty, \qquad (4.2.18)$$

and by the Fubini–Tonelli's theorem (Theorem 1.2.4) we may interchange the orders of integrations over K and α in (4.2.7) and integrate over K first if we wish.

To this end we assume that the matrix $a = [a_{lj}]$ defined in (4.2.4) is of rank n.[5] Consider the situation when $\alpha_1 > 0, \ldots, \alpha_L > 0$. Then in this case the matrix $[A_{jj'}]$ with elements defined in (4.2.13) is nonsingular. Let k' be the solution of the linear equations

$$\sum_{j=1}^{n} A_{jj'} k_j'^\mu = -B_j^\mu, \qquad (4.2.19)$$

where the B_j^μ have been defined in (4.2.10). The solution (4.2.19) corresponds to the stationary value of the expression $\sum_{l=1}^{L} \alpha_l Q_l^2 \equiv D(k)$ in (4.2.8), i.e., $(\partial/\partial k^\mu) D(k)|_{k=k'} = 0$. From (4.2.19) we may then write

$$D(k) \equiv \sum_{l=1}^{L} \alpha_l Q_l^2 = \sum_{j, j'=1}^{n} (k_j - k_j') A_{jj'} (k_{j'} - k_{j'}') + D(k'), \qquad (4.2.20)$$

$$D_0(\mathbf{k}) \equiv \sum_{l=1}^{L} \alpha_l \mathbf{Q}_l^2 = \sum_{j, j'=1}^{n} (\mathbf{k}_j - \mathbf{k}_j') \cdot A_{jj'} (\mathbf{k}_{j'} - \mathbf{k}_{j'}') + D_0(\mathbf{k}'). \qquad (4.2.21)$$

---

[5] In Chapter 5 we shall see that n of the k^l, for example, could be consistently chosen to coincide with the integration variables $k_1, \ldots, k_n$, i.e., $k^l = k_l$, $1 \leq l \leq n$, when properly labeled, and hence $a_{lj} = \delta_{lj}$ for $1 \leq l, j \leq n$. More generally, n of the k^l may be chosen to be of the form $k^l = \sum_{j=1}^{n} a_{lj} k_j$, for $1 \leq l \leq n$, with $[a_{lj}]$, $1 \leq l, j \leq n$, a nonsingular matrix, which reduces to the case just mentioned if we choose $a_{lj} = \delta_{lj}$. The remaining k^l would be of the form $k^l = \sum_{j=1}^{n} a_{lj} k_j$ with $l > n$, for some a_{lj}. The rank of the matrix $[a_{lj}]$, $1 \leq l \leq L$, $1 \leq j \leq n$, is then clearly of rank n. Note that we always have $L > n$ in (4.2.1).

We note, in particular, from the very definition of $D_0(\mathbf{k})$, that

$$D_0(\mathbf{k}) \geq 0. \tag{4.2.22}$$

Let $O = [O_{jj'}]$ be a matrix such that

$$k_j^\mu - k_j'^\mu = \sum_{j'} O_{jj'} \tilde{\mathbf{k}}_{j'}^\mu, \tag{4.2.23}$$

$$O^{\mathrm{T}} A O = \mathbb{1}, \tag{4.2.24}$$

where $\tilde{K} = (\tilde{k}_1^0, \ldots, \tilde{\mathbf{k}}_n^3)$ will define new integration variables.

We have established, through (4.2.17) and (4.2.18), by the application of the Fubini–Tonelli theorem, that we may interchange the orders of integration over K and $\alpha \in \mathscr{D}$ in (4.2.7); hence with $\tilde{K}$ as new integration variables we obtain from (4.2.7)

$$\mathscr{F}_\varepsilon(P, \mu) = (L - 1)! \int_\mathscr{D} d\alpha \int_{\mathbb{R}^{4n}} dK \, \mathscr{P}(P, K, \mu, \varepsilon) \left(\sum_{l=1}^L \alpha_l D_l \right)^{-L}$$

$$= (L - 1)! \int_{\mathscr{D}-\mathscr{E}} \frac{d\alpha}{(\det A)^2} \int_{\mathbb{R}^{4n}} d\tilde{K} \, \frac{\mathscr{P}(P, K' + O\tilde{K}, \mu, \varepsilon)}{V^L}, \tag{4.2.25}$$

where $\tilde{K} = (\tilde{k}_1^0, \ldots, \tilde{k}_n^3)$, $\mathscr{E}$ (a set of measure zero) is the set of all $\alpha \in \mathscr{D}$ for which at least one of the $\alpha_i \in \mathscr{D}$, $1 \leq i \leq L$, vanishes, and from (4.2.2), (4.2.5), (4.2.20)–(4.2.24)

$$V = (1 - i\varepsilon) \sum_{j=1}^n \tilde{\mathbf{K}}_j^2 - \sum_{j=1}^n (K_j^0)^2 + D(k') - i\varepsilon D_0(\mathbf{k}') + (1 - i\varepsilon) \sum_{l=1}^L \alpha_l \mu_l^2. \tag{4.2.26}$$

From (4.2.19), (4.2.8), (4.2.9), and definitions (4.2.10)–(4.2.13), we see that $D(k')$ and $D_0(\mathbf{k}')$ are some quadratic forms in p and $\mathbf{p}$ respectively,[6] i.e.,

$$D(k') = \sum_{j,j'=1}^m p_j U_{jj'}(\alpha) p_{j'} \equiv p U p, \tag{4.2.27}$$

$$D_0(\mathbf{k}') = \sum_{j,j'=1}^m \mathbf{p}_j \cdot (U_{jj'}(\alpha) \mathbf{p}_{j'}) \equiv \mathbf{p} \cdot (U\mathbf{p}), \tag{4.2.28}$$

where the $U_{jj'}(\alpha)$ are rational in α. We note, in particular, directly from (4.2.22) that

$$\mathbf{p} \cdot (U\mathbf{p}) \geq 0. \tag{4.2.29}$$

---

[6] There is a very long history in the literature on quantum field theory of matrices of the form in (4.2.27) and (4.2.28); cf. Todorov (1971), and references therein.

The matrix $U = [U_{jj'}]$, with $U_{jj'}$ as a rational function in α, originally defined for $\alpha \in \mathscr{D} - \mathscr{E}$ may be extended to all $\alpha \in \mathscr{D}$. To this end we introduce real variables $x_1, \ldots, x_n, y_1, \ldots, y_m$, and by analogy with Q_l in (4.2.3) define

$$z_l = \sum_{j=1}^{n} a_{lj} x_j + \sum_{j=1}^{m} b_{lj} y_j \qquad (4.2.30)$$

with the a_{lj}, b_{lj} introduced in (4.2.4). Consider the quadratic form

$$G(x, y, \alpha) = \sum_{l=1}^{L} \alpha_l z_l^2 \qquad (4.2.31)$$

defined by analogy with $D_0(\mathbf{k})$ in (4.2.21). We may define a matrix $U = [U_{ij}(\alpha)]$ by

$$\inf_x G(x, y, \alpha) = \sum y_i U_{ij}(\alpha) y_j, \qquad (4.2.32)$$

which is continuous in α for all $\alpha \in \mathscr{D}$ and coincides with the coefficients in (4.2.27) and (4.2.28) for $\alpha_i > 0$, $i = 1, \ldots, L$. Accordingly, the matrix $U = [U_{ij}(\alpha)]$ in (4.2.27) and (4.2.28) may be extended to one with continuous elements for all $\alpha \in \mathscr{D}$.

Since $\mathscr{P}(P, K' + O\tilde{K}, \mu, \varepsilon)$ is, in particular, a polynomial in the $\tilde{k}_j^\mu$, the integration over $\tilde{K}$ in (4.2.25) may be explicitly carried out in a standard manner,[7] which yields an expression of the form (up to overall multiplicative constants):

$$\int_{\mathbb{R}^{4n}} dK\, \mathscr{P}(P, K, \mu, \varepsilon) \Big/ \left(\sum_{l=1}^{L} \alpha_l D_l \right)^L = \int_{\mathbb{R}^{4n}} d\tilde{K}\, \frac{\mathscr{P}(P, K' + O\tilde{K}, \mu, \varepsilon)}{(\det A)^2 V^L}$$

$$= \text{const}\ (1 - i\varepsilon)^{-(3n/2)+r_0} N(\alpha, P, \mu, \varepsilon) F_\varepsilon(\alpha, P)^{-t}, \qquad (4.2.33)$$

or

$$\mathscr{F}_\varepsilon(P, \mu) = \text{const}\ (1 - i\varepsilon)^{-(3n/2+r_0)} \int_{\mathscr{D}} d\alpha\, N(\alpha, P, \mu, \varepsilon) F_\varepsilon(\alpha, P)^{-t}, \qquad (4.2.34)$$

where

$$F_\varepsilon(\alpha, P) = pUp + M^2 - i\varepsilon(\mathbf{p} \cdot U\mathbf{p} + M^2), \qquad (4.2.35)$$

$$M^2 = \sum_{l=1}^{L} \alpha_l \mu_l^2, \qquad (4.2.36)$$

r_0 is the largest integer $\leq d_P/2$, where d_P is the degree of the polynomial $\mathscr{P}$ in $\tilde{K}$, and

$$t = L - 2n - r_0. \qquad (4.2.37)$$

---

[7] Cf. Jauch and Rohrlich (1976), Appendix A5.

$N(\alpha, P, \mu, \varepsilon)$ is a polynomial in P, μ, ε and in the μ_l^{-1}, in general, as well and is rational in α. We note, in particular, that $\mathscr{P}(P, K, \mu, 0)$ is Lorentz covariant, and hence so is $N(\alpha, P, \mu, 0)$.

Lemma 4.2.1

(i) $\displaystyle \int_{\mathscr{D}} d\alpha \, N(\alpha, P, \mu, \varepsilon)$ (4.2.38)

and

(ii) $\displaystyle \int_{\mathbb{R}^{4m}} dP \, f(P) \int_{\mathscr{D}} d\alpha \, N(\alpha, P, \mu, \varepsilon) F_\varepsilon(\alpha, P)^{-t},$ (4.2.39)

for all $f(P) \in \mathscr{S}(\mathbb{R}^{4m})$, are absolutely convergent for $\varepsilon > 0$.

From the elementary inequality (2.2.10), with $x = (\mathbf{p} \cdot U\mathbf{p} + M^2)/(p^0 U p^0)$, we have from (4.2.35)

$$|F_\varepsilon(\alpha, P)| \leq \sqrt{1 + \varepsilon^2} \, (p_E U p_E + M^2),$$ (4.2.40)

where

$$p_E U p_E = \sum_{j, j' = 1}^{m} p_{jE} U_{jj'}(\alpha) p_{j'E} \geq 0.$$ (4.2.41)

Since the $U_{jj'}(\alpha)$ are continuous on $\mathscr{D}$, we may find a strictly positive constant H (cf. Lemma 1.1.1), independent of α, such that

$$(p_E U p_E + M^2) \leq H\left(1 + \sum_{i=1}^{m} \left(\frac{p_{iE}^2}{\bar{\mu}^2}\right)\right),$$ (4.2.42)

where

$$\bar{\mu} = \max_l \mu_l.$$ (4.2.43)

From (4.2.40) and (4.2.42), we then have

$$1 \leq H\left[1 + \sum_{i=1}^{m} \left(\frac{p_{iE}^2}{\bar{\mu}^2}\right)\right]\sqrt{1 + \varepsilon^2} |F_\varepsilon(\alpha, P)|^{-1}.$$ (4.2.44)

Accordingly, we have for (i)

$$\int_{\mathscr{D}} d\alpha \, |N(\alpha, P, \mu, \varepsilon)| \leq H^t\left[1 + \sum_{i=1}^{n} \frac{p_{iE}^2}{\bar{\mu}^2}\right]^t (1 + \varepsilon^2)^{t/2}$$

$$\times \int_{\mathscr{D}} d\alpha \, |N(\alpha, P, \mu, \varepsilon)| \, |F_\varepsilon(\alpha, P)|^{-t}.   (4.2.45)$$

The integral on the right-hand side of the inequality (4.3.45) may be majorized from (4.2.33) as

$$\int_{\mathscr{D}} d\alpha \, |N(\alpha, P, \mu, \varepsilon)| \, |F_\varepsilon(\alpha, P)|^{-t} \leq \text{const} \int_{\mathscr{D}} d\alpha \int_{\mathbb{R}^{4n}} dK \, |\mathscr{P}(P, K, \mu, \varepsilon)|$$

$$\times \left(\sum_{l=1}^{L} \alpha_l D_{lE} \right)^{-L} < \infty, \qquad (4.2.46)$$

as the absolute convergence of (4.2.1) implies, from (2.2.15), the absolute convergence of the integral on the right-hand side of (4.2.46). Equations (4.2.45) and (4.2.46) then establish the absolute convergence of the integral in (i).

To establish the absolute convergence of the integral in (ii) we write

$$N(\alpha, P, \mu, \varepsilon) = \sum_{a, m'} \varepsilon^a p^{m'} N_{a, m'}(\alpha, \mu), \qquad (4.2.47)$$

in a notation similar to the one in (2.2.18). The generalized Lagrange interpolating formula, in particular, inequality (4.1.12), then establishes from (i) that $\int_{\mathscr{D}} d\alpha \, N_{a, m'}(\alpha, \mu)$ is absolutely convergent. Finally, we use the inequality

$$|F_\varepsilon(\alpha, P)| \geq \varepsilon(\mathbf{p} \cdot U\mathbf{p} + M^2) \geq \varepsilon \underline{\mu}^2, \qquad (4.2.48)$$

where

$$\underline{\mu} = \min_l \mu_l > 0, \qquad (4.2.49)$$

to derive the following inequality for $\varepsilon > 0$:

$$\int_{R^{4m}} dP \, |f(P)| \int_{\mathscr{D}} d\alpha \, |N(\alpha, P, \mu, \varepsilon)| \, |F_\varepsilon(\alpha, P)|^{-t}$$

$$\leq G_\varepsilon \sum_{a, m'} \varepsilon^a \int_{\mathbb{R}^{4m}} dP \, |f(P)| \, |p^{m'}| \int_{\mathscr{D}} d\alpha \, |N_{a, m'}(\alpha, \mu)| < \infty, \quad (4.2.50)$$

where G_ε is a positive constant depending on ε and we have used the fact that $f(P)p^{m'} \in \mathscr{S}(\mathbb{R}^{4m})$ and that $\int d\alpha \, N_{a, m}(\alpha, \mu)$ is absolutely convergent.[8] This completes the proof of the lemma.

Inequality (4.2.50) establishes the absolute convergence of $T_\varepsilon(f)$ in (4.1.3) for $\varepsilon > 0$, and, in particular it implies that to study the $\varepsilon \to +0$ limit of $T_\varepsilon(f)$ it suffices to study the limit $\varepsilon \to +0$ of the integral

$$I_{\varepsilon am'}(h) = \int_{\mathbb{R}^{4m}} dP \int_{\mathscr{D}} d\alpha \, h(P) N_{a, m'}(\alpha, \mu) F_\varepsilon(\alpha, P)^{-t}, \qquad (4.2.51)$$

---

[8] That $\int_{\mathbb{R}^{4m}} dP \, |f(P)| \, |p^{m'}| < \infty$ follows from the very definition of $fp^{m'}$ as a function in $\mathscr{S}(\mathbb{R}^{4m})$. The latter property (see 4.1.1), in particular, states that we may find an arbitrary positive integer N and a constant $C > 0$ such that $|f(P)p^{m'}| < C[1 + |p|^2]^{-N}$. By choosing N sufficiently large, the integral $\int_{\mathbb{R}^{4m}} dP \, [1 + |p|^2]^{-N}$ then obviously exists.

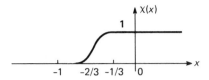

Fig. 4.1   Sketch of the function $\chi(x)$.

where $h(P) \in \mathscr{S}(\mathbb{R}^{4m})$ since $f(P)p^{m'} \in \mathscr{S}(R^{4m})$, as $T_\varepsilon(f)$ is a linear combination of integrals of the form in (4.2.51) with, in particular, coefficients that have well-defined limits for $\varepsilon \to +0$.

To study the $\varepsilon \to +0$ of (4.2.51) we introduce a function $0 \le \chi(x) \le 1$ defined by

$$\chi(x) = \begin{cases} 0 & \text{for } x < -\tfrac{2}{3} \\ 1 - \exp\{-(3x+2)^2/[1-(3x+2)^2]\} & \text{for } -\tfrac{2}{3} \le x < -\tfrac{1}{3} \\ 1 & \text{for } -\tfrac{1}{3} \le x, \end{cases}$$

(4.2.52)

which together with all its derivatives vanishes for $x < -\tfrac{2}{3}$. The function $\chi(x)$ is sketched in Fig. 4.1. By setting $x = pUp/\mu^2$, we may rewrite for (4.2.51),

$$I_{\varepsilon am'}(h) = I^1_{\varepsilon am'}(h) + I^2_{\varepsilon am'}(h),$$

(4.2.53)

where

$$I^1_{\varepsilon am'}(h) = \int_{\mathbb{R}^{4m}} dP \int_{\mathscr{D}} d\alpha \, h(P) \chi\left(\frac{pUp}{\mu^2}\right) N_{a,m'}(\alpha, \mu) F_\varepsilon(\alpha, P)^{-t},$$

(4.2.54)

$$I^2_{\varepsilon am'}(h) = \int_{\mathbb{R}^{4m}} dP \int_{\mathscr{D}} d\alpha \, h(P) \left[1 - \chi\left(\frac{pUp}{\mu^2}\right)\right] N_{a,m'}(\alpha, \mu) F_\varepsilon(\alpha, P)^{-t}.$$

(4.2.55)

Because of the χ factor multiplying the integrand in (4.2.54) we have, for P contributing to this integral,

$$pUp + M^2 \ge pUp + \underline{\mu}^2 \ge (-\tfrac{2}{3} + 1)\underline{\mu}^2 = \underline{\mu}^2/3,$$

(4.2.56)

i.e., $pUp + M^2$ never vanishes for the integral $I^1_{\varepsilon am'}(h)$. Also

$$|F_\varepsilon(\alpha, P)|^{-1} \le |F_0(\alpha, P)|^{-1} \le (3/\underline{\mu}^2),$$

(4.2.57)

where we have used (4.2.56). Accordingly we have

$$|I^1_{\varepsilon am'}(h)| \le \int_{\mathbb{R}^{4m}} dP \int_{\mathscr{D}} d\alpha \, |h(P)| \chi\left(\frac{pUp}{\mu^2}\right) |N_{a,m'}(\alpha, \mu)| |F_0(\alpha, P)|^{-t}$$

$$\le \text{const} \int_{\mathbb{R}^{4m}} dP |h(P)| \int_{\mathscr{D}} d\alpha \, |N_{a,m'}(\alpha, \mu)| < \infty,$$

(4.2.58)

where we have used the fact that $\chi(x) \leq 1$ and Lemma 4.2.1(i). Accordingly, from the Lebesgue dominated convergence theorem [Theorem 1.2.2(ii)],

$$\lim_{\varepsilon \to +0} I^1_{\varepsilon am'}(h) = I^1_{0am'}(h) = \int_{\mathbb{R}^{4m}} dP \int_{\mathscr{D}} d\alpha \, h(P)\chi\left(\frac{pUp}{\mu^2}\right)N_{a,m'}(\alpha, \mu)F_0(\alpha, P)^{-t},$$

(4.2.59)

and the latter exists.

The study of $I^2_{\varepsilon am'}$, for $\varepsilon \to +0$, is more difficult. To this end we note that we may write

$$F_\varepsilon(\alpha, P)^{-t} = -\frac{(-\frac{1}{2})^t}{(t-1)!}\left[(pUp)_\varepsilon^{-1}\sum_{j=1}^m p_j^\mu \frac{\partial}{\partial p_j^\mu}\right]^t \ln F_\varepsilon(\alpha, P),$$

(4.2.60)

whose validity is easily verified, where $(pUp)_\varepsilon \equiv pUp - i\varepsilon \mathbf{p} \cdot U\mathbf{p}$. Substituting the expression (4.2.60) into $I^2_{\varepsilon am'}(h)$, integrating by parts, and using the vanishing property of $h(p)$. together with all its derivatives, at infinity, we obtain for $I^2_{\varepsilon am'}(h)$ explicitly

$$I^2_{\varepsilon am'}(h) = -\frac{(\frac{1}{2})^t}{(t-1)!}\int_{\mathbb{R}^{4m}} dP \int_{\mathscr{D}} d\alpha \, N_{am'}(\alpha, \mu)(\ln F_\varepsilon(\alpha, P))$$

$$\times \left\{\left[\sum_{j=1}^m \frac{\partial}{\partial p_j^\mu}\frac{p_j^\mu}{(pUp)_\varepsilon}\right]^t h(P)\left[1 - \chi\left(\frac{pUp}{\mu^2}\right)\right]\right\}.$$

(4.2.61)

It is easy to see that the expression in the curly brackets may be written in the form of a finite sum of terms each of which is of the form of some non-negative power of ε multiplied by a term of a finite sum of the form

$$(pUp)_\varepsilon^{-t}\sum_i h_i(P)\chi_i(\alpha, P),$$

(4.2.62)

where $h_i(P) \in \mathscr{S}(\mathbb{R}^{4m})$ and the $\chi_i(\alpha, P)$ are bounded functions and are vanishing for (α, P) with values outside the set

$$S = \{(\alpha, P) : pUp/\mu^2 \leq -\tfrac{1}{3}\},$$

(4.2.63)

(recall the definition of $[1 - \chi(pUp/\mu^2)]$). Accordingly, to consider the $\varepsilon \to +0$ limit of $I^2_{\varepsilon am'}(h)$ it suffices to consider the limit of integrals of the form

$$\tilde{I}^2_{\varepsilon am'}(h) = -\frac{(\frac{1}{2})^t}{(t-1)!}\int_{\mathbb{R}^{4m}} dP \int_{\mathscr{D}} d\alpha \, \frac{N_{a,m'}(\alpha, \mu)}{(pUp)_\varepsilon^t}(\ln F_\varepsilon(\alpha, P))\sum_i h_i(P)\chi_i(\alpha, P).$$

(4.2.64)

For $(\alpha, P) \in S$, $F_0(\alpha, P)$ may vanish. The factor $(pUp)_\varepsilon$, however, does not vanish in (4.2.64) for all ε, as

$$\tfrac{1}{3}\mu^2 \leq |pUp| \leq |(pUp)_\varepsilon|.$$

(4.2.65)

Accordingly, the integral (4.2.64) without the $\ln F_\varepsilon(\alpha, P)$ factor in the integrand, i.e.,

$$\hat{I}^2_{\varepsilon am'}(h) \equiv -\frac{(\frac{1}{2})^t}{(t-1)!} \int_{\mathbb{R}^{4m}} dP \int_{\mathcal{D}} d\alpha \frac{N_{a,m'}(\alpha, \mu)}{(pUp)^t_\varepsilon} \sum_i h_i(P)\chi_i(\alpha, P), \qquad (4.2.66)$$

is absolutely convergent, and from (4.2.65) and the Lebesgue dominated convergence theorem $\hat{I}^2_{0am'}$ exists.

Now we consider the expression $\tilde{I}^2_{\varepsilon am'}(h)$ containing the $\ln F_\varepsilon(\alpha, P)$ factor. To this end we note that for any positive x, y and $y_0, y_0 > y > 0$,

$$|\ln(x + y)| \leq |\ln x| + |\ln y_0| + 1. \qquad (4.2.67)$$

The proof of (4.2.67) is elementary. We consider all possible cases. First suppose that $x + y > 1$. If $x \geq y$, then $2x \geq x + y > 1$, and

$$\ln(x + y) \leq \ln 2x \leq |\ln x| + \ln 2 < |\ln x| + 1 + |\ln y_0|. \qquad (4.2.68)$$

If $x < y$, then $1 < x + y < 2y \leq 2y_0$, and

$$\ln(x + y) \leq |\ln y_0| + \ln 2 < |\ln y_0| + 1 + |\ln x|. \qquad (4.2.69)$$

Now suppose $x + y < 1$; then

$$|\ln(x + y)| \leq |\ln x| < |\ln x| + |\ln y_0| + 1. \qquad (4.2.70)$$

If $x + y = 1$, then (4.2.67) is trivially true. We now use (4.2.67) to obtain

$$|\ln F_\varepsilon(\alpha, P)| \leq \tfrac{1}{2}[\ln |F_\varepsilon(\alpha, P)|^2 + 2\pi]$$
$$\leq |\ln |pUp + M^2|| + |\ln \varepsilon_0(\mathbf{p} \cdot U\mathbf{p} + M^2)| + \pi + \tfrac{1}{2}, \qquad (4.2.71)$$

where $\varepsilon_0 > \varepsilon > 0$. Therefore to establish the limit $\varepsilon \to +0$ of $\tilde{I}^2_{\varepsilon am'}(h)$, it suffices from (4.2.71) and (4.2.65) to establish the absolute convergence of the following integrals:

$$J^1_{am'} = \int_{\mathbb{R}^{4m}} dP \int_{\mathcal{D}} d\alpha \frac{N_{a,m'}(\alpha, \mu)}{(pUp)^t} h_i(P)\chi_i(\alpha, P), \qquad (4.2.72)$$

$$J^2_{am'} = \int_{\mathbb{R}^{4m}} dP \int_{\mathcal{D}} d\alpha \frac{N_{a,m'}(\alpha, \mu)}{(pUp)^t} h_i(P)\chi_i(\alpha, P)\ln(\mathbf{p} \cdot U\mathbf{p} + M^2), \qquad (4.2.73)$$

$$J^3_{am'} = \int_{\mathbb{R}^{4m}} dP \int_{\mathcal{D}} d\alpha \frac{N_{a,m'}(\alpha, \mu)}{(pUp)^t} h_i(P)\chi_i(\alpha, P) \ln|pUp + M^2|. \qquad (4.2.74)$$

From (4.2.58), $J^1_{am'}$ is obviously absolutely convergent. Also, $(\mathbf{p} \cdot U\mathbf{p} + M^2) > 0$ and U is continuous in $\mathcal{D}$, and hence $J^2_{am'}$ is also absolutely convergent. On the other hand, we may, for example, use a very powerful theorem in distribution theory (the Hironaka–Atiyah–Bernstein–Gel'fand theorem),[9] which may be tailored to the problem at hand; i.e., it may applied to the integral $J^3_{am'}$. It states that if the integral

$$\hat{J}^3_{am'} = \int_{\mathbb{R}^{4m}} dP \int_{\mathscr{D}} d\alpha \, \frac{N_{am'}(\alpha, \mu)}{(pUp)^t} \, h_i(P)\chi_i(\alpha, P), \tag{4.2.75}$$

is absolutely convergent, which obviously it is since it coincides with $J^1_{am'}$ in (4.2.72), then so is the integral $J^3_{am'}$ containing the $\ln |(pUp + M^2)|$ factor. This means that the $\ln |(pUp + M^2)|$ "singularity" is too mild to break the absolute convergence of $J^3_{am'}$. Therefore by the Lebesgue dominated convergence theorem we have

$$\lim_{\varepsilon \to +0} \tilde{I}^2_{\varepsilon am'}(h) = \tilde{I}^2_{0am'}(h)$$

$$= -\frac{(\tfrac{1}{2})^t}{(t-1)!} \int_{\mathbb{R}^{4m}} dP \int_{\mathscr{D}} d\alpha \, \frac{N_{a,m'}(\alpha, \mu)}{(pUp)^t} \ln |F_0(\alpha, P)| \sum_i h_i(P)\chi_i(\alpha, P). \tag{4.2.76}$$

Accordingly, we may state

Theorem 4.2.1: $\lim_{\varepsilon \to +0} T_\varepsilon(f)$, with $T_\varepsilon(f)$ *as given in* (4.1.3), *exists and defines a Lorentz covariant distribution.*

NOTES

The $\varepsilon \to +0$ limit of absolutely convergent Feynman integrals, starting in momentum space, is worked out in the elegant papers by Hahn and Zimmermann (1968), where, in particular, the generalized Lagrange's interpolating formula is derived; Zimmermann (1968; see also Hepp, 1966); and Lowenstein and Speer (1976). In the latter reference one will find how the Hironaka–Atiyah–Bernstein–Gel'fand (HABG) theorem is tailored to the problem at hand. For the proof of the HABG theorem see Hironaka (1964), Atiyah (1970), and Bernstein and Gel'fand (1969). This theorem is quite powerful and goes beyond standard results in distribution theory.

---

[9] The proof of this theorem is extremely long and is beyond the scope of the present book, as it requires very special tools.

Chapter 5 / THE SUBTRACTION FORMALISM

This chapter is devoted to the subtraction formalism. In Section 5.1 we present basic definitions and work out examples concerning subgraphs, subdiagrams, canonical variables, Taylor operations, and the associated subtractions. In Section 5.2 we introduce the subtraction scheme, and some examples are then worked out. Here the reader will learn how to write down explicitly the expression for a renormalized Feynman integrand. A recursive formula for the subtraction scheme is also given. The convergence proof of the subtraction scheme is given in Section 5.3. In Section 5.4 we establish the "unifying theorem of renormalization," which demonstrates essentially of the equivalence of the paths taken in the ingenious approaches of Salam and Bogoliubov.

5.1 BASIC DEFINITIONS AND EXAMPLES

5.1.1 Subgraphs and Subdiagrams

We define a graph G by specifying a set of vertices $\mathscr{V} = \{v_1, \ldots, v_r\}$ and a set of lines $\mathscr{L} = \{\ell_1, \ldots, \ell_s\}$ joining these vertices as obtained from the so-called Feynman rules.[1] Given a graph G, a subgraph G' of G, which we

---

[1] In an abstract way we may, for our purposes, define a graph G by specifying two *universal* sets of vertices $\mathscr{V}$ and lines $\mathscr{L}$ with the lines joining the vertices according to given rules. Subgraphs and subdiagrams, as discussed below, are then defined in reference to these universal sets and the given rules.

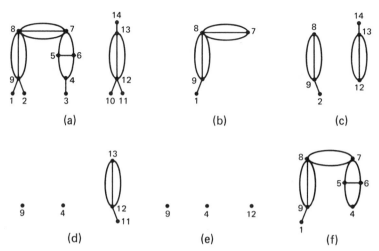

Fig. 5.1   Parts (b), (c), (d), and (e) represent subgraphs of the graph G defined in (a). Part (f) does not represent a subgraph of G but a subdiagram of G.

write $G' \subset G$, is defined by specifying a subset of vertices $\mathscr{V}' \subset \mathscr{V}$ and all those lines in $\mathscr{L}$ of G that join these vertices in the original graph G. To make this definition more transparent we give some examples. Given a graph G as shown in Fig. 5.1a, some examples of subgraphs are given in Figs. 5.1b–e specified, respectively, by the vertices $v_1, v_7, v_8, v_9; v_2, v_8, v_9,$ $v_{12}, v_{13}, v_{14}; v_4, v_9, v_{11}, v_{12}, v_{13};$ and v_4, v_9, v_{12}. Note that Fig. 5.1e represents a subgraph of G because there are no lines joining the vertices $v_4, v_9,$ and v_{12} in Fig. 5.1a in the original graph G. Clearly, Fig. 5.1f does not represent a subgraph of G, as one of the lines joining the vertices v_7 and v_8 in G is missing in Fig. 5.1f. Quite generally, by specifying a subset of the vertices $\mathscr{V}' \subset \mathscr{V}$ and some, not necessarily all, of the lines in $\mathscr{L}$ of G joining the vertices in $\mathscr{V}'$, we define a subdiagram. Accordingly, a subdiagram is a more general concept than a subgraph and reduces to the latter if it contains all those lines in G that join its vertices in $\mathscr{V}'$. Figure 5.1f, as mentioned, is not a subgraph, but it represents a subdiagram of G of 5.1a. In the sequel when we say a subdiagram we may also mean a subgraph. We do not allow a line joining a vertex to itself.

A subdiagram is said to be disconnected if it is constructed out of two or more subdiagrams that have no vertices in common. Otherwise a subdiagram is said to be connected. Figure 5.2a represents a subdiagram g of G of Fig. 5.1a that is disconnected, as it is constructed out of two subdiagrams g_1 and g_2 that have no vertices in common. Figure 5.2b provides another example of a disconnected subdiagram. Figure 5.1f represents a connected subdiagram. In this terminology the graph G of Fig. 5.1a is itself disconnected,

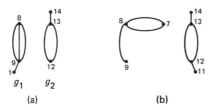

Fig. 5.2 (a) A disconnected subdiagram g of G of Fig. 5.1a as it is constructed out of g_1 (a subgraph) and g_2 (a subdiagram) which have no vertices in common. (b) Another example of a disconnected subdiagram g of G.

as it is constructed out of two subgraphs specified, respectively, by the two disjoint set of vertices $\{v_1, v_2, \ldots, v_g\}$ and $\{v_{10}, v_{11}, \ldots, v_{14}\}$. We say that a subdiagram g has n connected parts if it is constructed out of n subdiagrams each being connected. Trivially, if $n = 1$, then g is connected.

A vertex of a subdiagram g to which is attached just one and only one line or no lines at all will be called an extral vertex of the subdiagram g. Figure 5.3a represents a subdiagram g of G where the vertices v_9 and v_{11} are extral vertices of g. The extral vertices of the graph G itself of Fig. 5.1a are v_1, v_2, v_3, v_{10}, v_{11}, and v_{14}. By removing the extral vertices from a subdiagram and the lines attached to them, we obtain a new subdiagram, say, g', which may or may not have extral vertices. Figure 5.3b represents the subdiagram g' of g of Fig. 5.3a as obtained from the latter by removing the extral vertices v_9 and v_{11} and the lines attached to these vertices, i.e., the lines joining the vertices v_9 to v_8 and v_{11} to v_{12}. The new subdiagram g' of g has an extral vertex v_8. By continuing the process of removing the extral vertices (and the lines attached to them) of the resulting new subdiagrams repeatedly, we finally obtain either a nontrivial (i.e., nonempty) subdiagram with no extral vertices or no subdiagram at all (i.e., no lines and no vertices). We call a subdiagram with no extral vertices an amputated subdiagram. Figures 5.3a–c show the process of obtaining the amputated version, say, g'' of g. The

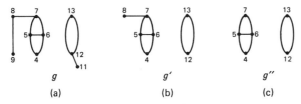

Fig. 5.3 (a) A subdiagram g of G of Fig. 5.1a where the vertices v_9 and v_{11} are extral vertices. By removing the extral vertices v_9 and v_{11} of g and the lines attached to them, we obtain a new subdiagram g' in (b) having v_8 as an extral vertex. Finally, (c) represents the amputated version g'' of g. Vertices v_4, v_7, v_{12}, and v_{13} are external vertices of g'' in (c). The vertices v_5 and v_6 are internal vertices of all the subdiagrams g, g', and g''.

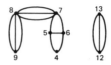

Fig. 5.4   The amputated version of the graph G of Fig. 5.1a.

lines deleted in the process of amputating a subdiagram g are called the external lines of g. That is, the amputated version g'' of a subdiagram g is obtained by removing the external lines (and the relevant vertices) of g. The external lines of g will be called external lines *to* g'' (the amputated version of g). A line that is not an external line of a subdiagram g is called an internal line. Accordingly, an amputated subdiagram may have only internal lines. In Fig. 5.4 we represent the amputated version of the graph G of Fig. 5.1a.

A subdiagram is proper, or called a proper subdiagram, if it is amputated and its number of connected parts does not increase upon removing any one of its (internal!) lines. Note, by this definition that a proper subdiagram is not necessarily connected. A line in an amputated subdiagram g is called an improper line of g if upon removing it, the number of the connected parts of g is increased. The proper part of a subdiagram g is a subdiagram obtained from g by amputating the latter and removing the improper lines (if any) in the resulting subdiagram. Figure 5.5a represents a subdiagram (with two connected parts) and is not proper, since the line joining the vertex v_7 to v_8 is an improper line and upon removing it we introduce three, not two, connected parts, as shown in Fig. 5.5b. On the other hand, the subdiagram in 5.5c is proper. The subdiagram g' in Fig. 5.5b is the proper part of g of Fig. 5.5a. The proper part of the graph G of Fig. 5.1a was given in Fig. 5.4. In the sequel, the proper part of a graph will be referred to as a proper graph. A subdiagram that is not proper is also called an improper subdiagram.

A vertex of the graph G that is not an extral vertex and to which is attached an external line will be called an external vertex of G. A vertex of the graph G that is neither an extral vertex nor an external vertex is called an internal vertex. Now consider a subdiagram $g \nsubseteq G$. A vertex v_i of g that is not an extral vertex of g, and if v_i is an external vertex of the graph G and/or there

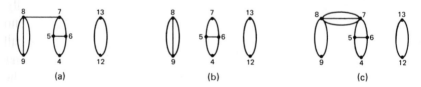

Fig. 5.5   The subdiagram g in (a) is not proper, as it contains an improper line joining the vertex v_7 to v_8. The subdiagram g' in (b) is the proper version of g in (a). Part (c) represents a proper subdiagram.

is a line ℓ belonging to G, but not to g, attached to v_i, then the latter is called an external vertex of g. A vertex of g that is neither an extral vertex nor an external vertex is called an internal vertex of g.

Consider the subdiagram $g \subset G$ in Fig. 5.3a. The vertices $v_4, v_7, v_8, v_{12}, v_{13}$ are external vertices of g and v_5, v_6 are its internal vertices. For $g' \subset G$ in Fig. 5.3b, v_4, v_7, v_{12}, v_{13} are its external vertices and v_5, v_6 are its internal vertices. For the graph G, its external vertices are v_4, v_9, v_{12}, v_{13}, as they are attached to external lines and they are not extral vertices of G. Its internal vertices are v_5, v_6, v_7, v_8. For the proper part of the graph G of 5.1a, as given in Fig. 5.4, the external vertices of the latter are also v_4, v_9, v_{12}, v_{13} and its internal vertices are v_5, v_6, v_7, v_8.

In this book we require that each proper and connected graph together with all its proper connected *sub*diagrams have each *at least two* external vertices.

Let g_1 and g_2 be two subdiagrams defined by the pairs $(\mathscr{V}_1, \mathscr{L}_1)$ and $(\mathscr{V}_2, \mathscr{L}_2)$. If $\mathscr{V}_1 \subset \mathscr{V}_2$ and $\mathscr{L}_1 \subset \mathscr{L}_2$, then we write $g_1 \subset g_2$. On the other hand, if $\mathscr{V}_1 \cap \mathscr{V}_2 = \varnothing$ (empty), then we write $g_1 \cap g_2 = \varnothing$ and we say that g_1 and g_2 are disjoint.

Several lines may connect any two vertices of a subdiagram. For example, there are three lines connecting the vertices v_8 and v_9 of the subdiagram g_1 in Fig. 5.2a. We may number these lines. An *l*th line joining a vertex v_i to a vertex v_j will carry a momentum Q_{ijl} depending on the indices i, j, and l. The appearance of the indices i and j in the order in which they appear in Q_{ijl} is meant, as a convention, to show that the momentum is flowing from v_i to v_j. Accordingly, we take $Q_{ijl} = -Q_{jil}$ when interchanging the indices i and j, i.e., when "reversing" the direction of flow.

An object of interest in quantum field theory is a proper and connected graph. Consider a *proper* and connected graph G. With each line ℓ_l joining a vertex v_i to a vertex v_j of G carrying a momentum Q_{ijl} and a mass μ_{ijl} we associate a propagator as a function of Q_{ijl} and μ_{ijl}, defined by a polynomial in Q_{ijl}, μ_{ijl}, and, in general, in $(\mu_{ijl})^{-1}$ as well: $P_{ijl}(Q_{ijl}, \mu_{ijl})$, multiplied by (see Chapter 2)

$$[Q_{ijl}^2 + \mu_{ijl}^2 - i\varepsilon(\mathbf{Q}_{ijl}^2 + \mu_{ijl}^2)]^{-1}.$$

Accordingly a line will be represented by[2]

$$D_{ijl}^+(Q_{ijl}, \mu_{ijl}) = P_{ijl}(Q_{ijl}, \mu_{ijl})[Q_{ijl}^2 + \mu_{ijl}^2 - i\varepsilon(\mathbf{Q}_{ijl}^2 + \mu_{ijl}^2)]^{-1}. \qquad (5.1.1)$$

We choose $\mu_{ijl}^2 > 0$,[3] and we do not allow a line connecting a vertex to itself.

---

[2] The polynomial dependence of P_{ijl} on $(\mu_{ijl})^{-1}$, in general, is well known for higher-spin fields.

[3] Zero-mass behavior of renormalized Feynman amplitudes will be studied in Chapter 6.

Let j be fixed and consider $\{v_{i(j)}\}_{1 \leq i \leq r_j}$, the set of vertices attached by lines to v_j in G, and consider the set $\mathscr{L}^G(v_j)$ of all the lines joining the vertices $v_{i(j)}$, $1 \leq i \leq r_j$, to the vertex v_j. Let $\{Q_{ijl}\}_{1 \leq l \leq s_{ij}}$ be the set of momenta carried by these lines. Then we assign to the vertex v_j a polynomial $\mathscr{P}_j = \mathscr{P}_j(Q_{1j1}, \ldots, Q_{r_j j s_{r_j j}})$. The *unrenormalized* Feynman *integrand* associated with the proper and connected graph G, up to an overall multiplicative constant (involving couplings, etc.), is of the form

$$I_G = \prod_{\substack{ijl \\ i<j}}^{G} \mathscr{P}_j D_{ijl}^{+}, \tag{5.1.2}$$

consisting the product over all the polynomials $\mathscr{P}_j$ assigned to the vertices $\{v_j\}$ of G and the propogators D_{ijl}^{+} (5.1.1) joining the vertices of G.

Let $P = \{p_1^0, \ldots, p_m^3\}$ denote the set of the components of the external independent momenta of the graph G carried by the external lines *to* G, and let $K = \{k_1^0, \ldots, k_n^3\}$ be the set of the $4n$ integration variables associated with the graph G. The latter will be called the set of the components of the internal momenta of G.

A momentum Q_{ijl} flowing in an lth line from a vertex v_i to a vertex v_j of G is of the form

$$Q_{ijl} = k_{ijl} + q_{ijl}, \tag{5.1.3}$$

where

$$k_{ijl} = \sum_{s=1}^{n} a_{ijl}^{s} k_s, \tag{5.1.4}$$

$$q_{ijl} = \sum_{r=1}^{m} b_{ijl}^{r} p_r, \tag{5.1.5}$$

$$a_{ijl}^{s} = -a_{jil}^{s},$$
$$b_{ijl}^{r} = -b_{jil}^{r}. \tag{5.1.6}$$

That is, k_{ijl} is a linear combination of the internal momenta of G and q_{ijl} is a linear combination of its external momenta. The sets $\{k_{ijl}\} \equiv k$ and $\{q_{ijl}\} \equiv \tilde{q}$ will be called the sets of internal and external variables of G, respectively. Let q_j be the *total* momentum carried by the external lines *to* G attached to the vertex v_j in a direction of flow *away*[4] from the vertex v_j. If a vertex v_j of G is an internal vertex, we simply set $q_j = 0$, by definition, as there are no external lines attached to such a vertex. With such a convention we then have at each vertex v_j of G

$$q_j = \sum_{il}^{G} Q_{ijl}, \tag{5.1.7}$$

---

[4] The direction of flow of the momentum q_j away from the vertex v_j is taken as a convention. Clearly, q_j is a linear combination of the external independent momenta in P of G.

by momentum conservation. The sum is over all i corresponding to the vertices v_i attached to the vertex v_j all belonging to G, and over all the lines ℓ_l of G joining the vertices v_i to the vertex v_j. As mentioned, if v_j is an internal vertex, then we simply have $q_j = 0$. Finally, by overall momentum conservation we have, by summing over all j corresponding to all the vertices of G, the expression

$$\sum_j^G q_j = 0. \tag{5.1.8}$$

5.1.2   Canonical Variables

In this subsection we write down the expression for the Q_{ijl} in the unrenormalized Feynman integrand I_G (5.1.2) in a form suitable for carrying out the so-called subtractions of renormalization. This is important for constructing the renormalized Feynman integrand R. The particular way of writing the expressions for the Q_{ijl} in a form suitable for carrying out the subtractions, as will be given, will be called canonical. Once the renormalized Feynman integrand R has been constructed (Section 5.2) one may, of course, freely make a change of variables in R for the purpose of evaluating the renormalized Feynman amplitude $\mathscr{A} = \int dK\, R$, once its (absolute) convergence (Section 5.3) has been established.

We say that the set $\{q_{ijl}\} \equiv \tilde{q}$ is a canonical set of external variables of a proper and connected graph G if the q_{ijl} are chosen in the following manner:

$$\sum_{il}^G q_{ijl} = q_j \tag{5.1.9}$$

at each vertex v_j of G, and where these are chosen of the form:

$$q_{ijl} = u_i - u_j, \tag{5.1.10}$$

where u_i, u_j are four-vectors. Equation (5.1.10) shows, in particular, that the external variables of the lines joining a vertex v_i to a vertex v_j are all chosen to be equal. If G has $\#\mathscr{V}^G$ vertices, then (5.1.9) together with the constraint (5.1.8) gives $\#\mathscr{V}^G - 1$ independent equations for the $\#\mathscr{V}^G - 1$ independent differences $u_i - u_j$, and hence the external variables q_{ijl} of G are then uniquely determined.

Equations (5.1.3) and (5.1.7)–(5.1.9) imply

$$\sum_{ijl}^G k_{ijl} = 0, \qquad \sum_{il}^G k_{ijl} = 0 \tag{5.1.11}$$

at all the vertices v_j of G for the latter equation. Finally, we note that as we have $\#\mathscr{V}^G - 1$ independent external variables q_{ijl} [see (5.1.10)], we may write the variables in $\{q_{ijl}\} \equiv \tilde{q}$ as functions (linear combinations) of $\# \mathscr{V}^G - 1$ independent variables in $\{q_j\} \equiv q: q_{ijl} = q_{ijl}(q)$ [see (5.1.8) and (5.1.9)].

The choice of the Q_{ijl} of G such that the q_{ijl} are canonical external variables [(5.1.8)–(5.1.10)] and the k_{ijl} are *any* consistent solutions of (5.1.11) will be called a canonical choice of variables. A property of a proper and connected graph G (or for a proper and connected subdiagram, in general) is that if $4n$ denotes the number of independent integration variables associated with G, then $n = \# \mathscr{L}^G - \# \mathscr{V}^G + 1$, where $\# \mathscr{L}^G$ is the number of lines of G and $\# \mathscr{V}^G$, already introduced, is its number of vertices. To this end note that the second of the $\# \mathscr{V}^G$ equations in (5.1.11) with the first equation in the latter as a constraint allows only n of the $\# \mathscr{L}^G$ k_{ijl} in k to be independent. If $(k_1^0, \ldots, k_n^3)$ are the $4n$ integration variables, then n of the internal variables k_{ijl} may be, consistently, chosen to be the internal momenta $k_1, \ldots, k_n$. More generally, by relabeling the elements in the set $\{k_{ijl}\} \equiv \{k(i)\}$, we may, consistently, choose n of the $k(i)$ in the form $k(i) = \sum_{j=1}^n a_{ij} k_j, i = 1, \ldots, n$, with $[a_{ij}]$, for $1 \le i, j \le n$, a nonsingular matrix. The remaining $k(i)$ would be of the form $k(i) = \sum_{j=1}^n a_{ij} k_j, i > n$ for some a_{ij}. The *rank* of the matrix $[a_{ij}]$ in $k(i) = \sum_{j=1}^n a_{ij} k_j, i = 1, \ldots, n$, is then equal to n.[5] This reduces to the case mentioned earlier if we choose $a_{ij} = \delta_{ij}$ for $1 \le i, j \le n$.

We work out a few examples for the illustration of the above rules.

Example 5.1:   Consider the graph in Fig. 5.6a with its external vertices defined to be v_1 and v_2 with $q_1 + q_2 = 0$. Let $q_1 = q$; then $q_2 = -q$. Equation (5.1.9) states

$$q_{211} + q_{212} = q, \qquad q_{121} + q_{122} = -q \qquad (5.1.12)$$

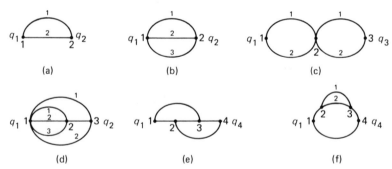

(a)   (b)   (c)

(d)   (e)   (f)

Fig. 5.6   Some proper and connected graphs. The vertices v_1, v_2 are taken as the external vertices of the graphs in (a) and (b). The external vertices of the graphs in (c) and (d) are taken to be v_1 and v_3. The external vertices of the graphs in (e) and (f) are taken to be v_1 and v_4. The overall momentum conservations are $q_1 + q_2 = 0$, $q_1 + q_3 = 0$, and $q_1 + q_4 = 0$, respectively.

[5] The $k(i)$, for $i > n$, may be also rewritten as linear combinations of the $k(i)$ with $1 \le i \le n$.

at the vertices v_1 and v_2, respectively, which are, of course, not independent. Using the definition (5.1.10) and choosing any one of the equations in (5.1.12), we obtain $2(u_2 - u_1) = q$, or

$$u_2 - u_1 = \tfrac{1}{2}q,$$

i.e.,

$$q_{211} = \tfrac{1}{2}q, \qquad l = 1, 2. \tag{5.1.13}$$

We also have from (5.1.11)

$$k_{121} + k_{122} = 0$$

at the vertex v_2. If we choose k_{121} as the internal momentum k (the integration variable), we then have

$$k_{121} = k = -k_{122}. \tag{5.1.14}$$

Equations (5.1.13) and (5.1.14) imply that a canonical choice of the variables Q_{121}, Q_{122} of G is

$$Q_{121} = k + \tfrac{1}{2}q, \qquad Q_{122} = -k + \tfrac{1}{2}q. \tag{5.1.15}$$

Example 5.2: For the subdiagram g in Fig. 5.6b, with external vertices v_1 and v_2, let $q_1 = q, q_2 = -q$. At the vertex v_1,

$$q = q_{211} + q_{212} + q_{213}, \tag{5.1.16}$$

or

$$u_2 - u_1 = \tfrac{1}{3}q, \tag{5.1.17}$$

i.e.,

$$q_{211} = \tfrac{1}{3}q, \qquad l = 1, 2, 3. \tag{5.1.18}$$

At the vertex v_1 we also have

$$k_{211} + k_{212} + k_{213} = 0. \tag{5.1.19}$$

Let $k_{211} = k_1$ and $k_{212} = k_2$ be the integration variables; then $k_{213} = -k_1 - k_2$, and a canonical choice of variables would be

$$Q_{211} = k_1 + \tfrac{1}{3}q, \quad Q_{212} = k_2 + \tfrac{1}{3}q, \quad Q_{213} = -k_1 - k_2 + \tfrac{1}{3}q. \tag{5.1.20}$$

Example 5.3: Consider the graph in Fig. 5.6c with v_2 as the only internal vertex. Let $q_1 = -q$ and $q_3 = q$. At v_1,

$$-q = q_{211} + q_{212},$$

or

$$u_2 - u_1 = -\tfrac{1}{2}q. \tag{5.1.21}$$

At v_2 (with $q_2 = 0$, since v_2 is an internal vertex),

$$0 = q_{121} + q_{122} + q_{321} + q_{322} = 2(u_1 - u_2) + 2(u_3 - u_2), \quad (5.1.22)$$

which together (5.1.21) implies

$$u_3 - u_2 = -\tfrac{1}{2}q.$$

That is,

$$q_{121} = \tfrac{1}{2}q, \qquad l = 1, 2,$$
$$q_{231} = \tfrac{1}{2}q, \qquad l = 1, 2. \qquad (5.1.23)$$

Let $k_{121} = k_1$ and $k_{231} = k_2$ be the integration variables. Using the conservation laws at v_1 and v_3,

$$k_{211} + k_{212} = 0, \qquad k_{231} + k_{232} = 0, \qquad (5.1.24)$$

we obtain $k_{122} = -k_1, k_{232} = -k_2$. Therefore a canonical choice of variables is

$$Q_{121} = k_1 + \tfrac{1}{2}q, \qquad Q_{122} = -k_1 + \tfrac{1}{2}q,$$
$$Q_{231} = k_2 + \tfrac{1}{2}q, \qquad Q_{232} = -k_2 + \tfrac{1}{2}q. \qquad (5.1.25)$$

Example 5.4:   For the graph in Fig. 5.6d, v_1 and v_3 are defined to be its external vertices, v_2 its internal vertex, and hence $q_2 = 0$. At the vertices v_1, v_2, and v_3, respectively,

$$
\begin{aligned}
v_1: \quad q_1 &= q_{211} + q_{212} + q_{213} + q_{311} + q_{312} \\
&= 3(u_2 - u_1) + 2(u_3 - u_1), \qquad (5.1.26)
\end{aligned}
$$

$$
\begin{aligned}
v_2: \quad 0 &= q_{121} + q_{122} + q_{123} + q_{32} \\
&= 3(u_1 - u_2) + (u_3 - u_2), \qquad (5.1.27)
\end{aligned}
$$

$$
\begin{aligned}
v_3: \quad q_3 &= q_{131} + q_{132} + q_{23} \\
&= 2(u_1 - u_3) + (u_2 - u_3). \qquad (5.1.28)
\end{aligned}
$$

With the constraint $q_1 + q_3 = 0$, we have at our disposal only two independent equations, and we choose (5.1.26) and (5.1.28) with $q_1 = q$ and $q_3 = -q$ and $(u_1 - u_2)$, $(u_1 - u_3)$ as two independent differences. These lead to

$$-q = 3(u_1 - u_2) + 2(u_1 - u_3), \qquad (5.1.29)$$

$$-q = -(u_1 - u_2) + 3(u_1 - u_3), \qquad (5.1.30)$$

where in writing (5.1.30) we have used the identity $u_2 - u_3 = (u_2 - u_1) + (u_1 - u_3)$. Equations (5.1.29) and (5.1.30) lead to the unique solution

$$u_1 - u_2 = -\tfrac{1}{11}q, \; u_1 - u_3 = -\tfrac{4}{11}q, \text{ or}$$

$$
\begin{aligned}
q_{12l} &= -\tfrac{1}{11}q, & l &= 1, 2, 3, \\
q_{13l} &= -\tfrac{4}{11}q, & l &= 1, 2, \\
q_{23} &= -\tfrac{3}{11}q.
\end{aligned}
\tag{5.1.31}
$$

For the internal variables we may choose

$$k_{121} = k_1, \qquad k_{122} = k_2, \qquad k_{123} = k_3, \qquad k_{131} = k_4, \tag{5.1.32}$$

corresponding to the 16 integration variables. For the remaining internal variables, we have at v_2 and v_3, respectively,

$$k_{121} + k_{122} + k_{123} + k_{32} = 0, \qquad k_{131} + k_{132} + k_{23} = 0, \tag{5.1.33}$$

leading to $k_{32} = -k_1 - k_2 - k_3$ and $k_{132} = -k_1 - k_2 - k_3 - k_4$.

A canonical choice of the variables is then

$$
\begin{aligned}
&Q_{121} = k_1 - \tfrac{1}{11}q, & &Q_{122} = k_2 - \tfrac{1}{11}q, \\
&Q_{123} = k_3 - \tfrac{1}{11}q, & &Q_{131} = k_4 - \tfrac{4}{11}q, \\
&Q_{132} = -k_1 - k_2 - k_3 - k_4 - \tfrac{4}{11}q, & &Q_{32} = -k_1 - k_2 - k_3 + \tfrac{3}{11}q.
\end{aligned}
\tag{5.1.34}
$$

Example 5.5: Now we consider the classic example of the graph given in Fig. 5.6e, where the external vertices are v_1 and v_4. At the vertices v_1, v_2, v_3, and v_4 we have, respectively,

$$
\begin{aligned}
q_1 &= q_{21} + q_{31}, & 0 &= q_{12} + q_{32} + q_{42}, \\
0 &= q_{23} + q_{43} + q_{13}, & q_4 &= q_{34} + q_{24},
\end{aligned}
\tag{5.1.35}
$$

or

$$
\begin{aligned}
q_1 &= (u_2 - u_1) + (u_3 - u_1), & 0 &= (u_1 - u_2) + (u_3 - u_2) \\
& & &\quad + (u_4 - u_2), \\
0 &= (u_2 - u_3) + (u_4 - u_3) & q_4 &= (u_3 - u_4) + (u_2 - u_4). \\
&\quad + (u_1 - u_3),
\end{aligned}
\tag{5.1.36}
$$

We also have $q_1 + q_4 = 0$. Let $q_1 = -q$, $q_4 = q$, and choose the first three equations in (5.1.36) with $u_1 - u_3$, $u_2 - u_3$, and $u_4 - u_3$ as the three independent differences. These equations lead uniquely to

$$u_1 - u_3 = \tfrac{1}{2}q, \qquad u_2 - u_3 = 0, \qquad u_4 - u_3 = -\tfrac{1}{2}q, \tag{5.1.37}$$

i.e.,

$$q_{21} = -\tfrac{1}{2}q, \quad q_{31} = -\tfrac{1}{2}q, \quad q_{32} = 0, \quad q_{24} = \tfrac{1}{2}q, \quad q_{34} = \tfrac{1}{2}q. \tag{5.1.38}$$

For the internal variables we may choose $k_{24} = k_1$ and $k_{13} = k_2$, corresponding to the eight integration variables. At the vertex v_1 we have $k_{21} + k_{31} = 0$ or $k_{21} = k_2$. At the vertex v_2 we have $k_{12} + k_{32} + k_{42} = 0$ or $k_{32} = k_1 + k_2$. Finally at the vertex v_3 we have $k_{23} + k_{13} + k_{43} = 0$ or $k_{43} = k_1$. Therefore a canonical choice of variables is

$$Q_{12} = -k_2 + \tfrac{1}{2}q, \qquad Q_{13} = k_2 + \tfrac{1}{2}q,$$

$$Q_{32} = k_1 + k_2, \qquad Q_{34} = -k_1 + \tfrac{1}{2}q, \qquad Q_{24} = k_1 + \tfrac{1}{2}q. \qquad (5.1.39)$$

Since the internal variables k_{ijl} may be chosen as any consistent solution of (5.1.11), another choice of canonical variables Q_{ij} is

$$Q_{12} = \tfrac{3}{8}k_1 + \tfrac{9}{8}k_2 + \tfrac{1}{2}q, \qquad Q_{13} = -\tfrac{3}{8}k_1 - \tfrac{9}{8}k_2 + \tfrac{1}{2}q,$$

$$Q_{32} = \tfrac{3}{4}k_1 - \tfrac{3}{4}k_2, \qquad Q_{34} = -\tfrac{3}{8}k_1 - \tfrac{3}{8}k_2 + \tfrac{1}{2}q, \qquad (5.1.40)$$

$$Q_{24} = \tfrac{9}{8}k_1 + \tfrac{3}{8}k_2 + \tfrac{1}{2}q.$$

Here we have chosen $k_{12} = \tfrac{3}{8}(k_1 + 3k_2)$ and $k_{34} = -\tfrac{3}{8}(3k_1 + k_2)$, and the coefficient matrix $\tfrac{3}{8}\begin{pmatrix} -\tfrac{1}{3} & -3 \end{pmatrix}$ is nonsingular.

Example 5.6:   We finally consider another classic example with the graph given in Fig. 5.6f with v_1 and v_4 as external vertices. At the vertices v_1, v_2, v_3, and v_4 we have, respectively,

$$q_1 = q_{21} + q_{41}, \qquad\qquad 0 = q_{12} + q_{322} + q_{321},$$

$$0 = q_{231} + q_{232} + q_{43}, \qquad q_4 = q_{34} + q_{14}. \qquad (5.1.41)$$

Let $q_1 = q, q_4 = -q$; then (5.1.41) gives

$$q = (u_2 - u_1) + (u_4 - u_1), \qquad 0 = 2(u_3 - u_2) + (u_1 - u_2),$$

$$-q = (u_3 - u_4) + (u_1 - u_4), \qquad (5.1.42)$$

as only three of the former equations are independent. We choose $u_1 - u_2$, $u_1 - u_4$, $u_2 - u_3$ as the three independent differences. From (5.1.42), rewritten in terms of these differences, we obtain

$$-q = (u_1 - u_2) + (u_1 - u_4), \qquad 0 = (u_1 - u_2) - 2(u_2 - u_3),$$

$$-q = 2(u_1 - u_4) - (u_1 - u_3),$$

giving the unique solution $u_1 - u_3 = -\tfrac{3}{7}q, u_1 - u_2 = -\tfrac{2}{7}q, u_1 - u_4 = -\tfrac{5}{7}q$, $u_2 - u_3 = -\tfrac{1}{7}q, u_4 - u_3 = \tfrac{2}{7}q$. For the internal variables we have

$$v_1: \quad k_{21} + k_{41} = 0,$$

$$v_2: \quad k_{12} + k_{321} + k_{322} = 0,$$

$$v_3: \quad k_{231} + k_{232} + k_{43} = 0, \qquad (5.1.43)$$

$$v_4: \quad k_{34} + k_{14} = 0.$$

A consistent solution of (5.1.43) is $k_{12} = \frac{2}{7}k_1 - \frac{8}{7}k_2 = -k_{14}$, $k_{231} = \frac{8}{7}k_1 - \frac{4}{7}k_2$, $k_{232} = -\frac{6}{7}k_1 - \frac{4}{7}k_2$, $k_{34} = \frac{2}{7}k_1 - \frac{8}{7}k_2$, where k_1 and k_2 denote the eight integration variables. Therefore a canonical choice of the variables Q_{ijl} is

$$Q_{12} = \tfrac{2}{7}k_1 - \tfrac{8}{7}k_2 - \tfrac{2}{7}q, \qquad Q_{14} = -\tfrac{2}{7}k_1 + \tfrac{8}{7}k_2 - \tfrac{5}{7}q,$$

$$Q_{231} = \tfrac{8}{7}k_1 - \tfrac{4}{7}k_2 - \tfrac{1}{7}q, \qquad Q_{232} = -\tfrac{6}{7}k_1 - \tfrac{4}{7}k_2 - \tfrac{1}{7}q, \qquad (5.1.44)$$

$$Q_{34} = \tfrac{2}{7}k_1 - \tfrac{8}{7}k_2 - \tfrac{2}{7}q.$$

5.1.3 Canonical Decomposition of the Q_{ijl} of Subdiagrams $g \subset G$

Let G be a proper and connected graph. Its variables Q_{ijl} have been defined through (5.1.3)–(5.1.11), where $\{k_{ijl}\} \equiv k$ and $\{q_{ijl}\} = \tilde{q}$ are its sets of the internal and external variables. It is convenient to rewrite

$$Q_{ijl} = k_{ijl}^G + q_{ijl}^G, \qquad (5.1.45)$$

with a superscript G to emphasize that the k_{ijl}^G and the q_{ijl}^G are the internal and the external variables of the graph G. We may also rewrite $\{k_{ijl}^G\} \equiv k^G$ and $\{q_{ijl}^G\} \equiv \tilde{q}^G$. Once the sets k^G and $\tilde{q}^G$ have been constructed, we proceed to construct the sets k^g and $\tilde{q}^g$ of the internal and external variables of any proper subdiagram g of G. The construction of the sets k^G and $\tilde{q}^G$ will be important when carrying out the subtractions of renormalization (Sections 5.1.4 and 5.2).

Suppose first that g is a proper and connected subdiagram of G. Let $\{Q_{ijl}\}^G \equiv Q^G$ and $\{Q_{ijl}\}^g \equiv Q^g$ be the sets of variables of G and g, respectively. If $Q_{ijl} \in Q^g$, then obviously the same $Q_{ijl} \in Q^G$. For any $Q_{ijl} \in Q^g$, we write

$$Q_{ijl} = k_{ijl}^g + q_{ijl}^g, \qquad (5.1.46)$$

and since the k_{ijl}^G and the q_{ijl}^G may be written as linear combinations of independent variables in k^G and q^G, respectively, we expect from (5.1.45) and (5.1.46) that, in general, we may express k_{ijl}^g and q_{ijl}^g as functions (linear combinations) of independent variables in k^G and q^G; i.e., we may write $k_{ijl}^g = k_{ijl}^g(k^G, q^G)$ and $q_{ijl}^g = q_{ijl}^g(k^G, q^G)$. We will show below how this may be done. We recall that $q^G \equiv \{q_j^G\}$ and $k^G \equiv \{k_{ijl}^G\}$. The k_{ijl}^g and q_{ijl}^g will be defined in the same manner as the k_{ijl}^G and q_{ijl}^G, once the Q_{ijl}, pertaining also to g, are chosen in a canonical way as described in Section 5.1.2.

At each external vertex v_j of g we have, as before,

$$\sum_{il}^g Q_{ijl} = q_j^g(k^G, q^G), \qquad (5.1.47)$$

where $q_j^g(k^G, q^G)$ is the total momentum carried *away* from the external vertex v_j of g and is, in general, a function of the elements in k^G and q^G.[6] The sum in (5.1.47) is over all i corresponding to the vertices v_i of g attached to the vertex v_j and over all the l corresponding to all the lines ℓ_l of g attaching the vertices v_i to the vertex v_j in question. If v_j is an internal vertex of g, then we define $q_j^g = 0$, and (5.1.47) then holds for all vertices v_j, external and internal, of the proper and connected subdiagram g of G.

In the same way as for G, the external variables q_{ijl}^g of g are chosen in a canonical way as the unique solution of the equations

$$\sum_{il}^g q_{ijl}^g(k^G, q^G) = q_j^g(k^G, q^G),\tag{5.1.48}$$

$$\sum_{j}^g q_j^g(k^G, q^G) = 0,\tag{5.1.49}$$

with

$$q_{ijl}^g = w_i - w_j.\tag{5.1.50}$$

The sum over j, in particular, in (5.1.49) corresponds to all the vertices v_j of g. If g has $\#\mathscr{V}^G$ vertices, then (5.1.48)–(5.1.50) provide $\#\mathscr{V}^g - 1$ independent equations for the $\#\mathscr{V}^g - 1$ independent differences $w_i - w_j$. Once the Q_{ijl} are determined in a canonical way, as described in Section 5.1.2, and the q_{ijl}^g uniquely determined from (5.1.48)–(5.1.50), then the k_{ijl}^g are also uniquely determined from (5.1.46) by

$$k_{ijl}^g = Q_{ijl} - q_{ijl}^g.\tag{5.1.51}$$

We now show the important fact that the k_{ijl}^g are linear combinations of the k_{ijl}^G only, and hence are *independent* of the q_j^G (and also of the q_{ijl}^G). To this end set the k_{ijl}^G in (5.1.48)–(5.1.50) equal to zero. We then have

$$\sum_{il}^g q_{ijl}^g(0, q^G) = q_j^g(0, q^G),\tag{5.1.52}$$

$$\sum_{j}^g \left(\sum_{i,l}^g q_{ijl}^g(0, q^G) \right) = 0,\tag{5.1.53}$$

$$q_{ijl}^g(0, q^G) = w_i' - w_j',\tag{5.1.54}$$

where $w_i' - w_j' = w_i - w_j|_{k_{ijl}^G = 0}$. Again these equations determine the $q_{ijl}^g(0, q^G)$, i.e., they determine the differences $w_i' - w_j'$, *uniquely*. However, when the k_{ijl}^G are set equal to zero, we have from (5.1.45) $Q_{ijl} = q_{ijl}^G$, and from

---

[6] The functions are, actually, in the simple forms as linear combinations of elements from k^G and q^G.

(5.1.47), (5.1.49), and (5.1.10), we have, for i, j, and l pertaining to the subdiagram g,

$$\sum_{il}^{g} q_{ijl}^{G} = q_{j}^{g}(0, q^{G}),$$ (5.1.55)

$$\sum_{j}^{g} \left(\sum_{i,l}^{g} q_{ijl}^{G} \right) = 0,$$ (5.1.56)

$$q_{ijl}^{G} = u_{i}' - u_{j}',$$ (5.1.57)

where $u_{i}' - u_{j}' = u_{i} - u_{j}|_{k_{jl}^{G}=0}$ in (5.1.10). Therefore q_{ijl}^{G}, with i, j, and l pertaining to the subdiagram g, satisfy the same equations as those of q_{ijl}^{g} in (5.1.52)–(5.1.54). Since (5.1.52)–(5.1.54) and (5.1.55)–(5.1.57), respectively, determine the $\#\mathscr{V}^{g} - 1$ independent differences $w_{i}' - w_{j}'$ and $u_{i}' - u_{j}'$ uniquely,

$$w_{i}' - w_{j}' = u_{i}' - u_{j}',$$ (5.1.58)

or

$$q_{ijl}^{g}(0, q^{G}) = q_{ijl}^{G},$$ (5.1.59)

for i, j, and l pertaining to the subdiagram $g \subset G$. Finally, we use (5.1.51) and (5.1.59) to obtain

$$k_{ijl}^{g}(0, q^{G}) = q_{ijl}^{G} - q_{ijl}^{g}(0, q^{G}) = 0,$$ (5.1.60)

and hence we conclude that

$$k_{ijl}^{g} = k_{ijl}^{g}(k^{G}).$$ (5.1.61)

Therefore the k_{ijl}^{g} depend *only* on the elements in k^{G}, and are *independent* of the elements in q^{G} (and also in $\tilde{q}^{G}$).

Therefore for any proper and connected subdiagram $g \subset G$, we may summarize by saying that the internal k_{ijl}^{g} and the external q_{ijl}^{g} variables of g are uniquely determined by:

$$\sum_{il}^{g} q_{ijl}^{g}(k^{G}, q^{G}) = \sum_{il}^{g} Q_{ijl},$$ (5.1.62)

$$\sum_{j}^{g} \left(\sum_{il}^{g} q_{ijl}^{g}(k^{G}, q^{G}) \right) = 0,$$ (5.1.63)

$$q_{ijl}^{g}(k^{G}, q^{G}) = w_{i} - w_{j},$$ (5.1.64)

$$k_{ijl}^{g}(k^{G}) = Q_{ijl} - q_{ijl}^{g}(k^{G}, q^{G}),$$ (5.1.65)

once the Q_{ijl} are determined in a canonical way for the graph G as described in Section 5.1.2. The decomposition $Q_{ijl} = k_{ijl}^{g} + q_{ijl}^{g}$ in (5.1.46) will be called a canonical decomposition of the Q_{ijl} in reference to the subdiagram $g \subset G$, with i, j, and l pertaining to the subdiagram g. We note that (5.1.62), (5.1.63), and (5.1.65) imply, by analogy with (5.1.11) for G,

$$\sum_{ijl}^{g} k_{ijl}^{g} = 0, \qquad \sum_{il}^{g} k_{ijl}^{g} = 0 \qquad (5.1.66)$$

at each vertex v_j of g for the latter equation. Again (5.1.66) implies that $L(g) \equiv \# \mathscr{L}^{g} - \# \mathscr{V}^{g} + 1$ of the k_{ijl}^{g} may be independent. Since the k_{ijl}^{G} are linear combinations of the integration variables [see (5.1.4)], it follows from (5.1.61) that the k_{ijl}^{g} also are linear combinations of the integration variables *only*. Finally, if we set $\{q_j^{g}\} \equiv q^{g}$, where q_j^{g} is the total momentum carried away from the external vertex v_j of g [see (5.1.47) and (5.1.48)], then from (5.1.48)–(5.1.50), we may write the q_{ijl}^{g} as linear combinations of the q_j^{g}, i.e., $q_{ijl}^{g} = q_{ijl}^{g}(q^{g})$.

In general, if g is a proper and not necessarily connected subdiagram of g, then we apply the above construction for each of the connected parts of g.

Similarly, if g' is a proper subdiagram of g ($g' \subset g$), then we carry out the canonical decomposition of the Q_{ijl} in the form $Q_{ijl} = k_{ijl}^{g} + q_{ijl}^{g} = k_{ijl}^{g} + q_{ijl}^{g}$, and repeat the above analysis word for word to determine the internal $k_{ijl}^{g'}$ and external $q_{ijl}^{g'}$ variables uniquely by equations of the form (5.1.62)–(5.1.65) with G and g replaced by g and g', respectively, in the latter, and thus obtaining, in particular,

$$k_{ijl}^{g'} = k_{ijl}^{g'}(k^{g}), \qquad (5.1.67)$$

$$q_{ijl}^{g'} = q_{ijl}^{g'}(k^{g}, q^{g}). \qquad (5.1.68)$$

Also if we denote by $q_j^{g'}$ the total momentum carried away from the external vertex v_j of g and set $\{q_j^{g'}\} \equiv q^{g'}$, then we may also write $q_{ijl}^{g'} = q_{ijl}^{g'}(q^{g'})$.

We give a few examples to illustrate the canonical decomposition of the Q_{ijl} in reference to a proper subdiagram $g \subset G$.

Example 5.7:   Consider the subdiagram g_2 in Fig. 5.7a as a subdiagram of the graph of Fig. 5.6b. From (5.1.62), at the vertex v_1

$$q_{211}^{g_2} + q_{212}^{g_2} = Q_{211} + Q_{212}, \qquad (5.1.69)$$

and hence from (5.1.20) and (5.1.64),

$$2(w_2 - w_1) = k_1 + k_2 + \tfrac{2}{3}q$$

or

$$q_{21l}^{g_2} = \tfrac{1}{2}(k_1 + k_2) + \tfrac{1}{3}q, \qquad l = 1, 2. \qquad (5.1.70)$$

From (5.1.65) and (5.1.20),

$$k_{211}^{g_2} = k_1 + \tfrac{1}{3}q - \tfrac{1}{2}(k_1 + k_2) - \tfrac{1}{3}q = \tfrac{1}{2}(k_1 - k_2)$$

and   (5.1.71)

$$k_{212}^{g_2} = \tfrac{1}{2}(k_2 - k_1).$$

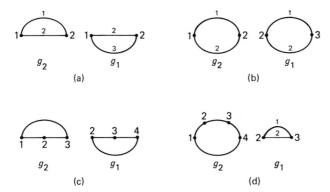

Fig. 5.7 Parts (a), (b), (c), and (d) represent subdiagrams of the graphs in Figs. 6b, 6c, 6e and 6f, respectively.

As shown in general in (5.1.61), we see that the $k_{ijl}^{g_2}$ are independent of q^G, i.e., of q.

Similarly, for g_1 in Fig. 5.7a,

$$q_{211}^{g_1} = -\tfrac{1}{2}k_1 + \tfrac{1}{3}q, \qquad l = 2, 3, \tag{5.1.72}$$

$$k_{212}^{g_1} = k_2 + \tfrac{1}{2}k_1, \tag{5.1.73}$$

$$k_{213}^{g_1} = -\tfrac{1}{2}k_1 - k_2. \tag{5.1.74}$$

Example 5.8:   For the subdiagrams g_1 and g_2 in Fig. 5.7b as subdiagrams of the graph of Fig. 5.6c, we have at the vertices v_3 and v_2, respectively,

$$q_{231}^{g_1} + q_{232}^{g_1} = Q_{231} + Q_{232} = q, \tag{5.1.75}$$

$$q_{121}^{g_2} + q_{122}^{g_2} = Q_{121} + Q_{122} = q \tag{5.1.76}$$

from (5.1.25), or

$$q_{231}^{g_1} = \tfrac{1}{2}q, \qquad l = 1, 2, \tag{5.1.77}$$

$$q_{121}^{g_2} = \tfrac{1}{2}q, \qquad l = 1, 2. \tag{5.1.78}$$

From (5.1.65) and (5.1.25), we then obtain,

$$\begin{aligned}
k_{231}^{g_1} = k_2, \qquad k_{232}^{g_1} = -k_2, \\
k_{121}^{g_2} = k_1, \qquad k_{122}^{g_2} = -k_1.
\end{aligned} \tag{5.1.79}$$

Example 5.9:   We consider the subdiagram g_1 in Fig. 5.7c as a subdiagram of the graph in Fig. 5.6e. We may, for example, choose the canonical variables Q_{ijl} as given in (5.1.40). At the vertices v_2 and v_3 we have, respectively,

$$q_{42}^{g_1} + q_{32}^{g_1} = Q_{42} + Q_{32}, \qquad q_{43}^{g_1} + q_{23}^{g_1} = Q_{43} + Q_{23},$$

or

$$w_4 - w_2 + w_3 - w_2 = -\tfrac{3}{8}k_1 - \tfrac{9}{8}k_2 - \tfrac{1}{2}q,$$
$$w_4 - w_3 + w_2 - w_3 = \tfrac{3}{8}k_1 + \tfrac{9}{8}k_2 - \tfrac{1}{2}q. \tag{5.1.80}$$

Choosing $w_2 - w_3$ and $w_4 - w_3$ as the independent differences, we obtain

$$w_4 - w_3 + 2(w_3 - w_2) = -\tfrac{3}{8}k_1 - \tfrac{9}{8}k_2 - \tfrac{1}{2}q,$$
$$w_4 - w_3 - (w_3 - w_2) = \tfrac{3}{8}k_1 + \tfrac{9}{8}k_2 - \tfrac{1}{2}q,$$

leading to

$$q_{34}^{g_1} = -\tfrac{1}{8}k_1 - \tfrac{3}{8}k_2 + \tfrac{1}{2}q, \qquad q_{24}^{g_1} = \tfrac{1}{8}k_1 + \tfrac{3}{8}k_2 + \tfrac{1}{2}q. \tag{5.1.81}$$

Finally, for the internal variables of g_1,

$$k_{34}^{g_1} = Q_{34} - q_{34}^{g_1} = -k_1, \qquad k_{24}^{g_1} = Q_{24} - q_{24}^{g_1} = k_1. \tag{5.1.82}$$

Similarly, for g_2,

$$q_{32}^{g_2} = \tfrac{3}{4}k_1 + \tfrac{1}{4}k_2, \qquad q_{13}^{g_2} = -\tfrac{3}{8}k_1 - \tfrac{1}{8}k_2 + \tfrac{1}{2}q, \tag{5.1.83}$$

$$k_{32}^{g_2} = -k_2, \qquad k_{13}^{g_2} = -k_2. \tag{5.1.84}$$

The reader may wish to work out the canonical decomposition of the Q_{ijl} in reference to the subdiagrams g_1 and g_2 with the canonical choice in (5.1.39).

Example 5.10: Now we consider the classic example of the graph in Fig. 5.6f with proper subdiagrams g_1 and g_2 as depicted in Fig. 5.7d. At vertex v_2 of subdiagram g_1,

$$q_{321}^{g_1} + q_{322}^{g_1} = Q_{321} + Q_{322},$$

or from (5.1.44) and (5.1.64),

$$q_{32l}^{g} = -\tfrac{1}{7}k_1 + \tfrac{4}{7}k_2 + \tfrac{1}{7}q, \qquad l = 1, 2. \tag{5.1.85}$$

From (5.1.65) and (5.1.44), we then obtain

$$k_{321}^{g_1} = -k_1 = -k_{322}^{g_1}. \tag{5.1.86}$$

Similarly, for g_2,

$$q_{14}^{g_2} = -\tfrac{2}{7}k_1 + \tfrac{1}{7}k_2 - \tfrac{5}{7}q,$$
$$q_{12}^{g_2} = \tfrac{2}{7}k_1 - \tfrac{1}{7}k_2 - \tfrac{2}{7}q,$$
$$q_{34}^{g_2} = \tfrac{2}{7}k_1 - \tfrac{1}{7}k_2 - \tfrac{2}{7}q, \tag{5.1.87}$$
$$q_{23}^{g_2} \equiv q_{232}^{g_2} = -\tfrac{6}{7}k_1 + \tfrac{3}{7}k_2 - \tfrac{1}{7}q,$$

and

$$k_{14}^{g_2} = k_2 = -k_{12}^{g_2} = -k_{34}^{g_2} = -k_{232}^{g_2}. \tag{5.1.88}$$

5.1.4 Taylor Operations

Let G be a proper and connected graph. Let us scale its Q_{ijl} variables in the expression for the unrenormalized integrand I_G in (5.1.2) by a parameter λ. If $d(P_{ijl})$ denotes the degree of a polynomial P_{ijl} with respect to λ in (5.1.1), we then define, as in Chapter 2, the degree of D_{ijl}^+ in (5.1.1) by $d(D_{ijl}^+) = d(P_{ijl}) - 2$. Similarly, let $d(\mathscr{P}_j)$ be the degree, with respect to λ, of the polynomial $\mathscr{P}_j$ associated with the vertex v_j of G. We then define the dimensionality $d(G)$ of the graph G by

$$d(G) = \sum_{\substack{ijl \\ i<j}}^{G} d(D_{ijl}^+) + \sum_{j}^{G} d(\mathscr{P}_j) + 4L(G),$$
$$\tag{5.1.89}$$
$$L(G) \equiv \#\mathscr{L}^G - \#\mathscr{V}^G + 1,$$

where $L(G)$ coincides with the number of the independent internal momenta $k_1, \ldots, k_n$ and hence also with the number of independent internal variables k_{ijl}^G associated with G. As before, $\mathscr{L}^G$ and $\mathscr{V}^G$ denote the set of lines and vertices of G, and $\#\mathscr{L}^G$, $\#\mathscr{V}^G$ denote the number of elements in them. Similarly, if g is a proper and connected subdiagram of G, then a similar expression to (5.1.89) may be written for the dimensionality $d(g)$ of g, with G replaced by g in the latter, i.e., in particular, with $L(g) = \#\mathscr{L}^g - \#\mathscr{V}^g + 1$.

If a proper but disconnected subdiagram g has m proper and connected parts $g_1, \ldots, g_m$, we write $g = \bigcup_{i=1}^{m} g_i$. The dimensionality $d(g)$ of g is then given by

$$d(g) = \sum_{i=1}^{m} d(g_i). \tag{5.1.90}$$

If I_{g_i} is the unrenormalized Feynman integrand associated with the subdiagram g_i, we write $I_g = \prod_{i=1}^{m} I_{g_i}$. Suppose g' is a subdiagram of g with unrenormalized integrand $I_{g'}$; we then define the expression $I_{g/g'}$ by

$$I_g = I_{g/g'} I_{g'}. \tag{5.1.91}$$

In other words, $I_{g/g'}$ represents the unrenormalized integrand of g with $I_{g'}$, corresponding to $g' \nsubseteq g$, replaced by unity. We use the notation $I_\varnothing = 1$. By definition, for $g' \nsubseteq g$, a line belonging to g' does not belong to g/g'. We may interpret g/g' diagrammatically as the subdiagram obtained from g by shrinking g' in it to a point. A few examples of this process are given in Fig. 5.8. Accordingly, the subdiagram g' is considered as a "vertex" of g/g' with which is associated the analytical expression $\mathscr{P}_{g'} = 1$. If g' is a proper subdiagram with s connected parts $g_1', \ldots, g_s'$, we may then write

$$I_g = I_{g/g'} \prod_{i=1}^{s} I_{g_i'}. \tag{5.1.92}$$

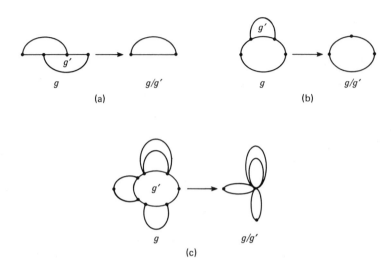

Fig. 5.8   The subdiagrams on the right-hand side of each subdiagram g is obtained by shrinking the subdiagram $g' \not\subseteq g$ in it to a point.

Finally, we note that if g_1 and g_2, $g_1 \not\supseteq g_2$, are proper and connected subdiagrams having a certain number X of vertices in common, then if g_2 in g_1 is shrunk to a point, these X vertices "merge" into a single vertex and we may write $\#\mathcal{V}^{g_1} = \#\mathcal{V}^{g_1/g_2} + \#\mathcal{V}^{g_2} - 1$. Also, $\#\mathcal{L}^{g_1} = \#\mathcal{L}^{g_1/g_2} + \#\mathcal{L}^{g_2}$. Hence $L(g_1) = L(g_1/g_2) + L(g_2)$ and we note that $d(g_1) = d(g_1/g_2) + d(g_2)$. If g' is a proper subdiagram with s connected parts, then $L(g') = \sum_{i=1}^{s} L(g_i')$ and we may write for the dimensionality of g in (5.1.92) $d(g) = d(g/g') + \sum_{i=1}^{s} d(g_i')$.

As before, let $\{q_j^G\} \equiv q^G$ be the set of the total external momenta carried away from the external vertices of G. We define by T_G the Taylor operation on I_G in the external independent momenta in q^G, about the origin, up to the order $d(G)$. If the dimensionality $d(G) < 0$, then we set $T_G = 0$.[7] Accordingly, $T_G I_G$ is of the form

$$T_G I_G = \sum_{r=0}^{d(G)} \frac{1}{r!} \sum_{j_1,\dots,j_r} q_{j_1}^{\mu_1} \cdots q_{j_r}^{\mu_r} \left[\frac{\partial^r}{\partial q_{j_1}^{\mu_1} \cdots q_{j_r}^{\mu_r}} I_G \right]_{q^G = 0}. \qquad (5.1.93)$$

To this end, note that the external variables of G in $\{q_{ijl}^G\} \equiv \tilde{q}^G$ may be written as linear combinations of the external total momenta of G in $\{q_j^G\} \equiv q^G$: $q_{ijl}^G = q_{ijl}^G(q^G)$ [see following Eq. (5.1.11)].

---

[7] In general, one may carry out a Taylor operation to any order $d'(G)$ as long as $d'(G) \geq d(G)$. Without loss of generality, we shall take the order associated with T_G to coincide with the dimensionality $d(G)$ of G. This will also apply to the Taylor operations in reference to proper subdiagrams $g \subset G$ to be discussed.

Similarly, if g is proper (not necessarily connected) subdiagram g of G, $T_g I_g$ is defined with reference to the external independent total momenta in $\{q_j^g\} \equiv q^g$, up to the order $d(g)$. If $d(g) < 0$, then we set $T_g = 0$. Again the external variables of g in $\{q_{ijl}^g\} = \tilde{q}^g$ may be written as linear combinations of the external total momenta of g in $\{q_j^g\} \equiv q^g$ [see following Eq. (5.1.66)].

We are particularly interested in the consecutive application of two or more Taylor operations as follows. Suppose g' is a proper subdiagram of g $(g' \not\subseteq g)$; then we are interested in the operation in $T_g T_{g'} I_g$. This is precisely defined in the following manner. We write I_g in the form (5.1.92) and carry out the Taylor operation $T_{g'}$ on $I_{g'}$ with reference to the external independent total momenta in $\{q_j^{g'}\} \equiv q^{g'}$. The expression $I_{g/g'} T_{g'} I_{g'}$ is of the form

$$I_{g/g'} T_{g'} I_{g'} = F(Q, k^{g'}, q^{g'}), \tag{5.1.94}$$

where $F(Q, k^{g'}, q^{g'})$ is a function of the variables in $\{Q_{ijl}^g\} \equiv Q^g$, the variables in $\{k_{ijl}^{g'}\} \equiv k^{g'}$, and the variables in $\{q_j^{g'}\} \equiv q^{g'}$. *Before* carrying out the Taylor operation T_g on $I_{g/g'} T_{g'} I_{g'}$, we express this F as a function of the internal and external variables of g, i.e., of k^g and q^g. This is easy to do. We use (5.1.46) to write $Q_{ijl} = k_{ijl}^g + q_{ijl}^g$. Similarly, we use (5.1.67) and (5.1.68) to write $k_{ijl}^{g'} = k_{ijl}^{g'}(k^g)$ and $q_{ijl}^{g'} = q_{ijl}^{g'}(k^g, q^g)$, $q_j^{g'} = \sum_{il}^{g'} q_{ijl}^{g'}(k^g, q^g)$, with the latter by analogy with (5.1.48), or directly from the expression corresponding to (5.1.47) for g': $\sum_{il}^{g'} Q_{ijl} = q_j^{g'}$, with $Q_{ijl} = k_{ijl}^g + q_{ijl}^g$, to write

$$F(Q, k^{g'}, q^{g'}) = \tilde{F}(k^g, q^g). \tag{5.1.95}$$

Now we may finally carry out the Taylor operation T_g with respect to the independent components in q^g by applying it to $\tilde{F}$ in (5.1.95). This gives the precise procedure of carrying out the consecutive Taylor operations and may be summarized through the following equation:

$$T_g T_{g'} I_g = T_g[I_{g/g'} T_{g'} I_{g'}]$$
$$= T_g[F(Q, k^{g'}, q^{g'})] = T_g \tilde{F}(k^g, q^g), \qquad g' \not\subseteq g. \tag{5.1.96}$$

The meaning of the consecutive Taylor operations of the form in $T_{g_1} T_{g_2} \cdots T_{g_n} I_g$, where $g_1 \not\supseteq g_2 \not\supseteq \cdots \not\supseteq g_n$ are proper subdiagrams of g, is now obvious.

We work out a few examples. For the purpose of illustrations we omit the $i\varepsilon$ factor in the denominators only to simplify the notation.

Example 5.11: Consider the graph g of Fig. 5.6b and the proper subdiagram $g_1 \subset g$ in Fig. 5.7a. We take

$$I_g = \prod_{l=1}^{3} (Q_{121}^2 + \mu^2)^{-1}, \qquad I_{g_1} = \prod_{l=2}^{3} (Q_{121}^2 + \mu^2)^{-1},$$

where $d(g) = 2$ and $d(g_1) = 0$. Then

$$T_{g_1}I_{g_1} = \prod_{l=2}^{3} [(k_{12l}^{g_1})^2 + \mu^2]^{-1}, \tag{5.1.97}$$

where the $k_{12l}^{g_1}$ are given by (5.1.73) and (5.1.74). Also,

$$I_{g/g_1}T_{g_1}I_{g_1} = (Q_{121}^2 + \mu^2)^{-1} \prod_{l=2}^{3} [(k_{12l}^{g_1})^2 + \mu^2]^{-1}, \tag{5.1.98}$$

where we write [see (5.1.20)] $Q_{121} = k_{121}^g + q_{121}^g$, with $k_{121}^g = -k_1$, $q_{121}^g = -\frac{1}{3}q$. $T_{g_1}I_{g_1}$ is independent of q_{ijl}^g. We scale q_{121}^g in I_{g/g_1} by λ and carry out the Taylor operation T_g, with respect to λ, up to the order $d(g) = 2$, and then set $\lambda = 1$, to obtain

$$T_g T_{g_1}I_g = \{[(k_{121}^g)^2 + \mu^2]^{-1} - [2q_{121}^g k_{121}^g + (q_{121}^g)^2][(k_{121}^g)^2 + \mu^2]^{-2}$$
$$+ (2q_{121}^g k_{121}^g)^2[(k_{121}^g)^2 + \mu^2]^{-3}\} \prod_{l=2}^{3} [(k_{12l}^{g_1})^2 + \mu^2]^{-1}, \tag{5.1.99}$$

and the latter may be expressed as a function of the integration variables k_1, k_2 and the external momentum q [see (5.1.72)–(5.1.74) and (5.1.20)]. The expression for $T_g T_{g_2} I_g$ may be similarly obtained.

Example 5.12:   Consider the graph g of Fig. 5.6c and the proper subdiagrams $g_1, g_2 \subset g$ of Fig. 5.7b. For simplicity, we take

$$I_g = \prod_{l'=1}^{2} (Q_{12l'}^2 + \mu^2)^{-1} \prod_{l=1}^{2} (Q_{23l}^2 + \mu^2)^{-1}, \tag{5.1.100}$$

and hence

$$I_{g_1} = \prod_{l=1}^{2} (Q_{23l}^2 + \mu^2)^{-1}, \qquad I_{g_2} = \prod_{l=1}^{2} (Q_{12l}^2 + \mu^2)^{-1}. \tag{5.1.101}$$

We have

$$T_{g_1}I_{g_1} = \prod_{l=1}^{2} [(k_{23l}^{g_1})^2 + \mu^2]^{-1}, \qquad T_{g_2}I_{g_2} = \prod_{l=1}^{2} [(k_{12l}^{g_2})^2 + \mu^2]^{-1}, \tag{5.1.102}$$

where the $k_{23l}^{g_1}$ and $k_{12l}^{g_2}$ are given in (5.1.79). We may write $Q_{12l} = k_{12l}^g + q_{12l}^g$, where $k_{121}^g = k_1 = -k_{122}^g$ and $q_{12l}^g = \frac{1}{2}q, l = 1, 2$ [see (5.1.25)]. Accordingly,

$$T_g T_{g_1}I_g = \prod_{l'=1}^{2} [(k_{12l'}^g)^2 + \mu^2]^{-1} \prod_{l=1}^{2} [(k_{23l}^{g_1})^2 + \mu^2]^{-1} \tag{5.1.103}$$

and may be written as a function of the integration variables k_1, k_2, and the external momentum q.

Similarly,

$$T_g T_{g_2} I_g = \prod_{l'=1}^{2} [(k^g_{23l'})^2 + \mu^2]^{-1} \prod_{l=1}^{2} [(k^{g_2}_{12l})^2 + \mu^2]^{-1}; \quad (5.1.104)$$

also,

$$T_g I_g = \prod_{l'=1}^{2} [(k^g_{23l'})^2 + \mu^2]^{-1} \prod_{l=1}^{2} [(k^g_{12l})^2 + \mu^2]^{-1}, \quad (5.1.105)$$

when the k^g_{ijl} are given by (5.1.25) and the $k^{g_2}_{12l}$ are given in (5.1.79).

Now we consider a slightly more complicated example.

Example 5.13:   Consider the graph in Fig. 5.9a with external vertices v_1 and v_2 with $q_1 + q_2 = 0$, $q_1 = q$, and hence $q_2 = -q$. g_1 in Fig. 5.9b is a proper subdiagram of g. A canonical choice of the variables Q_{ijl} with respect to the graph g is

$$\begin{aligned} Q_{121} = k_1 - \tfrac{1}{4}q, \qquad Q_{122} = k_2 - \tfrac{1}{4}q, \\ Q_{123} = k_3 - \tfrac{1}{4}q, \qquad Q_{124} = -k_1 - k_2 - k_3 - \tfrac{1}{4}q. \end{aligned} \quad (5.1.106)$$

The canonical decomposition of the Q_{ijl} with respect to the subdiagram g_1 is then $Q_{ijl} = k^{g_1}_{ijl} + q^{g_1}_{ijl}$, with

$$q^{g_1}_{12l} = \tfrac{1}{3}(k_1 + k_2 + k_3) - \tfrac{1}{4}q, \qquad l = 1, 2, 3, \quad (5.1.107)$$

$$k^{g_1}_{121} = \tfrac{2}{3}k_1 - \tfrac{1}{3}(k_2 + k_3),$$

$$k^{g_1}_{122} = \tfrac{2}{3}k_2 - \tfrac{1}{3}(k_1 + k_3), \quad (5.1.108)$$

$$k^{g_1}_{123} = \tfrac{2}{3}k_3 - \tfrac{1}{3}(k_1 + k_2),$$

where k_1, k_2, and k_3 are integration variables. The dimensionalities of g and g_1 are, respectively, $d(g) = 4$ and $d(g_1) = 2$, where

$$I_g = \prod_{l=1}^{4} (Q^2_{12l} + \mu^2)^{-1}, \qquad I_{g_1} = \prod_{l=1}^{3} (Q^2_{12l} + \mu^2)^{-1}, \quad (5.1.109)$$

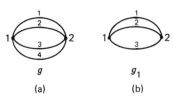

$$g \qquad\qquad g_1$$

(a) \qquad\qquad\qquad (b)

Fig. 5.9   A graph g involving scalar particles for Example 5.13. g_1 is a proper subdiagram of g with nonnegative dimensionality.

and the expression for I_{g_1} follows from that of I_g. A straightforward application of T_{g_1} on I_{g_1} gives

$$
T_{g_1} I_{g_1} = \left[1 - \sum_{l=1}^{3} \frac{[2q_{12l}^{g_1} k_{12l}^{g_1} + (q_{12l}^{g_1})^2]}{[(k_{12l}^{g_1})^2 + \mu^2]} + 2 \sum_{l=1}^{3} \frac{(q_{12l}^{g_1} k_{12l}^{g_1})^2}{[(k_{12l}^{g_1})^2 + \mu^2]^2} \right.
$$
$$
\left. + 2 \left(\sum_{l=1}^{3} \frac{q_{12l}^{g_1} k_{12l}^{g_1}}{[(k_{12l}^{g_1})^2 + \mu^2]} \right)^2 \right] I_{g_1}^0,
\tag{5.1.110}
$$

where

$$
I_{g_1}^0 = \prod_{l=1}^{3} [(k_{12l}^{g_1})^2 + \mu^2]^{-1}.
\tag{5.1.111}
$$

We also have $I_{g/g_1} = (Q_{124}^2 + \mu^2)^{-1}$ with Q_{124} as given in (5.1.106). In order to apply the Taylor operation T_g on $I_{g/g_1} T_{g_1} I_{g_1}$, we have to write the $q_{12l}^{g_1}$ as functions of the q_{12l}^g. We note that

$$
q_{12l}^{g_1} = Q_{12l} - k_{12l}^{g_1} = q_{12l}^g + (k_{12l}^g - k_{12l}^{g_1}),
\tag{5.1.112}
$$

with

$$
q_{12l}^g = -\tfrac{1}{4}q,
\tag{5.1.113}
$$

where k_{12l}^g, $k_{12l}^{g_1}$ are functions of k_1, k_2, and k_3 as given by (5.1.106) and (5.1.108). Hence we may express $I_{g/g_1} T_{g_1} I_g$ as a function of k_1, k_2, k_3 and the external variables q_{12l}^g of g. From (5.1.112) we may express the $q_{12l}^{g_1}$ in the numerators of (5.1.110) in terms of q_{ijl}^g and k_{12l}^g, $k_{12l}^{g_1}$. We may also write $Q_{124} = k_{124}^g + q_{124}^g$ in $I_{g/g_1} = (Q_{124}^2 + \mu^2)^{-1}$, and hence we may finally carry out the Taylor operation T_g on $I_{g/g_1} T_{g_1} I_{g_1}$ up to the fourth order in q in a straightforward, though tedious, manner.

Similar examples may be also given when different types of propagators and different masses are involved, and they may be treated in the same manner.

Let g and g' be two proper and connected subdiagrams with $g' \not\supseteq g$. For future reference we show that the actual dependence of the q_j^g and the q_{ijl}^g on the $k_{ijl}^{g'}$ is only on the $k_{ijl}^{g'}$ corresponding to the lines of the subdiagram g'/g, in the notation (5.1.91), i.e., of g' but not of g.

To this end we consider the expression (5.1.47) with G in it replaced by g', which gives

$$
\sum_{il}^{g} Q_{ijl} = q_j^g(k^{g'}, q^{g'}),
\tag{5.1.114}
$$

at each external vertex v_j of g. We also write

$$
Q_{ijl} = k_{ijl}^{g'} + q_{ijl}^{g'},
\tag{5.1.115}
$$

and use (5.1.7), with G replaced by g', to obtain

$$\sum_{il}^{g'} Q_{ijl} = q_j^{g'}$$

or

$$\sum_{il}^{g} Q_{ijl} + \sum_{il}^{'g'/g} Q_{ijl} = q_j^{g'}, \qquad (5.1.116)$$

where the second term in (5.1.116) is *defined* as a sum over all the i and l corresponding to all the vertices v_i and lines ℓ_l, belonging to g' *but not* to g, such that the lines ℓ_l join the vertices v_i to the external vertex v_j of g. From (5.1.114), (5.1.115), and (5.1.116), we then have

$$q_j^g(k^{g'}, q^{g'}) = q_j^{g'} - \sum_{il}^{'g'/g} (k_{ijl}^{g'} + q_{ijl}^{g'}). \qquad (5.1.117)$$

Also, the q_{ijl}^g may be written as linear combinations of the q_j^g [see following Eqs. (5.1.66) and (5.1.11)]. Accordingly, we see from (5.1.117) that the q_j^g and the q_{ijl}^g are linear combinations of the $k_{ijl}^{g'}$ in g'/g, i.e., $q_j^g(k^{g'}, q^{g'})$, $q_{ijl}^g(k^{g'}, q^{g'})$, with $k_{ijl}^{g'}$ in g'/g. This result will be quite useful later on. The same conclusion may be reached if $g' \not\supseteq g$, with g' and g proper but not necessarily connected subdiagrams, by replacing g' in (5.1.117) in turn by each of its connected components g_i' and by replacing g by those connected components of g falling in each g_i'.

Finally let $\mathcal{V}_1^{g'}$ be the set of vertices in $\mathcal{V}^{g'}$ but not in $\mathcal{V}^g$. Then for $v_j \in \mathcal{V}_1^{g'}$,

$$\sum_{il}^{'g'/g} k_{ijl}^{g'} = 0. \qquad (5.1.118)$$

Also, we denote by $\mathcal{V}_2^{g'}$ the set of vertices in $\mathcal{V}^g$ such that if $v_j \in \mathcal{V}_2^{g'}$, then we may find at least one line in g'/g that joins the vertex v_j. By summing over all the $v_j \in \mathcal{V}_2^{g'}$, we have by momentum conservation,

$$\sum_j^{\mathcal{V}_2^{g'}} \left(\sum_{il}^{'g'/g} k_{ijl}^{g'} \right) = 0.^8 \qquad (5.1.119)$$

The $\#\mathcal{V}^{g'/g}$ equations in (5.1.118) and (5.1.119) together with the constraint[9]

$$\sum_j^{\mathcal{V}_1^{g'}} \left(\sum_{il}^{'g'/g} k_{ijl}^{g'} \right) + \sum_j^{\mathcal{V}_2^{g'}} \left(\sum_{il}^{'g'/g} k_{ijl}^{g'} \right) = 0, \qquad (5.1.120)$$

---

[8] Note that, by definition, the vertices v_i and the lines ℓ_l joining these vertices to the vertices in $\mathcal{V}_2^{g'}$, do not belong to g.

[9] The constraint in (5.1.120) simply means that one of the $\#\mathcal{V}^{g'/g}$ expressions on the left-hand sides of (5.1.118)–(5.1.119) is a linear combination of the remaining $\#\mathcal{V}^{g'/g} - 1$ expressions.

then imply that $\# \mathscr{L}^{g'/g} - (\# \mathscr{V}^{g'/g} - 1) \equiv L(g'/g)$ of the $\# \mathscr{L}^{g'/g} \, k^{g'}_{ijl}$ in g'/g may be independent.

5.2   THE SUBTRACTION SCHEME

Let I_G be the unrenormalized Feynman integrand associated with a proper and connected graph G. We recall that a subdiagram is proper if it is amputated and its number of connected parts does not increase upon cutting any one of its lines. We define the renormalized Feynman integrand R, associated with G, involving subtractions as follows:

$$R = \left[1 + \sum_D \prod_{g \in D} (-T_g) \right] I_G, \qquad (5.2.1)$$

where the sum is over *all* nonempty sets D such that

   (i)   If $g \in D$, then g is a proper (but not necessarily connected) sub-diagram of G with $d(g) \geq 0$. If $d(G) \geq 0$, then one of the elements of D may be G itself.
   (ii)   If $g_1, g_2 \in D$, then either $g_1 \not\subseteq g_2$ or $g_2 \not\subseteq g_1$. If $g_1 \not\subseteq g_2$, then the ordering of the Taylor operations in (5.2.1) is as $\cdots T_{g_2} \cdots T_{g_1} \cdots$, as defined in Section 5.1.4.

Remarks

   1.   The Taylor operations are directly applied to the integrand I_G (directly in *momentum space*), and hence no questions of divergences arise in (5.2.1). The sets D in (5.2.1) will be called renormalization sets.
   2.   In Corollary 5.4.1 we shall prove that in obtaining the *final* expression for R, one may restrict the summation in (5.2.1) over renormalization sets D such that for each connected part g_i of a $g \in D$ we have $d(g_i) \geq 0$, as the other D sets will not contribute to the sum in (5.2.1), and hence to the final expression for R. The more general form for the D sets given will be useful in providing the convergence proof of renormalization in Section 5.3.

   The structure in (5.2.1) is very simple, as will be seen in the examples to follow. The simplicity of this structure will also be reflected in the convergence proof (Section 5.3) of the subtraction scheme and in later work in the appendix.

The renormalized Feynman amplitude associated with G is then

$$\mathscr{A}_\varepsilon(P, \mu) = \int_{\mathbb{R}^{4n}} dK \, R(P, K, \mu, \varepsilon), \qquad (5.2.2)$$

where $K = (k_1^0, \ldots, k_n^3)$ and $P = (p_1^0, \ldots, p_m^3)$ denote, respectively, the integration variables (components of the internal momenta) and the components of the independent external momenta associated with G. The absolute convergence of (5.2.2) will be given in Section 5.3 for $\varepsilon > 0$. The existence of the limit $\varepsilon \to +0$ of $\mathscr{A}_\varepsilon$ in the sense of distributions then follows from the work in Chapter 4.

We give a few examples concerning our subtraction scheme (5.2.1) that also illustrate the simplicity of the structure of $[1 + \sum_D \prod_{g \in D} (-T_g)]$.

Example 5.14:   For Example 5.11, the graph g (Fig. 5.6b) and the proper subdiagrams g_1 and g_2 (Fig. 5.7a) all have nonnegative dimensionalities; hence the renormalization sets D are

$$\{g\}, \quad \{g, g_1\}, \quad \{g, g_2\}, \quad \{g_1\}, \quad \{g_2\}. \qquad (5.2.3)$$

Note that $\{g_1, g_2\}$ is *not* a D set since neither $g_1 \nsubseteq g_2$ nor $g_2 \nsubseteq g_1$. The renormalized integrand is then given by

$$R = (1 - T_g)[1 - T_{g_1} - T_{g_2}]I_g. \qquad (5.2.4)$$

Example 5.15:   For Example 5.12, the graph g (Fig. 5.6c) and the proper subdiagrams g_1 and g_2 (Fig. 5.7b) all have nonnegative dimensionalities. Therefore the renormalization sets D are as given in (5.2.3) and the renormalized integrand is of the form (5.2.4). Note again that $\{g_1, g_2\}$ is not a D set.

Example 5.16:   For the graph g of Fig. 5.10a, the dotted lines denote scalar particles and the solid lines denote spin-$\frac{1}{2}$ particles with, for example, a $\bar\psi\psi\phi$ interaction. The degrees of the corresponding propagators are, respectively, -2 and -1. $g, g_1, \ldots, g_5$ denote all proper subdiagrams (Figs. 5.10a–f) of g with nonnegative dimensionalities. Accordingly, the renormalization D sets are

$$
\begin{aligned}
&\{g, g_3, g_1\}, \quad \{g, g_3\}, \quad \{g, g_4, g_2\}, \quad \{g, g_4\}, \quad \{g, g_5, g_1\}, \\
&\{g, g_5, g_2\}, \quad \{g, g_5\}, \quad \{g, g_2\}, \quad \{g, g_1\}, \quad \{g\}, \\
&\{g_3, g_1\}, \quad \{g_3\}, \quad \{g_4, g_2\}, \quad \{g_4\}, \quad \{g_5, g_1\}, \\
&\{g_5, g_2\}, \quad \{g_5\}, \quad \{g_2\}, \quad \{g_1\}.
\end{aligned} \qquad (5.2.5)
$$

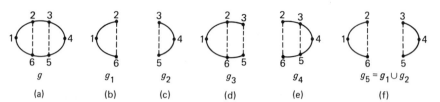

Fig. 5.10   A graph g in (a) where the dotted lines denote scalar particles and the solid lines denote spin-$\frac{1}{2}$ particles. The degree of the corresponding propagators are, respectively, -2 and -1. The subdiagrams $g_1, \ldots, g_5$ and g itself denote the proper subdiagrams of g with nonnegative dimensionalities.

The renormalized integrand associated with g is then

$$R = [1 - T_g]\{1 - T_{g_3}[1 - T_{g_1}] - T_{g_4}[1 - T_{g_2}] - T_{g_5}[1 - T_{g_1} - T_{g_2}]$$
$$- T_{g_1} - T_{g_2}\} I_g. \tag{5.2.6}$$

It is easy to write down the expression (5.2.6) immediately, by inspection, without writing out first the sets in (5.2.5) in detail. For example, the proper subdiagrams of g, not equal to g, with nonnegative dimensionalities are g_3, g_4, g_5, g_1, and g_2. Hence in the curly brackets we have

$$\{1 - T_{g_3}[\cdot] - T_{g_4}[\cdot] - T_{g_5}[\cdot] - T_{g_1}[\cdot] - T_{g_2}[\cdot]\}. \tag{5.2.7}$$

We may now consider each term in (5.2.7). For the T_{g_3}, the only proper subdiagram of g_3, not equal to g_3, with the nonnegative dimensionality is g_1, accordingly the expression, for the square brackets multiplying T_{g_3} (from the right!) is $[1 - T_{g_1}]$. The 1 factor in $[1 - T_{g_1}]$ occurs, of course, because $\{g_3, g_1\}$ as well as $\{g_3\}$ are renormalization sets. Here we note that g_1 does not contain a proper subdiagram, not equal to g_1, with nonnegative dimensionality; hence we may write $[1 - T_{g_1}[\cdot]] = [1 - T_{g_1}]$. Otherwise, we would have repeated the above procedure for filling in the square brackets multiplying T_{g_1} in the same manner as for $T_{g_3}[\cdot]$. The square brackets in $T_{g_4}[\cdot]$ and $T_{g_5}[\cdot]$ are filled in the same manner. The square brackets multiplying $T_{g_1}[\cdot]$ and $T_{g_2}[\cdot]$ in (5.2.7) are simply replaced by 1 since g_1 and g_2 do not contain proper subdiagrams, not equal to g_1 and g_2, respectively, with nonnegative dimensionalities, as just mentioned.

The method of writing the expressions for R, as exemplified in (5.2.4) and (5.2.6), is thus straightforward.

Example 5.17:   We explicitly work out the Taylor operations in (5.2.1) for Example 5.12 corresponding to the graph in Fig. 5.6c to construct R.

According to (5.2.4), corresponding to Example 5.15,

$$R = (1 - T_g)[1 - T_{g_1} - T_{g_2}]I_g, \tag{5.2.8}$$

where from (5.1.100) and (5.1.102)–(5.1.105),

$$I_g = \prod_{l'=1}^{2} [Q_{12l'}^2 + \mu^2]^{-1} \prod_{l=1}^{2} [Q_{23l}^2 + \mu^2]^{-1},$$

$$T_{g_1} I_g = \prod_{l'=1}^{2} [Q_{12l'}^2 + \mu^2]^{-1} \prod_{l=1}^{2} [(k_{23l}^{g_1})^2 + \mu^2]^{-1},$$

$$T_{g_2} I_g = \prod_{l'=1}^{2} [(k_{12l'}^{g_2})^2 + \mu^2]^{-1} \prod_{l=1}^{2} [Q_{23l}^2 + \mu^2]^{-1},$$

$$T_g T_{g_1} I_g = \prod_{l'=1}^{2} [(k_{12l'}^{g})^2 + \mu^2]^{-1} \prod_{l=1}^{2} [(k_{23l}^{g_1})^2 + \mu^2]^{-1}, \qquad (5.2.9)$$

$$T_g T_{g_2} I_g = \prod_{l'=1}^{2} [(k_{12l'}^{g_2})^2 + \mu^2]^{-1} \prod_{l=1}^{2} [(k_{23l}^{g})^2 + \mu^2]^{-1},$$

$$T_g I_g = \prod_{l'=1}^{2} [(k_{12l'}^{g})^2 + \mu^2]^{-1} \prod_{l=1}^{2} [(k_{23l}^{g})^2 + \mu^2]^{-1},$$

with [see (5.1.25), (5.1.79)]

$$Q_{121} = k_1 + \tfrac{1}{2}q, \qquad Q_{122} = -k_1 + \tfrac{1}{2}q,$$
$$Q_{231} = k_2 + \tfrac{1}{2}q, \qquad Q_{232} = -k_2 + \tfrac{1}{2}q, \qquad (5.2.10)$$

$$k_{231}^{g_2} = k_2 = -k_{232}^{g_1}, \qquad k_{121}^{g_2} = k_1 = -k_{122}^{g_2}. \qquad (5.2.11)$$

Also, from (5.2.10),

$$k_{121}^{g} = k_1 = -k_{122}^{g}, \qquad k_{231}^{g} = k_2 = -k_{232}^{g}. \qquad (5.2.12)$$

Putting (5.2.9)–(5.2.12) in (5.2.8), we obtain, after some simplification,

$$R = \frac{[\tfrac{1}{16}q^4 - (k_1 q)^2 + \tfrac{1}{2}q^2(k_1^2 + \mu^2)][\tfrac{1}{16}q^4 - (k_2 q)^2 + \tfrac{1}{2}q^2(k_2^2 + \mu^2)]}{(k_1^2 + \mu^2)^2(k_2^2 + \mu^2)^2(k_{1+}^2 + \mu^2)(k_{1-}^2 + \mu^2)(k_{2+}^2 + \mu^2)(k_{2-}^2 + \mu^2)}, \qquad (5.2.13)$$

where we have defined

$$k_{1+} \equiv Q_{121} = k_1 + \tfrac{1}{2}q, \qquad k_{1-} \equiv Q_{122} = -k_1 + \tfrac{1}{2}q,$$
$$k_{2+} \equiv Q_{231} = k_2 + \tfrac{1}{2}q, \qquad k_{2-} \equiv Q_{232} = -k_2 + \tfrac{1}{2}q. \qquad (5.2.14)$$

It is instructive at this stage for the reader to compare the degrees of R with respect to k_1, k_2 and (k_1, k_2) with the corresponding ones of the unrenormalized integrand I_g defined in (5.2.9).

A *recursion* formula for determining the renormalized integrand R in (5.2.1) may be also given that will be useful later on. To this end let $D_{G'}$

denote a renormalization set D such that the largest subdiagram in $D_{G'}$ is G', i.e., if $g \in D_{G'}$, then necessarily $g \subset G'$. Let $D_\varnothing = \varnothing$ denote the empty set and define $(-T_\varnothing) = 1$. Then (5.2.1) may be rewritten in the form[10]

$$R = \sum_{\varnothing \subset G' \subset G} \sum_{D_{G'}} \prod_{g \in D_{G'}} (-T_g) I_G, \qquad (5.2.15)$$

where $\sum_{D_{G'}}$ denotes a summation over all renormalization sets $D_{G'}$ with largest subdiagram G' in $D_{G'}$, and $\sum_{\varnothing \subset G' \subset G}$ denotes a summation over all proper (but not necessarily connected) subdiagrams of G with $d(G') \geq 0$ (for $G' \neq \varnothing$) and, by definition, also includes the term with $G' = \varnothing$. The expression (5.2.15) may be rewritten in the equivalent form

$$R = \sum_{\varnothing \subset G' \nsubseteq G} \sum_{D_{G'}} \prod_{g \in D_{G'}} (-T_g) I_G + \sum_{D_G} \prod_{g \in D_G} (-T_g) I_G, \qquad (5.2.16)$$

where $\sum_{D_G}$ is the sum over all renormalization sets having G as their largest subdiagram (graph), and the corresponding second term in (5.2.16) reduces to zero if $d(G) < 0$ since

$$\sum_{D_G} \prod_{g \in D_G} (-T_g) I_G = (-T_G) \sum_{\varnothing \subset G' \nsubseteq G} \sum_{D_{G'}} \prod_{g \in D_{G'}} (-T_g) I_G, \qquad (5.2.17)$$

and $T_G = 0$ for the case $d(G) < 0$, by definition. Using (5.2.16) and (5.2.17), we may rewrite (5.2.15):

$$R = (1 - T_G) \left[1 + \sum_{\varnothing \nsubseteq G' \nsubseteq G} \sum_{D_{G'}} \prod_{g \in D_{G'}} (-T_g) \right] I_G, \qquad (5.2.18)$$

where we have used the definition $(-T_\varnothing) = 1$. We define the expression

$$M_{G'} = \sum_{D_{G'}} \prod_{g \in D_{G'}} (-T_g) I_{G'}. \qquad (5.2.19)$$

The latter may be rewritten

$$M_{G'} = (-T_{G'}) \sum_{D_{G'}} \prod_{\substack{g \in D_{G'} \\ g \nsubseteq G'}} (-T_g) I_{G'}$$

$$= (-T_{G'}) \left[1 + \sum_{D_{G'}} \prod_{\substack{g \in D_{G'} \\ \varnothing \nsubseteq g \nsubseteq G'}} (-T_g) \right] I_{G'}$$

$$= (-T_{G'}) \left[I_{G'} + \sum_{\varnothing \nsubseteq G'' \nsubseteq G'} I_{G'/G''} \sum_{D_{G''}} \prod_{g \in D_{G''}} (-T_g) I_{G''} \right], \qquad (5.2.20)$$

where we have used in the process of writing (5.2.20) the relation $I_{G'} = I_{G'/G''} I_{G''}$ for $G'' \nsubseteq G'$. Again using the definition (5.2.19) and noting that G'

---

[10] Recall that in this book the symbol $\subset$ in $\varnothing \subset G' \subset G$ may include equality as well. We use the symbol $\nsubseteq$ to exclude equality. $G' = \varnothing$ means no lines and no vertices.

is an arbitrary proper subdiagram of G (with $d(G') \geq 0$), we obtain from the last equality in (5.2.20)

$$M_{G'} = (-T_{G'})\left[I_{G'} + \sum_{\varnothing \not\subseteq G'' \not\subseteq G'} I_{G'/G''} M_{G''} \right]. \tag{5.2.21}$$

Therefore from (5.2.18) and the definition (5.2.19) we have the following equivalent expression for R:

$$R = (1 - T_G)\left[I_G + \sum_{\varnothing \not\subseteq G' \not\subseteq G} I_{G/G'} M_{G'} \right], \tag{5.2.22}$$

where $T_G = 0$ if $d(G) < 0$, and $M_{G'}$ is defined recursively by (5.2.21).

5.3   CONVERGENCE OF THE SUBTRACTION SCHEME

The renormalized Feynman amplitude associated with a proper and connected graph G is from (5.2.2) given by

$$\mathscr{A}_\varepsilon(P, \mu) = \int_{\mathbb{R}^{4n}} dK \, R(P, K, \mu, \varepsilon), \qquad \varepsilon > 0, \tag{5.3.1}$$

where R is given by (5.2.1) with the unrenormalized integrand I_G having the form (5.1.2). Obviously, R has the very general structure given (2.2.3). Let R_E be the Euclidean version of R by replacing the Minkowski metric $g_{\mu\nu}$ by the Euclidean metric $\eta_{\mu\nu}$ and setting $\varepsilon = 0$ in the latter as defined in (2.2.83). As in (2.2.84) we define

$$\mathscr{A}_E(P, \mu) = \int_{\mathbb{R}^{4n}} dK \, R_E(P, K, \mu). \tag{5.3.2}$$

In Chapter 2 we have established, in particular, that R_E and R, with $\varepsilon > 0$ in the latter, belong to class $B_{4n+4m+\rho}(I)$. In this section we prove the absolute convergence of the integral (5.3.2) and the absolute convergence of (5.3.1) for $\varepsilon > 0$. The limit $\varepsilon \to +0$ of $\mathscr{A}_\varepsilon(P, \mu)$ may be then seen to exist, in the sense of distributions, from Chapter 4.

In Section 5.3.1 some basic properties of Taylor operations and the remainder terms are obtained in view of applications to subtracted-out Feynman integrands. In Section 5.3.2 a basic grouping of the Taylor operations is carried out that is indispensable for the convergence proof that is finally completed in Section 5.3.3 by making use of the power-counting theorem established in Chapter 3 coupled to the fact that R_E and R belong to class $B_{4n+4m+\rho}(I)$.

5.3.1   Some Properties of Taylor Operations and Their Remainders

Lemma 5.3.1[11]: *Let $f(\zeta)$ be a function that is differentiable an arbitrary number of times in $\zeta \in \mathbb{R}^1$. Define recursively*

$$f_n(\zeta) = [f_{n-1}(\zeta) - f_{n-1}(0)]/\zeta, \tag{5.3.3}$$

$$f_0(\zeta) = f(\zeta); \tag{5.3.4}$$

then

$$f(\zeta) = \sum_{j=0}^{n-1} \zeta^j f_j(0) + \zeta^n f_n(\zeta) \tag{5.3.5}$$

for arbitrary positive integer n.

We prove (5.3.5) by induction. For $n = 1$, $f_1(\zeta) = [f(\zeta) - f(0)]/\zeta$, by definition, i.e.,

$$f(\zeta) = f(0) + \zeta f_1(\zeta), \tag{5.3.6}$$

which establishes (5.3.5) for $n = 1$. Suppose (5.3.5) is true for some $n = k > 1$, i.e.,

$$f(\zeta) = \sum_{j=0}^{k-1} \zeta^j f_j(0) + \zeta^k f_k(\zeta). \tag{5.3.7}$$

By definition, $f_{k+1}(\zeta) = [f_k(\zeta) - f_k(0)]/\zeta$, or

$$f_k(\zeta) = f_k(0) + \zeta f_{k+1}(\zeta). \tag{5.3.8}$$

Substituting (5.3.8) in (5.3.7), we obtain

$$f(\zeta) = \sum_{j=0}^{k-1} \zeta^j f_j(0) + \zeta^k f_k(0) + \zeta^{k+1} f_{k+1}(\zeta)$$

$$= \sum_{j=0}^{k} \zeta^j f_j(0) + \zeta^{k+1} f_{k+1}(\zeta); \tag{5.3.9}$$

i.e., the lemma is also true for $n = k + 1$. This completes the proof of the lemma by induction.

Lemma 5.3.2:   *Let*

$$f(\lambda, \zeta, \mu) = \prod_{j=1}^{L} [(\lambda k_j + \zeta q_j)^2 + \mu_j^2 - i\varepsilon\{(\lambda \mathbf{k}_j + \zeta \mathbf{q}_j)^2 + \mu_j^2\}]^{-1}. \tag{5.3.10}$$

---

[11] The classic formula in this lemma is known in the literature as Newton's divided difference interpolating formula (cf. Isaacson and Keller, 1966).

As in Lemma 5.3.1, we define recursively

$$f_n(\lambda, \zeta, \mu) = [f_{n-1}(\lambda, \zeta, \mu) - f_{n-1}(\lambda, 0, \mu)]/\zeta. \tag{5.3.11}$$

Define a set of coefficients $\{F_a(\lambda, \mu)\}$ by

$$\prod_{j=1}^{L} [(\lambda k_j)^2 + \mu_j^2 - i\varepsilon\{(\lambda \mathbf{k}_j)^2 + \mu_j^2\}]$$

$$- \prod_{j=1}^{L} [(\lambda k_j + \zeta q_j)^2 + \mu_j^2 - i\varepsilon\{(\lambda \mathbf{k}_j + \zeta \mathbf{q}_j)^2 + \mu_j^2\}]$$

$$= \sum_{a=1}^{2L} \zeta^a F_a(\lambda, \mu). \tag{5.3.12}$$

Then

$$f_n(\lambda, \zeta, \mu) = f(\lambda, 0, \mu) \sum_{a=1}^{n} f_{n-a}(\lambda, \zeta, \mu)F_a(\lambda, \mu)$$

$$+ f(\lambda, 0, \mu)f(\lambda, \zeta, \mu) \sum_{a=n+1}^{2L} \zeta^{a-n}F_a(\lambda, \mu)$$

$$\text{for } 1 \leq n < 2L \tag{5.3.13a}$$

and

$$f_n(\lambda, \zeta, \mu) = f(\lambda, 0, \mu) \sum_{a=1}^{2L} f_{n-a}(\lambda, \zeta, \mu)F_a(\lambda, \mu) \qquad \text{for } n \geq 2L. \tag{5.3.13b}$$

Let T_ζ^d denote the Taylor operation in ζ, about the origin, up to the order d; then

$$[1 - T_\zeta^d]f(\lambda, \zeta, \mu) = \zeta^{d+1}f_{d+1}(\lambda, \zeta, \mu), \tag{5.3.14}$$

where $f_{d+1}(\lambda, \zeta, \mu)$ is given recursively by (5.3.13) with $n = d + 1$, and from the latter, or by inspection, it is the ratio of two polynomials in λ, ζ, μ and is of the form

$$f_{d+1}(\lambda, \zeta, \mu) = \hat{P}_\varepsilon(\zeta q, q, \lambda k, \mu)G^{-1}(\zeta q, \lambda k, \mu), \tag{5.3.15}$$

where $\hat{P}_\varepsilon$ is a polynomial in its arguments in (5.3.15) and ε and

$$G(\zeta q, \lambda k, \mu) = \prod_{j=1}^{L} [(\lambda k_j + \zeta q_j)^2 + \mu_j^2 - i\varepsilon\{(\lambda \mathbf{k}_j + \zeta \mathbf{q}_j)^2 + \mu_j^2\}]$$

$$\times [(\lambda k_j)^2 + \mu_j^2 - i\varepsilon\{(\lambda \mathbf{k}_j)^2 + \mu_j^2\}]^{\beta_j}, \tag{5.3.16}$$

where β_j are some strictly positive integers.

To prove (5.3.13a) we proceed by induction. For $n = 1$, from (5.3.11),

$$f_1(\lambda, \zeta, \mu) = \frac{f(\lambda, \zeta, \mu) - f(\lambda, 0, \mu)}{\zeta}$$

$$= f(\lambda, \zeta, \mu)f(\lambda, 0, \mu)\left[\frac{f^{-1}(\lambda, 0, \mu) - f^{-1}(\lambda, \zeta, \mu)}{\zeta}\right]$$

$$= f(\lambda, \zeta, \mu)f(\lambda, 0, \mu) \sum_{a=1}^{2L} \zeta^{a-1}F_a(\lambda, \mu), \qquad (5.3.17)$$

where we have used (5.3.12). Equation (5.3.17) coincides with (5.3.13a) for $n = 1$. Now suppose that (5.3.13a) is true for some $n = k$ with $1 < k < 2L - 1$, i.e.,

$$f_k(\lambda, \zeta, \mu) = f(\lambda, 0, \mu) \sum_{a=1}^{k} f_{k-a}(\lambda, \zeta, \mu)F_a(\lambda, \mu)$$

$$+ f(\lambda, 0, \mu)f(\lambda, \zeta, \mu) \sum_{a=k+1}^{2L} \zeta^{a-k}F_a(\lambda, \mu). \qquad (5.3.18)$$

Using the definition (5.3.11) with $n = k + 1$, we obtain

$$f_{k+1}(\lambda, \zeta, \mu) = [f_k(\lambda, \zeta, \mu) - f_k(\lambda, 0, \mu)]/\zeta, \qquad (5.3.19)$$

and hence from (5.3.18),

$$f_{k+1}(\lambda, \zeta, \mu) = f(\lambda, 0, \mu) \sum_{a=1}^{k} F_a(\lambda, \mu)\left[\frac{f_{k-a}(\lambda, \zeta, \mu) - f_{k-a}(\lambda, 0, \mu)}{\zeta}\right]$$

$$+ f(\lambda, 0, \mu)f(\lambda, \zeta, \mu) \sum_{a=k+1}^{2L} \zeta^{a-k-1}F_a(\lambda, \mu) \qquad (5.3.20)$$

or

$$f_{k+1}(\lambda, \zeta, \mu) = f(\lambda, 0, \mu) \sum_{a=1}^{k+1} F_a(\lambda, \mu)f_{k+1-a}(\lambda, \zeta, \mu)$$

$$+ f(\lambda, 0, \mu)f(\lambda, \zeta, \mu) \sum_{a=k+2}^{2L} \zeta^{a-(k+1)}F_a(\lambda, \mu), \qquad (5.3.21)$$

where we have used definition (5.3.11) again, now with $n = k + 1 - a$, and have absorbed the first term of the second sum of (5.3.20) in the first sum; hence (5.3.21) verifies (5.3.13a) for all $n < 2L$. Using definition (5.3.11) and the expression (5.3.13a), we readily verify that (5.3.13b) is true for $n = 2L$, and by induction we easily see that (5.3.13b) is true for all $n \geq 2L$. This completes the proof of (5.3.13).

It is easy to see from definition (5.3.11) that

$$f_n(\lambda, 0, \mu) = \frac{1}{n!} \left(\frac{\partial}{\partial \zeta} \right)^n f(\lambda, \zeta, \mu) \Big|_{\zeta = 0} \equiv \frac{1}{n!} f^{(n)}(\lambda, 0, \mu). \qquad (5.3.22)$$

Thus if we choose $n = d + 1$, then (5.3.5) in Lemma 5.3.1 implies that

$$f(\lambda, \zeta, \mu) = \sum_{j=0}^{d} \frac{\zeta^j}{j!} f^{(j)}(\lambda, 0, \mu) + \zeta^{d+1} f_{d+1}(\lambda, \zeta, \mu), \qquad (5.3.23)$$

which in turn implies the result in (5.3.14) since the sum in (5.3.23) is the result of the Taylor operation on $f(\lambda, \zeta, \mu)$ up to the order d in ζ about the origin.

The final lemma in this subsection is the following:

Lemma 5.3.3:   *Consider an expression of the form*

$$H(\lambda, \zeta, \mu) = \frac{P(\zeta q, \lambda k, \mu)}{\prod_{j=1}^{L} [(\lambda k_j + \zeta q_j)^2 + \mu_j^2 - i\varepsilon\{(\lambda \mathbf{k}_j + \zeta \mathbf{q}_j)^2 + \mu_j^2\}]}, \qquad (5.3.24)$$

where $P(\zeta q, \lambda k, \mu)$ is a polynomial in the elements in ζq, λk, μ, and, in general, in the $(\mu_j)^{-1}$ as well. For all $j = 1, \ldots, L$ we assume $k_j^\sigma \neq 0$, for at least one $\sigma \in [0, 1, 2, 3]$ (corresponding to each j), in the denominators in (5.3.24). As before, let T_ζ^d denote the Taylor operation in ζ about the origin, up to an order $d \geq 0$. Then

$$\operatorname*{degr}_{\lambda} [1 - T_\zeta^d] H(\lambda, \zeta, \mu) \leq \operatorname*{degr}_{\lambda, \zeta} H(\lambda, \zeta, \mu) - d - 1. \qquad (5.3.25)$$

Also, if we scale any subset of the masses by a parameter η, then

$$\operatorname*{degr}_{\lambda, \eta} [1 - T_\zeta^d] H(\lambda, \zeta, \mu) \leq \operatorname*{degr}_{\lambda, \zeta, \eta} H(\lambda, \zeta, \mu) - d - 1. \qquad (5.3.26)$$

To prove (5.3.25) we write

$$P(\zeta q, \lambda k, \mu) = \sum_a \zeta^{|a|} q^a P_a(\lambda k, \mu), \qquad (5.3.27)$$

where[12]

$$a = (a_{01}, \ldots, a_{3m}), \qquad |a| = a_{01} + \cdots + a_{3m}, \qquad (5.3.28)$$

$$q^a \equiv (q_1^0)^{a_{01}} \cdots (q_m^3)^{a_{3m}}. \qquad (5.3.29)$$

By using the elementary property that for a differentiable function $f(\zeta)$

$$T_\zeta^d \zeta^{|a|} f(\zeta) = \zeta^{|a|} T_\zeta^{d-|a|} f(\zeta), \qquad (5.3.30)$$

---

[12] For $a = (a_{01}, \ldots, a_{3m})$ and $b = (b_{01}, \ldots, b_{3m})$, we define $a + b = (a_{01} + b_{01}, \ldots, a_{3m} + b_{3m})$.

where $T_\zeta^{d-|a|} = 0$ if $|a| > d$, we obtain

$$[1 - T_\zeta^d]H(\lambda, \zeta, \mu) = \sum_a \zeta^{|a|}q^a P_a(\lambda k, \mu)(1 - T_\zeta^{d-|a|})f(\lambda, \zeta, \mu), \quad (5.3.31)$$

where

$$f(\lambda, \zeta, \mu) = \prod_{j=1}^{L} [(\lambda k_j + \zeta q_j)^2 + \mu_j^2 - i\varepsilon\{(\lambda \mathbf{k}_j + \zeta \mathbf{q}_j)^2 + \mu_j^2\}]^{-1}. \quad (5.3.32)$$

Hence from (5.3.14) in Lemma (5.3.2) we have in an obvious notation

$$[1 - T_\zeta^d]H(\lambda, \zeta, \mu) = \zeta^{d+1} \sum_a q^a P_a(\lambda k, \mu)f_{d+1-|a|}(\lambda, \zeta, \mu), \quad (5.3.33)$$

where $f_{d+1-|a|}(\lambda, \zeta, \mu)$ is of the form (5.3.15) *and*

$$f_{d+1-|a|}(\lambda, 0, \mu) = \frac{1}{(d+1-|a|)!} \left(\frac{\partial}{\partial \zeta}\right)^{(d+1-|a|)} f(\lambda, \zeta, \mu)\Big|_{\zeta=0} \quad (5.3.34)$$

for $d + 1 - |a| \geq 0$. Let $h(\lambda, \zeta, \mu)$ be the sum multiplying ζ^{d+1} in (5.3.33). We use the facts that

$$\underset{\lambda}{\operatorname{degr}}[1 - T_\zeta^d]H(\lambda, \zeta, \mu) = \underset{\lambda}{\operatorname{degr}} h(\lambda, \zeta, \mu) \leq \underset{\lambda, \zeta}{\operatorname{degr}} h(\lambda, \zeta, \mu)$$

$$= \underset{\lambda, \zeta}{\operatorname{degr}}[1 - T_\zeta^d]H(\lambda, \zeta, \mu) - (d + 1), \quad (5.3.35)$$

where the last equality follows from the presence of the overall factor ζ^{d+1} multiplying $h(\lambda, \zeta, \mu)$ in (5.3.33). But

$$\underset{\lambda, \zeta}{\operatorname{degr}}[1 - T_\zeta^d]H(\lambda, \zeta, \mu) \leq \underset{\lambda, \zeta}{\operatorname{degr}} H(\lambda, \zeta, \mu), \quad (5.3.36)$$

and hence we conclude from (5.3.35)

$$\underset{\lambda}{\operatorname{degr}}[1 - T_\zeta^d]H(\lambda, \zeta, \mu) \leq \underset{\lambda, \zeta}{\operatorname{degr}} H(\lambda, \zeta, \mu) - (d + 1), \quad (5.3.37)$$

which is (5.3.25) in the lemma.

We rewrite $P(\zeta q, \lambda k, \mu)$ in (5.3.24) as $\prod_{j=1}^{\rho} (\mu_j)^{-\sigma_j} \tilde{P}(\zeta q, \lambda k, \mu)$, where the σ_j are positive integers and $\tilde{P}(\zeta q, \lambda k, \mu)$ is a polynomial in $\zeta q, \lambda k, \mu$. By repeating the proof leading to (5.3.35) with $\tilde{P}$ replacing P in (5.3.24) and H and h replaced by the corresponding $\tilde{H}$ and $\tilde{h}$, we obtain instead of Eq. (5.3.35),

$$\underset{\lambda, \eta}{\operatorname{degr}}[1 - T_\zeta^d]\tilde{H}(\lambda, \zeta, \mu) = \underset{\lambda, \eta}{\operatorname{degr}} \tilde{h}(\lambda, \zeta, \mu) \leq \underset{\lambda, \eta, \zeta}{\operatorname{degr}} \tilde{h}(\lambda, \zeta, \mu)$$

$$= \underset{\lambda, \eta, \zeta}{\operatorname{degr}}[1 - T_\zeta^d]\tilde{H}(\lambda, \zeta, \mu) - (d + 1)$$

$$\leq \underset{\lambda, \eta, \zeta}{\operatorname{degr}} \tilde{H}(\lambda, \zeta, \mu) - (d + 1). \quad (5.3.38)$$

Upon subtracting $\mathrm{degr}_n \prod_{j=1}^{\rho} (\mu_j)^{\sigma_j}$ on both sides of this inequality, we obtain

$$\underset{\lambda,\,\eta}{\mathrm{degr}}[1 - T_\zeta^d]H(\lambda, \zeta, \mu) \leq \underset{\lambda,\,\eta,\,\zeta}{\mathrm{degr}}\, H(\lambda, \zeta, \mu) - (d + 1), \qquad (5.3.39)$$

which completes the proof of the lemma.

5.3.2 Basic Grouping of the Taylor Operations in (5.2.1)

Let I_G be the unrenormalized Feynman integrand associated with a proper and connected graph G. Let I be a $4n$-dimensional subspace of $\mathbb{R}^{4n+4m+\rho}$ associated with the integration variables in I_G. As in (2.2.21), we introduce a vector $\mathbf{P}$ in $\mathbb{R}^{4n+4m+\rho}$ such that the integration variables, the components of the external independent momenta, and the masses in G may be written as some linear combinations of the components of $\mathbf{P}$. Let

$$\mathbf{P} = \mathbf{L}_1\eta_1\eta_2 \cdots \eta_k + \mathbf{L}_2\eta_2 \cdots \eta_k + \cdots + \mathbf{L}_k\eta_k + \mathbf{C}, \qquad (5.3.40)$$

where $k \leq 4n$ and $\mathbf{L}_1, \mathbf{L}_2, \ldots, \mathbf{L}_k$ are any k independent vectors in I. $\mathbf{L}_1$, $\mathbf{L}_2, \ldots, \mathbf{L}_r$, for all $r \leq k \leq 4n$, span a subspace $S_r \equiv \{\mathbf{L}_1, \mathbf{L}_2, \ldots, \mathbf{L}_r\} \subset I$. $\mathbf{C}$ is a vector confined to a finite region in $\mathbb{R}^{4n+4m+\rho}$ with $\mu^i \neq 0$ for all $i = 1, \ldots, \rho$.

Since, in particular, the integration variables may be written as some linear combinations of the components of $\mathbf{P}$, it follows that for any proper subdiagram $g \subset G$, the k_{ijl}^g of g may be also written as some linear combinations of the components of $\mathbf{P}$. Let r be a fixed integer in $1 \leq r \leq k$. The latter property then, in particular, means that a four-vector k_{ijl}^g may (or may not) depend on the parameter η_r in (5.3.40). For a four-vector k_{ijl}^g, if at least one of its four components depends on a parameter η_r in (5.3.40), then we say that the four-vector k_{ijl}^g depends on η_r. Otherwise (i.e., when all the four components of k_{ijl}^g are independent of η_r), we say that k_{ijl}^g is independent of η_r.

For convenience, in this subsection we extend the definition of D sets in (5.2.1) to ones including (proper) subdiagrams g with strictly negative dimensionalities as well by simply setting $(-T_g) = 0$ for $d(g) < 0$. This obviously does not change anything in the expression for R, and will simplify the construction to be given.

Choose a set D and consider the set $D \cup \{G\}$ obtained from D by adjoining to it the whole graph G. Obviously, if $G \in D$, then $D \cup \{G\} = D$. We arrange the subdiagrams in the set $D \cup \{G\}$ in increasing order, i.e., as $\{\ldots, g_i, g_{i+1}, \ldots\}$ with $g_{i+1} \not\supseteq g_i$.

As before, let r be a *fixed* integer in $1 \leq r \leq k$. Let $g \not\supseteq g'$ be any two consecutive (proper) subdiagrams in $D \cup \{G\}$. We denote by $c(g/g')$ the set of *all* the lines of g/g' that have the internal variables k_{ijl}^g in g/g' independent of η_r.[13] To simplify the notation we write $c(g/g') = g/g'$ if all the k_{ijl}^q of g/g' are independent of η_r, $c(g/g') \not\subseteq g/g'$ if not all the k_{ijl}^q of g/g' are independent of η_r, and $c(g/g') = \varnothing$ if all the k_{ijl}^q of g/g' are dependent on η_r. Finally, by the expression $c(g/g') \neq \varnothing$ it is meant that at least some (i.e., at least one) of the k_{ijl}^q of g/g' are independent of η_r.

For any two consecutive subdiagram g, g' in $D \cup \{G\}$ with $g \neq g'$, we write $g/g' = \bar{g}$. From the set $D \cup \{G\}$ we induce a set of subdiagrams by deleting, in general, some subdiagrams from $D \cup \{G\}$ and introducing, in general, some new subdiagrams not in $D \cup \{G\}$.

Let g be a subdiagram in the chosen set $D \cup \{G\}$.

(i)   Suppose $\varnothing \not\subseteq c(\bar{g}) \not\subseteq \bar{g}$, and *define* $g_0 = g - c(\bar{g})$ as the subdiagram consisting of the lines in $\mathscr{L}^g - c(\bar{g})$ and, of course, of the relevant vertices as their end points. We note in particular that $g' \not\subseteq g_0 \not\subseteq g$. By construction, all the k_{ijl}^q in g_0/g', where $\bar{g} = g/g'$, are dependent on η_r. We shall study later the nature of the subdiagram g_0. For this case, i.e., with $\varnothing \not\subseteq c(\bar{g}) \not\subseteq \bar{g}$, we keep g in $D \cup \{G\}$ and we induce a set $\{g_0\}$. (ii) Suppose $c(\bar{g}) = \varnothing$. Let g', g, g'' be the consecutive subdiagrams in $D \cup \{G\}$ with $g' \not\subseteq g \not\subseteq g''$, $\bar{g} = g/g', \bar{g}'' = g''/g$. If all the $k_{ijl}^{g''}$ in $\bar{g}''$ are independent of η_r, then we eventually delete the subdiagram g from the set $D \cup \{G\}$, thus inducing the set $\{g\}$. Otherwise (i.e., if at least some of the $k_{ijl}^{g''}$ in $\bar{g}''$ are dependent on η_r) we keep g in $D \cup \{G\}$. (iii) Suppose $c(\bar{g}) = \bar{g}$; then we keep g in $D \cup \{G\}$. We summarize the above process as follows:

(i) $\varnothing \not\subseteq c(\bar{g}) \not\subseteq \bar{g}$; keep g in $D \cup \{G\}$ and introduce the subdiagram $g_0 = g - c(\bar{g}) \not\subseteq g$, thus inducing the set $\{g_0\}$.

(ii) $c(\bar{g}) = \varnothing$; if all the $k_{ijl}^{g''}$ in $\bar{g}''$ are independent of η_r, then eventually delete g from $D \cup \{G\}$ and thus induce the set $\{g\}$.

$c(\bar{g}) = \varnothing$; if at least some of the $k_{ijl}^{g''}$ in $\bar{g}''$ are dependent on η_r, then keep g in $D \cup \{G\}$ and thus induce no set.

(iii) $c(\bar{g}) = \bar{g}$; keep g in $D \cup \{G\}$ and thus induce no set.

We carry out this analysis for all g in $D \cup \{G\}$; thus for each g, we either introduce a new subdiagram g_0 and keep g in $D \cup \{G\}$, induce the set $\{g\}$, or simply keep g in $D \cup \{G\}$ as discussed. We continue the above analysis starting from the smallest subdiagram in $D \cup \{G\}$ until we arrive to the graph G itself. If $\varnothing \not\subseteq c(\bar{G}) \not\subseteq \bar{G}$, then we introduce a subdiagram $G_0 = G - c(\bar{G}) \not\subseteq G$ and thus induce the set $\{G_0\}$. However, whether $c(\bar{G}) = \varnothing$ or $c(\bar{G}) \neq \varnothing$, we *delete* the graph G from $D \cup \{G\}$. The set of subdiagrams obtained from $D \cup \{G\}$ by deleting all those g in $D \cup \{G\}$ that were eventually

---

[13] Recall that g/g' denotes the subdiagram g with g' in it shrunk to a point. The k_{ijl}^q of g/g', then, mean k_{ijl}^q of g pertaining to the lines of g/g' but not of g'.

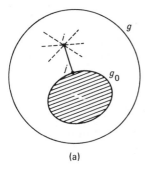 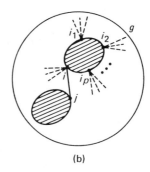

(a)   (b)

Fig. 5.11   The overall regions in (a) and (b) denote the proper subdiagram g. In (a) the line joining the vertex v_j to the vertex v_i is supposedly an external line of g_0, which is impossible. In (b) the line joining the vertices v_i and v_j is supposedly to be an improper line of g_0, which is again impossible.

to be deleted as well as deleting the graph G will be called a *nuclear* set and will be denoted D_N. By definition, $G \notin D_N$. The set of subdiagrams as induced from the set $D \cup \{G\}$ containing *all* those deleted subdiagrams g (including G) *and* those newly introduced subdiagrams g_0 as discussed in (i) will be denoted D_0.

It is easy to show that all the subdiagrams in D_0 are proper (but not necessarily connected) subdiagrams. First, G is in D_0 and it is proper. On the other hand, all those subdiagrams g deleted from $D \cup \{G\}$ and hence included in D_0 are, by definition, proper. Finally let $g_0 = g - c(\bar{g}) \in D_0$, with $\emptyset \not\subseteq c(\bar{g}) \not\subseteq \bar{g}$. Consider the subdiagram g depicted by the overall region in Fig. 5.11a. Suppose that $g_0 (\not\subseteq g)$ has an external line ℓ_l[14] joining some vertex v_j to some *extral* vertex v_i of g_0, as shown in Fig. 5.11a. The shaded region in this figure represents the remaining part of the subdiagram g_0. By momentum conservation [see (5.1.66)], we have at the vertex v_i

$$k_{ijl}^g = - \sum_{j'l'}{}' k_{ij'l'}^g , \qquad (5.3.41)$$

where the sum is over all those vertices $v_{j'}$ and lines $\ell_{l'}$ of $g/g_0 \subset \bar{g}$ with the lines $\ell_{l'}$ joining the vertices $v_{j'}$ to the vertex v_i. These lines are represented by the dotted lines in Fig. 5.11a. Since the $k_{ij'l'}^g$ in $\bar{g}$ are independent of η_r, it follows from (5.3.41) that k_{ijl}^g is also independent of η_r. Hence the extral vertex v_i and the line ℓ_l joining the vertex v_j to the vertex v_i cannot belong to $g - c(\bar{g}) = g_0$, by definition of $c(\bar{g})$. Therefore g_0 has no extral vertices and thus must be amputated. Now suppose that g_0 has an improper line ℓ_l joining some vertex v_i to some vertex v_j, and a situation as shown in Fig. 5.11b arises, with the dotted lines denoting lines belonging to g/g_0 and the

---

[14] That is, suppose that g_0 is not amputated.

shaded regions representing the remaining part of the subdiagram g_0, with external vertices $v_{i_1}, v_{i_2}, \ldots$. Again by momentum conservation we have in an obvious notation

$$k_{ijl}^g = - \sideset{}{'}\sum_{j_0 l_0} k_{ij_0 l_0}^g - \sideset{}{'}\sum_{j_1 l_1} k_{i_1 j_1 l_1}^g - \cdots - \sideset{}{'}\sum_{j_p l_p} k_{i_p j_p l_p}^g, \tag{5.3.42}$$

corresponding to the dotted lines in Fig. 5.11b, which represent some lines belonging to $g/g_0 \subset \bar{g}$. Accordingly, the $k_{ij_0 l_0}^g, k_{i_1 j_1 l_1}^g, \ldots, k_{i_p j_p l_p}^g$ are independent of η_r, and so, from (5.3.42), k_{ijl}^g is independent of η_r as well. Therefore the improper line ℓ_i cannot belong to $g - c(\bar{g}) = g_0$, by definition, and g_0 cannot contain improper lines. Therefore all the subdiagrams in D_0 are proper subdiagrams.

Finally, we show that *all* the $k_{ijl}^{g_0}$ in g_0/g', where $g_0 = g - c(\bar{g}) \in D_0$ $[c(\bar{g}) \neq \varnothing]$, with $\bar{g} = g/g'$, are dependent on η_r. From (5.1.67), we know that the $k_{ijl}^{g_0} = k_{ijl}^{g_0}(k^g)$, with $g_0 \subset g$, and they are independent of the elements in q^g. Thus, for convenience, we set the elements in q^g equal to zero to prove the result stated earlier. From a canonical decomposition as in (5.1.65) and (5.1.46),

$$k_{ijl}^{g_0}(k^g) = k_{ijl}^g - q_{ijl}^{g_0}(k^g, 0) \tag{5.3.43}$$

for i, j, and l pertaining to g_0/g'. Applying the result obtained in (5.1.117) to each of the connected parts of g, we arrive at the conclusion that the $q_{ijl}^{g_0}(k^g, 0)$ depend only on the k_{ijl}^g in g/g_0 and are, therefore, independent of η_r. On the other hand, we know, by construction, that all the k_{ijl}^g in g_0/g' are dependent on η_r. Therefore it follows from the (5.3.43) that all the $k_{ijl}^{g_0}$ in g_0/g' are dependent on η_r.

We have started out with a set $D \cup \{G\}$ and induced a set D_0. For $g \in D \cup \{G\}$, with $\varnothing \not\subset c(\bar{g}) \not\subset \bar{g}$ and $\bar{g} = g/g'$, we denote g_0/g' by $\bar{g}_0$. We have thus established that

all the subdiagrams $G' \in D_0 - \{G\}$ are proper and *all* the $k_{ijl}^{G'}$ in $\bar{G}'$ are dependent on η_r. $\tag{5.3.44}$

From the sets D_0 and D_N we define a new set

$$N = D_N \cup D_0 \equiv D \cup D_0. \tag{5.3.45}$$

We note, in particular, that for $c(\bar{g}) = \bar{g}$ or $c(\bar{g}) = \varnothing$, all the k_{ijl}^g in $\bar{g}$ are independent of η_r or dependent on η_r, respectively, by definition of $c(\bar{g})$. On the other hand, for $\varnothing \not\subset c(\bar{g}) \not\subset \bar{g}$, all the k_{ijl}^g in g/g_0, where $g_0 = g - c(\bar{g})$, are independent of η_r. Obviously the k_{ijl}^G in $\bar{G}$, with $G \in N$, are either all dependent on η_r or are all independent of η_r. Accordingly, if we arrange the subdiagrams in the set N in an increasing order, i.e., $\{\ldots, g', g, \ldots\}$ with $g \not\supseteq g'$, then we have, from (5.3.44) in particular, the following lemma.

Lemma 5.3.4:   *All the subdiagram in the set N, as defined in (5.3.45), are proper, and for any two consecutive subdiagrams $g, g' \in N$, with $g \not\supseteq g'$, the k_{ijl}^q in $g/g' \equiv \bar{g}$ are **either all** independent of η_r **or all** are dependent on η_r.*

Starting from a nuclear set D_N as obtained from a set $D \cup \{G\}$, we may generate *all* other possible sets that contain *in addition to* the subdiagrams in D_N any *one*, or any *two*, or ... or, finally *all* subdiagram(s) from D_0. The collection of *all* these sets, as just defined (as obtained from D_N and D_0), together with the set D_N will be denoted $\mathscr{D}$. Clearly the set $\mathscr{D}$ has the form $\mathscr{D} = \{D_N, \ldots, (D_N \cup D_0)\}$. The corresponding Taylor operations in (5.2.1) for the subdiagrams in the sets in $\mathscr{D}$ may be then readily combined, and the corresponding *sum* over the sets in $\mathscr{D}$ is then reduced to the expression

$$\mathscr{T}^{(N)} = \prod_{g \in N} (\delta_g^N - T_g) I_G, \tag{5.3.46}$$

where a set N is defined in (5.3.45) and

$$\delta_g^N = \begin{cases} 1 & \text{for} \quad g \in D_0 \\ 0 & \text{for} \quad g \in D_N. \end{cases} \tag{5.3.47}$$

The product in (5.3.46) must be taken in the correct order, that is, $\cdots (\delta_g^N - T_g) \cdots (\delta_{g'}^N - T_{g'}) \cdots I_G$, for $g \not\supseteq g'$. Clearly G is the largest subdiagram in N and $\delta_G^N = 1$. The sum over all D sets in (5.2.1) then reduces equivalently, for the *final* expression for R as a sum over all distinct sets N, with the latter as defined in (5.3.45), i.e.,

$$R = \sum_N \prod_{g \in N} (\delta_g^N - T_g) I_G. \tag{5.3.48}$$

For any $g \in N$, with $d(g) < 0$, we simply set $T_g = 0$, by definition.

Finally we introduce subsets of N: $H_1(N)$ and $H_2(N)$:

$$N = H_1(N) \cup H_2(N), \tag{5.3.49}$$

where $g \in H_1(N)$ if all the k_{ijl}^q in $\bar{g}$ are dependent on η_r and $g \in H_2(N)$ if all the k_{ijl}^q in $\bar{g}$ are dependent of η_r (see Lemma 5.3.4). We also introduce subsets $F_1(N)$ and $F_2(N)$ by

$$F_2(N) = H_1(N) \cap D_0 \tag{5.3.50}$$

and

$$F_1(N) = H_1(N) - F_2(N). \tag{5.3.51}$$

In particular,

$$H_1(N) = F_1(N) \cup F_2(N). \tag{5.3.52}$$

Also if $g \in F_2(N)$, then $\delta_g^N = 1$, and if $g \in F_1(N)$, then $\delta_g^N = 0$. We also remark that the whole graph $G \in H_2(N) \cup F_2(N)$, $\delta_G^N = 1$, by definition.

5.3.3   Completion of the Proof of Convergence

Now we apply Lemmas 5.3.2–5.3.4 and the grouping achieved in (5.3.48), with respect to the parameter η_r, to prove the following lemma, where r is fixed in $1 \le r \le k \le 4n$, as before.

Lemma 5.3.5

$$\mathop{\mathrm{degr}}_{\eta_r} R_{\mathrm{E}} < -\dim S_r, \qquad \mathop{\mathrm{degr}}_{\eta_r} R < -\dim S_r. \tag{5.3.53}$$

Consider a fixed set N in (5.3.48). To prove (5.3.53) we introduce the following recursion formula,

$$\mathcal{T}_g(N) = (\delta_g^N - T_g)I_{\bar{g}}\mathcal{T}_{g'}(N), \tag{5.3.54}$$

corresponding to a set N in (5.3.48), where g, g' are two consecutive sub-diagrams in N with $g \not\supseteq g'$. In the notation (5.3.54), we have from (5.3.46)

$$\mathcal{T}^{(N)} \equiv \mathcal{T}_G(N). \tag{5.3.55}$$

We also introduce the notation

$$\rho(g) = 4 \sum_{\substack{g' \in H_1(N) \\ g' \subset g}} L(\bar{g}'), \tag{5.3.56}$$

and we define $L(\varnothing) \equiv 0$. We note in particular

$$\rho(G) + 4 \sum_{\substack{g' \in H_2(N) \\ g' \subset G}} L(\bar{g}') = 4L(G). \tag{5.3.57}$$

For convenience, for any $g \in N$, we denote by $f_g^n(k^g, q^g, \mu)$ *any* function of the form

$$f_g^n(k^g, q^g, \mu) = \prod_{\substack{g' \in H_n(N) \\ g' \subset g}} \prod_{\substack{ijl \\ i<j}}^{\bar{g}'} [(\theta_{ijl}^{g'})^2 + \mu_{ijl}^2 - i\varepsilon\{(\theta_{ijl}^{g'})^2 + \mu_{ijl}^2\}]^{-\sigma_{ijl}^n} \tag{5.3.58}$$

for $n = 1, 2$, where $\theta_{ijl}^g \equiv Q_{ijl}^g$, and for the $g' \not\subset g$, $\theta_{ijl}^{g'} \equiv k_{ijl}^{g'}$. The σ_{ijl}^n are some strictly positive integers. Note that if $g \notin H_n(N)$ for $n = 1$ or $n = 2$, then the product in (5.3.58), for the corresponding n, is over the $g' \in H_n(N)$ with $g' \not\subset g$, by definition. If $g \notin H_n(N)$ for $n = 1$ or $n = 2$, and there is no $g' \in H_n(N)$, with $g' \not\subset g$, then we define $f_g^n \equiv 1$ for the corresponding n. Also by definition of $H_1(N)$ and $H_2(N)$, all the k_{ijl}^g in $\bar{g}'$ for all $g' \in H_1(N)$, $g' \subset g$, are dependent on η_r, and all the k_{ijl}^g in $\bar{g}'$ for all $g' \in H_2(N)$, $g' \subset g$, are independent of η_r in the denominators in (5.3.58). We also denote by $f^n(k^g, q^g, \mu)$ *any* function of the form

$$f^n(k^g, q^g, \mu) = \prod_{\substack{ijl \\ i<j}}^{\bar{g}} [(Q_{ijl}^g)^2 + \mu_{ijl}^2 - i\varepsilon\{(Q_{ijl}^g)^2 + \mu_{ijl}^2\}]^{-\sigma_{ijl}^n} \tag{5.3.59}$$

for $n = 1, 2$, with the σ_{ijl}^n some strictly positive integers. If the elements in q^g in (5.3.58) and (5.3.59) are set equal to zero, then the corresponding functions will be denoted by $f_g^{\prime n}(k^g, \mu)$ and $f^{\prime n}(k^g, \mu)$. Note that all the denominators in $f_g^1(k^g, \mu)$ and $f^1(k^g, \mu)$ are dependent on η_r and all the denominators in $f_g^2(k^g, \mu)$ and $f^2(k^g, \mu)$ are independent of η_r. Finally, we denote by $P(k^g, q^g, \mu)$ [or $P_g(k^g, q^g, \mu)$] any polynomial in the elements in k^g, q^g, μ and, in general, in the $(\mu_{ijl})^{-1}$ as well.

The sum over N in (5.3.48) is over a finite number of sets, and we establish (5.3.53) first for each $\mathscr{T}^{(N)}$, as defined in (5.3.46). As before, we arrange the sets N in an increasing order, i.e., as $\{\ldots, g', g, \ldots\}$, with $g \not\supseteq g'$. For additional clarity we shall deliberately give more details than is actually necessary.

Let g be the first element in N and note that $\bar{g} = g$. Suppose $g \in F_1(N)$. Then $\mathscr{T}_g(N)$ is of the form

$$\mathscr{T}_g(N) = -T_g I_g, \qquad (5.3.60)$$

where I_g is of the form

$$I_g = P(k^g, q^g, \mu) f^1(k^g, q^g, \mu). \qquad (5.3.61)$$

We explicitly have for $\mathscr{T}_g(N)$, in a convenient notation,

$$\mathscr{T}_g(N) = -\sum_{a,b} (q^g)^{a+b} P^a(k^g, 0, \mu) f_b^1(k^g, \mu), \qquad (5.3.62)$$

where

$$|a| + |b| \le d(g). \qquad (5.3.63)$$

The notation in (5.3.62) is similar to the ones in (5.3.27)–(5.3.29). We note that

$$|a| + \operatorname*{degr}_{\eta_r} P^a + |b| + \operatorname*{degr}_{\eta_r} f_b^1 \le d(\dot{g}) - 4L(g). \qquad (5.3.64)$$

Accordingly, if we scale η_r in P^a and f_b^1, and q^g as well, by a parameter α,

$$\operatorname*{degr}_{\alpha} \mathscr{T}_g(N) \le d(g) - \rho(g), \qquad (5.3.65)$$

where we have used the fact that $4L(g) = \rho(g)$ from the very definition in (5.3.56). We also note that

$$\operatorname*{degr}_{q^g} \mathscr{T}_g(N) \le d(g). \qquad (5.3.66)$$

Finally, we note from (5.3.62) that $\mathscr{T}_g(N)$ has the general structure

$$\mathscr{T}_g(N) = P_g(k^g, q^g, \mu) f_g^1(k^g, \mu). \qquad (5.3.67)$$

Suppose $g \in F_2(N)$. Then

$$\mathscr{T}_g(N) = (1 - T_g) I_g. \qquad (5.3.68)$$

If we scale the k^g_{ijl} in g, which, by definition, depend on η_r, by a parameter λ, and if $d(g) \geq 0$, then we may use the result (5.3.25) in Lemma 5.3.3 to conclude that

$$\underset{\lambda}{\operatorname{degr}} \, \mathcal{T}_g(N) < -\rho(g), \tag{5.3.69}$$

where $\rho(g) = 4L(g)$. If $d(g) < 0$, then $T_g = 0$,

$$\underset{\lambda}{\operatorname{degr}} \, \mathcal{T}_g(N) \leq d(g) - \rho(g) < -\rho(g). \tag{5.3.70}$$

For $d(g) \geq 0$, we see from (5.3.14), (5.3.15), and (5.3.33) that $\mathcal{T}_g(N)$ in (5.3.68) has the structure

$$\mathcal{T}_g(N) = P_g(k^g, q^g, \mu) f^1_g(k^g, q^g, \mu). \tag{5.3.71}$$

If $d(g) < 0$, then $\mathcal{T}_g(N) = I_g$, and I_g is again of the general form in (5.3.71). Finally, suppose $g \in H_2(N)$. Then

$$\mathcal{T}_g(N) = -T_g I_g, \tag{5.3.72}$$

where I_g is of the form

$$I_g = P(k^g, q^g, \mu) f^2(k^g, q^g, \mu). \tag{5.3.73}$$

We explicitly have

$$\mathcal{T}_g(N) = - \sum_{a,b} (q^g)^{a+b} P^a(k^g, 0, \mu) f^2_b(k^g, \mu), \tag{5.3.74}$$

where

$$|a| + |b| \leq d(g), \tag{5.3.75}$$

and hence if we scale q^g by a parameter α,

$$\underset{\alpha}{\operatorname{degr}} \, \mathcal{T}_g(N) \leq d(g), \tag{5.3.76}$$

Since $\rho(g) = 0$ [see (5.3.56)], in this case we may rewrite (5.3.76):

$$\underset{\alpha}{\operatorname{degr}} \, \mathcal{T}_g(N) \leq d(g) - \rho(g). \tag{5.3.77}$$

From (5.3.74) we also note that $\mathcal{T}_g(N)$ has the structure

$$\mathcal{T}_g(N) = P_g(k^g, q^g, \mu) f^2_g(k^g, \mu). \tag{5.3.78}$$

Since all the k^g_{ijl}, in this case, are independent of η_r, we may also trivially write in the notation (5.3.69)–(5.3.70)

$$\underset{\lambda}{\operatorname{degr}} \, \mathcal{T}_g(N) = 0. \tag{5.3.79}$$

Now as *induction* hypotheses, suppose that for some subdiagram $g \in N - \{G\}$, not necessarily the first element in it, the following are true:

(i)   If $g \in H_2(N)$, then $\mathcal{T}_g(N)$ has the structure

$$\mathcal{T}_g(N) = P_g(k^g, q^g, \mu) f_g^1(k^g, \mu) f_g^2(k^g, \mu). \qquad (5.3.80)$$

If we scale all those k_{ijl}^g in P_g, depending on η_r, and η_r in f_g^1 by a parameter λ, then[15]

$$\operatorname*{degr}_{\lambda} \mathcal{T}_g(N) < -\rho(g) \qquad (5.3.81)$$

if $\rho(g) \neq 0$, and

$$\operatorname*{degr}_{\lambda} \mathcal{T}_g(N) = 0 \qquad (5.3.82)$$

if $\rho(g) = 0$. Also,

$$\operatorname*{degr}_{q^g} \mathcal{T}_g(N) \leq d(g). \qquad (5.3.83)$$

Finally, if we scale λ and q^g by a parameter α (and set $\lambda = 1$), then

$$\operatorname*{degr}_{\alpha} \mathcal{T}_g(N) \leq d(g) - \rho(g). \qquad (5.3.84)$$

(ii)   If $g \in F_1(N)$, i.e., in particular, $\delta_g^N = 0$ [see Eq. (5.3.51) and following it), then $\mathcal{T}_g(N)$ has a structure as in (5.3.80). Equations (5.3.83) and (5.3.84) are also true.

(iii)   If $g \in F_2(N)$, i.e., in particular, that $\delta_g^N = 1$ [see (5.3.50)], then $\mathcal{T}_g(N)$ has a structure as in

$$\mathcal{T}_g(N) = P_g(k^g, q^g, \mu) f_g^1(k^g, q^g, \mu) f_g^2(k^g, \mu), \qquad (5.3.85)$$

and if we scale the $k_{ijl}^{g'}$ in all the $\bar{g}', g' \in H_1(N), g' \subset g$, in f_g^1, which, by definition, depend on η_r, and the k_{ijl}^g in P_g, depending on η_r, by a parameter λ, then

$$\operatorname*{degr}_{\lambda} \mathcal{T}_g(N) < -\rho(g). \qquad (5.3.86)$$

Suppose that the next subdiagram to g in N is g' and $g' \in N - \{G\}$; then we prove that (i), (ii), or (iii), as the case may be, is also true for g'. We finally treat the situation for G itself separately.

(i)   Suppose that $g \in H_2(N)$; then if $g' \in H_2(N), (g \nsubseteq g')$,

$$\mathcal{T}_{g'}(N) = (-T_{g'}) I_{\bar{g}'} \mathcal{T}_g(N), \qquad (5.3.87)$$

where $I_{\bar{g}'}$ is of the form

$$I_{\bar{g}'} = P_{\bar{g}'}(k^{g'}, q^{g'}, \mu) f^2(k^{g'}, q^{g'}, \mu). \qquad (5.3.88)$$

---

[15] Recall that $f_g^2(k^g, \mu)$ is independent of η_r [see definition (5.3.58)].

We write $k^g = k^g(k^{g'})$, $q^g = q^g(k^{g'}, q^{g'})$ in (5.3.80). Then with $g \in H_2(N)$, we explicitly have for $\mathcal{T}_{g'}(N)$ in (5.3.87)

$$\mathcal{T}_{g'}(N) = - \sum_{a,b,c} (q^{g'})^{a+b+c} P^a_{\bar{g}}(k^{g'}, 0, \mu) f^2_b(k^{g'}, \mu)$$

$$\times P^c_g(k^g(k^{g'}), q^g(k^{g'}, 0), \mu) f^1_g(k^g(k^{g'}), \mu) f^2_g(k^g(k^{g'}), \mu), \quad (5.3.89)$$

where

$$|a| + |b| + |c| \le d(g'). \tag{5.3.90}$$

If we scale η_r in f^1_g, and the $k^{g'}_{ijl}$ in P^c_g, $P^a_{\bar{g}}$, depending on η_r, by a parameter λ, we obtain according to the hypotheses in (5.3.81) and (5.3.82)

$$\underset{\lambda}{\operatorname{degr}} P^c_g + \underset{\lambda}{\operatorname{degr}} f^1_g < -\rho(g) \tag{5.3.91}$$

if $\rho(g) \ne 0$, and

$$\underset{\lambda}{\operatorname{degr}} P^c_g + \underset{\lambda}{\operatorname{degr}} f^1_g = 0 \tag{5.3.92}$$

if $\rho(g) = 0$ (in the latter case, we note that $f^1_g \equiv 1$). We note that $q^g(k^{g'}, 0)$ in P^c_g is independent of η_r since the $k^{g'}_{ijl}$ in g'/g, on which q^g depends [see (5.1.117)], are independent of η_r. Also note that $P^a_{\bar{g}'}, f^2_b$, and f^2_g are independent of η_r, i.e.,

$$\underset{\lambda}{\operatorname{degr}} P^a_{\bar{g}'} = \underset{\lambda}{\operatorname{degr}} f^2_b = \underset{\lambda}{\operatorname{degr}} f^2_g = 0.$$

Accordingly, we obtain from (5.3.91) and (5.3.92)

$$\underset{\lambda}{\operatorname{degr}} \mathcal{T}_{g'}(N) < -\rho(g') \tag{5.3.93}$$

if $\rho(g') \ne 0$, and

$$\underset{\lambda}{\operatorname{degr}} \mathcal{T}_{g'}(N) = 0, \tag{5.3.94}$$

if $\rho(g') = 0$, where

$$\rho(g') = \rho(g). \tag{5.3.95}$$

Also we note from (5.3.90) that

$$\underset{q^{g'}}{\operatorname{degr}} \mathcal{T}_{g'}(N) \le d(g'). \tag{5.3.96}$$

Finally, if we scale λ and $q^{g'}$ by a parameter α (and set $\lambda = 1$), we readily obtain

$$\underset{\alpha}{\operatorname{degr}} \mathcal{T}_{g'}(N) \le d(g') - \rho(g'). \tag{5.3.97}$$

We also note from (5.3.89) that $\mathcal{T}_{g'}(N)$ has a structure as in (5.3.80).

If $g' \in F_1(N)$ [see (5.3.51) and (5.3.52)], then

$$\mathcal{T}_{g'}(N) = (-T_{g'})I_{\bar{g}'}\mathcal{T}_g(N), \qquad (5.3.98)$$

where $I_{\bar{g}'}$ is of the form

$$I_{\bar{g}'} = P_{\bar{g}}(k^{g'}, q^{g'}, \mu)f^1(k^{g'}, q^{g'}, \mu). \qquad (5.3.99)$$

Accordingly, we may write

$$\mathcal{T}_{g'}(N) = -\sum_{a,b,c}(q^{g'})^{a+b+c}P^a_{\bar{g}}(k^{g'}, 0, \mu)f^1_b(k^{g'}, \mu)$$

$$\times\ P^c_g(k^g(k^{g'}), q^g(k^{g'}, 0), \mu)f^1_g(k^g(k^{g'}), \mu)f^2_g(k^g(k^{g'}), \mu), \quad (5.3.100)$$

where

$$|a| + |b| + |c| \le d(g'), \qquad (5.3.101)$$

i.e.,

$$\operatorname*{degr}_{q^{g'}}\mathcal{T}_{g'}(N) \le d(g'). \qquad (5.3.102)$$

If we scale η_r in f^1_b, f^1_g, the $k^{g'}_{ijl}$ in $P^a_{\bar{g}}, P^c_g$, depending on η_r, and scale as well $q^{g'}$ by a parameter α, then

$$|a| + |b| + \operatorname*{degr}_{\alpha} P^a_{\bar{g}} + \operatorname*{degr}_{\alpha} f^1_b \le d(\bar{g}') - 4L(\bar{g}'), \qquad (5.3.103)$$

and the induction hypothesis (5.3.84) for g we obtain

$$|c| + \operatorname*{degr}_{\alpha} P^c_g + \operatorname*{degr}_{\alpha} f^1_g \le d(g) - \rho(g). \qquad (5.3.104)$$

Accordingly,

$$\operatorname*{degr}_{\alpha}\mathcal{T}_{g'}(N) \le d(g') - \rho(g'), \qquad (5.3.105)$$

where we have used the facts that $d(g') = d(\bar{g}') + d(g)$ and $\rho(g') = 4L(\bar{g}') + \rho(g)$. Finally, we note from (5.3.100) that $\mathcal{T}_{g'}(N)$ has a structure as in (5.3.80).

If $g' \in F_2(N)$ [see (5.3.50)], then

$$\mathcal{T}_{g'}(N) = (1 - T_{g'})I_{\bar{g}'}\mathcal{T}_g(N), \qquad (5.3.106)$$

and $I_{\bar{g}'}$ is of the form in (5.3.99). We then have

$$\mathcal{T}_{g'}(N) = f^1_g(k^g, (k^{g'}), \mu)f^2_g(k^g(k^{g'}), \mu)\sum_c(q^{g'})^c$$

$$\times\ P^c_g(k^g(k^{g'}), q^g(k^{g'}, 0), \mu)(1 - T^{d(g')-|c|}_{g'})I_{\bar{g}'}, \quad (5.3.107)$$

where

$$|c| \le d(g). \tag{5.3.108}$$

From Lemma 5.3.3, by scaling the $k_{jkl}^{g'}$ in $\bar{g}'$ by λ, which by definition depend on η_r, we obtain for $d(g') \ge |c|$

$$\operatorname*{degr}_{\lambda}[1 - T_{g'}^{d(g')-|c|}]I_{\bar{g}'} \le d(\bar{g}') - 4L(\bar{g}') - d(g') + |c| - 1$$

$$= - d(g) - 4L(\bar{g}') + |c| - 1. \tag{5.3.109}$$

From the induction hypotheses in (5.3.81) or (5.3.82), we also have that (5.3.91) or (5.3.92) is true. Accordingly, from (5.3.109),

$$\operatorname*{degr}_{\lambda} \mathscr{T}_{g'}(N) < |c| - d(g) - \rho(g'), \tag{5.3.110}$$

where

$$\rho(g') = 4L(\bar{g}') + \rho(g). \tag{5.3.111}$$

Finally, from (5.3.108),

$$\operatorname*{degr}_{\lambda} \mathscr{T}_{g'}(N) < - \rho(g'). \tag{5.3.112}$$

If $d(g') < 0$, then

$$\operatorname*{degr}_{\lambda} I_{\bar{g}'} \le d(\bar{g}') - 4L(\bar{g}'), \tag{5.3.113}$$

which leads to

$$\operatorname*{degr}_{\lambda} \mathscr{T}_{g'}(N) \le d(g') - \rho(g') < - \rho(g'). \tag{5.3.114}$$

We also note from (5.3.107), with $d(g') \ge 0$, together from (5.3.14), (5.3.15), and (5.3.33), that $\mathscr{F}_{g'}(N)$ has a structure as in (5.3.85). The latter conclusion is also reached even if $d(g') < 0$.

(ii)   Suppose $g \in F_1(N)$. Then if $g' \in F_1(N)$,

$$\mathscr{T}_{g'}(N) = (- T_{g'})I_{\bar{g}'} \mathscr{T}_g(N), \tag{5.3.115}$$

where $I_{\bar{g}'}$ is of the form in (5.3.99). We explicitly have

$$\mathscr{T}_{g'}(N) = - \sum_{a,b,c} (q^{g'})^{a+b+c} P_{\bar{g}}^a(k^{g'}, 0, \mu) f_b^1(k^{g'}, \mu)$$

$$\times P_g^c(k^g(k^{g'}), q^g(k^{g'}, 0), \mu) f_g^1(k^g(k^{g'}), \mu) f_g^2(k^g(k^{g'}), \mu), \tag{5.3.116}$$

where

$$|a| + |b| + |c| \le d(g'), \tag{5.3.117}$$

i.e.,

$$\operatorname*{degr}_{q^{g'}} \mathscr{T}_{g'}(N) \le d(g'). \tag{5.3.118}$$

We scale η_r in f_g^1 and f_b^1, and all those $k_{ijl}^{g'}$ in $P_{\bar g}^a$ and P_g^c, depending on η_r, and scale $q^{g'}$ as well by a parameter α. By the induction hypotheses in (ii), we have from (5.3.84) as applied to g,

$$|c| + \underset{\alpha}{\operatorname{degr}} P_g^c + \underset{\alpha}{\operatorname{degr}} f_g^1 \le d(g) - \rho(g). \tag{5.3.119}$$

Also, by definition of $g' \in F_1(N)$,

$$|a| + |b| + \underset{\alpha}{\operatorname{degr}} P_{\bar g}^a + \underset{\alpha}{\operatorname{degr}} f_b^1 \le d(\bar g') - 4L(\bar g'). \tag{5.3.120}$$

Accordingly, from (5.3.119) and (5.3.120),

$$\underset{\alpha}{\operatorname{degr}} \mathscr{T}_{g'}(N) \le d(g') - \rho(g'). \tag{5.3.121}$$

We also note from (5.3.116) that $\mathscr{T}_{g'}(N)$ has a structure as in (5.3.80).

If $g' \in F_2(N)$, then

$$\mathscr{T}_{g'}(N) = (1 - T_{g'}) I_{\bar g'} \mathscr{T}_g(N), \tag{5.3.122}$$

where $I_{\bar g'}$ is again of the form in (5.3.99). By repeating a similar analysis as the one leading to (5.3.112) and (5.3.114), by using in the process the induction hypotheses in (ii), we obtain

$$\underset{\lambda}{\operatorname{degr}} \mathscr{T}_{g'}(N) < -\rho(g'), \tag{5.3.123}$$

and $\mathscr{T}_{g'}(N)$ has a structure as the one in (5.3.85).

(iii)   Suppose $g \in F_2(N)$. Then $g' \in H_2(N)$. In this case

$$\mathscr{T}_{g'}(N) = (-T_{g'}) I_{\bar g'} \mathscr{T}_g(N), \tag{5.3.124}$$

where $I_{\bar g'}$ is of the form in (5.3.88). We explicitly obtain

$$\mathscr{T}_{g'}(N) = - \sum_{a,b,c,d} (q^{g'})^{a+b+c+d} P_{\bar g}^a(k^{g'}, 0, \mu) f^2(k^{g'}, \mu)$$
$$\times P_g^c(k^g(k^{g'}), q^g(k^{g'}, 0), \mu) f_{g,d}^1(k^g(k^{g'}), q^g(k^{g'}, 0), \mu) f_g^2(k^g(k^{g'}), \mu), \tag{5.3.125}$$

where

$$|a| + |b| + |c| + |d| \le d(g'), \tag{5.3.126}$$

i.e.,

$$\underset{q^{g'}}{\operatorname{degr}} \mathscr{T}_{g'}(N) \le d(g'). \tag{5.3.127}$$

We make the important observation from (5.1.117), applied to each of the connected parts of g', that the dependence of the q^g of g on $k^{g'}$ is only on those in g'/g, i.e., $q^g(k^{g'}, 0)$ is *independent* of η_r. We scale η_r in $f_{g,d}^1$ [which by what has just been said about $q_{ijl}^g(k^{g'}, 0)$ has all its denominators, in (5.3.58) with $n = 1$, depend on η_r] by λ. We also scale the $k_{ijl}^{g'}$ in P_g^c, $P_{\bar{g}'}^a$, depending on η_r by λ. According to the induction hypothesis (5.3.86) in (iii),

$$\operatorname*{degr}_\lambda P_g^c + \operatorname*{degr}_\lambda f_{g,d}^1 < -\rho(g). \tag{5.3.128}$$

We also have $\operatorname*{degr}_\lambda P_{\bar{g}'}^a = 0$, which together (5.3.128) implies that

$$\operatorname*{degr}_\lambda \mathscr{T}_{g'}(N) < -\rho(g'), \tag{5.3.129}$$

where

$$\rho(g') = \rho(g). \tag{5.3.130}$$

If we scale $q^{g'}$ and λ, by a parameter α (and set $\lambda = 1$), we obtain by making use of (5.3.126)

$$\operatorname*{degr}_\alpha \mathscr{T}_{g'}(N) \le d(g') - \rho(g'). \tag{5.3.131}$$

Finally, we note from (5.3.125) that $\mathscr{T}_{g'}(N)$ has a structure as in (5.3.80).

Now we apply the above results to the whole graph G itself. If $G \in F_2(N)$, then (5.3.86) implies

$$\operatorname*{degr}_\lambda \mathscr{T}_G(N) < -\rho(G). \tag{5.3.132}$$

Since all the $q_{ijl} (\equiv q_{ijl}^G)$ of G are *independent* of η_r, we may simply replace $\operatorname{degr}_\lambda$ in (5.3.132) by $\operatorname{degr}_{\eta_r}$; i.e., we have

$$\operatorname*{degr}_{\eta_r} \mathscr{T}_G(N) < -\rho(G). \tag{5.3.133}$$

If $G \in H_2(N)$, then (5.3.81) implies, for $d(G) \ge 0$,

$$\operatorname*{degr}_{\eta_r}(-T_G)I_{\bar{G}}\mathscr{T}_g(N) < -\rho(G), \tag{5.3.134}$$

where g is the largest subdiagram in N contained in $G: g \nsubseteq G$, and we have used the same reasoning as in (5.3.132)–(5.3.133) to replace $\operatorname{degr}_\lambda$ by $\operatorname{degr}_{\eta_r}$. On the other hand, in this case

$$\operatorname*{degr}_{\eta_r} I_{\bar{G}} = 0, \tag{5.3.135}$$

or, from (5.3.81) or (5.3.86),

$$\operatorname*{degr}_{\eta_r} I_{\bar{G}}\mathscr{T}_g(N) < -\rho(G), \tag{5.3.136}$$

since $\rho(G) = \rho(g)$, where we have used (5.1.117) to conclude that $q^g(k^G, q^G)$ is independent of η_r, as the k_{ijl}^G in $\bar{G}$ are independent of η_r. From (5.3.134) and (5.3.136), if $d(G) \geq 0$, or just from (5.3.136) if $d(G) < 0$, we then conclude again that

$$\operatorname*{degr}_{\eta_r} \mathscr{T}_G(N) < -\rho(G). \tag{5.3.137}$$

Equations (5.3.133) and (5.3.137) then imply that we always have

$$\operatorname*{degr}_{\eta_r} \mathscr{T}^{(N)} < -\rho(G), \tag{5.3.138}$$

where we have used the identity in (5.3.55). Equation (5.3.56) also implies that

$$\rho(G) = 4 \sum_{\substack{g' \in H_1(N) \\ g' \subset G}} L(\bar{g}'). \tag{5.3.139}$$

From (5.1.4), (5.1.61), and the fact that the integration variables $k_1^0, \ldots, k_n^3$ are some linear combinations of the components of $\mathbf{P}$ in (5.3.40), we note that for any $g' \in H_1(N)$, we may write for a $k_{ijl}^{g'}$ in $\bar{g}'$,

$$k_{ijl}^{g'} = \sum_{t=1}^{r} (A_t(ijl)\eta_t \cdots \eta_{r-1}\eta_r)\eta_{r+1} \cdots \eta_k + \sum_{t=r+1}^{k} A_t(ijl)\eta_t \cdots \eta_k + c_{ijl}^{g'},$$

$$\tag{5.3.140}$$

where the following must be true, from the definition of the set $H_1(N)$:

$$\sum_{t=1}^{r} (A_t(ijl)\eta_t \cdots \eta_{r-1}\eta_r) \neq 0 \tag{5.3.141}$$

for at least one of the components of $k_{ijl}^{g'}$.[16] With $\eta_{r+1}, \ldots, \eta_k$ fixed, we see from (5.3.140)–(5.3.141) that the components of $k_{ijl}^{g'}$, depending on η_r, are functions of the independent parameters $\eta_1, \eta_2, \ldots, \eta_r$. The $A_t(ijl)$ and the $c_{ijl}^{g'}$ are independent of $\eta_1, \ldots, \eta_k$. On the other hand, for any $g' \in H_2(N)$, the $k_{ijl}^{g'}$ in $\bar{g}'$ are of the form

$$k_{ijl}^{g'} = \sum_{t=r+1}^{k} A_t'(ijl)\eta_t \cdots \eta_k + c_{ijl}^{g'}. \tag{5.3.142}$$

The parameters $\eta_1, \cdots, \eta_k$ (and, in particular, $\eta_1, \ldots, \eta_r$) are independent. Quite generally, we know from (5.1.118)–(5.1.120) that $\rho(G)$ of the $k_{ijl}^{g'}$ in the set

$$\{k_{ijl}^{g'} : g' \in H_1(N), k_{ijl}^{g'} \text{ in } \bar{g}'\}, \tag{5.3.143}$$

---

[16] Recall that, by definition, a four-vector depends on η_r if at least one of its components depends on η_r.

may be independent. In particular we note that not all of the four components of the four-vectors in the set in (5.3.143) are necessarily dependent on
η_r. Accordingly, we may write[17]

$$\dim S_r \leq \rho(G), \tag{5.3.144}$$

which, together with (5.3.138) and the fact that, in general, $\deg r_{\eta_r} R \leq \max_N \deg r_{\eta_r} \mathcal{T}^{(N)}$, as N runs over the sets in (5.3.48), establishes the statement of Lemma 5.3.5 for R. The same analysis with ε set equal to zero and
a Euclidean metric establishes the statement of the lemma for R_E.

The work of Chapter 2 implies, in particular, that we may find a constant
$b_r > 1$, and the power asymptotic coefficient $\alpha(S_r)$ of R, with $\eta_r \geq b_r$, may
be identified with $\deg r_{\eta_r} R$. We also note that (5.3.53) is true for *any* independent vectors $L_1, \ldots, L_r$ in I, as given in (5.3.40), with $S_r \equiv \{L_1, \ldots, L_r\}$,
and for *any* arbitrary r and k in $1 \leq r \leq k \leq 4n$. Therefore for any nonzero
subspace $S \subset I$, we have from Lemma 5.3.5 that

$$\alpha(S) < -\dim S. \tag{5.3.145}$$

What has just been said about R is also true for R_E. The condition in (5.3.145)
is nothing but criterion [A], Eq. (3.1.3) of Theorem 3.1.1. Hence we obtain.

Theorem 5.3.1: *The renormalized Feynman amplitude $\mathscr{A}_E(P, \mu)$, in
Euclidean space as given in (5.3.2), is absolutely convergent.*

Finally from Theorem 4.2.1 on the $\varepsilon \to +0$ limit of absolutely convergent
(Feynman) integrals, and from (5.3.145) we obtain.

Theorem 5.3.2: *The renormalized Feynman amplitude $\mathscr{A}_\varepsilon(P, \mu)$, in
Minkowski space as defined in (5.3.1), is absolutely convergent for $\varepsilon > 0$,
and the limit $\varepsilon \to +0$ defines a Lorentz covariant distribution.*

5.4   THE UNIFYING THEOREM OF
RENORMALIZATION AND BASIC IDENTITIES

We define

$$A_{G'} I_{G'} = \sum_{\varnothing \subset G'' \not\subseteq G'} \sum_{D_{G''}} \prod_{g \in D_{G''}} (-T_g) I_{G'} \tag{5.4.1}$$

and hence from (5.2.19),

---

[17] We give an elementary example where the equality in (5.3.144) does not hold. Suppose that
$k = 2$ in (5.3.40), with L_1 and L_2 independent vectors in I. Suppose that G is a graph with only one
four-dimensional integral, i.e., $K = (k_1^0, \ldots, k_1^3)$. We note that $\dim S_1 = 1$ and $\dim S_2 = 2$;
however, the corresponding $\rho_1(G)$ and $\rho_2(G)$ are simply $4L(G) = 4$ as a consequence of the four-
dimensional property of space–time.

$$A_{G'}I_{G'} = \sum_{\varnothing \subset G'' \nsubseteq G'} I_{G'/G''}M_{G''}. \tag{5.4.2}$$

Lemma 5.4.1:   *Let g be a proper subdiagram with n connected parts*

$g_1, \ldots, g_n$: $g = \bigcup_{i=1}^{n} g_i$. *Then*

$$-T_g A_g I_g = (-T_{g_1} A_{g_1} I_{g_1}) \cdots (-T_{g_n} A_{g_n} I_{g_n}). \tag{5.4.3}$$

In particular the identity (5.4.3) says that even if $d(g) \geq 0$ and one or more of the connected parts g_j are such that $d(g_i) < 0$, then $-T_g A_g I_g = 0$, by definition of the corresponding Taylor operations $(-T_{g_j}) = 0$ for such (proper and connected) subdiagrams g_j.

To prove (5.4.3) we note that the left-hand side of it may be written

$$-T_g A_g I_g = -T_g \left[\sum_{\varnothing \subset g^1 \nsubseteq g} (-T_{g^1}) \left[\sum_{\varnothing \subset g^2 \nsubseteq g^1} (-T_{g^2}) \left[\cdots (-T_{g^{i-1}}) \right. \right. \right.$$
$$\left. \left. \left. \times \left[\sum_{\varnothing \subset g^i \nsubseteq g^{i-1}} (-T_{g^i}) A_{g^i} \right] \cdots \right] \right] \right] I_g \tag{5.4.4}$$

for arbitrary i. For convenience we allow in the sums in (5.4.4) all proper subdiagrams by simply defining, as usual, $T_{g^j} = 0$ if $d(g^j) < 0$. Suppose i in (5.4.4) is chosen large enough so that the g^i (for $g^i \neq \varnothing$) are connected. The lemma is then trivially true for such g^i. Suppose as an induction hypothesis, that the statement of the lemma is true for all $g^1 \nsubseteq g$ corresponding to $i - 1$ square brackets in (5.4.4), i.e., suppose that the statement of the lemma is true for the subdiagrams $g^1 \nsubseteq g$; then we prove that the statement is also true for g itself.

According to the induction hypothesis,

$$(-T_g A_g) I_g = (-T_g) \sum_{\varnothing \subset g^1 \nsubseteq g} (-T_{\tilde{g}_1^1} A_{\tilde{g}_1^1}) \cdots (-T_{\tilde{g}_{m_1}^1} A_{\tilde{g}_{m_1}^1}) I_g, \tag{5.4.5}$$

where $\tilde{g}_1^1, \ldots, \tilde{g}_{m_1}^1$ are the connected parts of a g^1, i.e., in particular, $g^1 = \bigcup_{j=1}^{m_1} \tilde{g}_j^1$ and $m_1 = m_1(g^1)$. Clearly, each of the $\tilde{g}_j^1$ may fall in one and only one of the g_i $(i = 1, \ldots, n)$—the connected parts of g. Also, one or more of the $\tilde{g}_i^1$ may fall in one of the connected parts of g. Accordingly, for a fixed g^1 in (5.4.5), let $g_1^1, \ldots, g_{n_1}^1$ be all proper, though not necessarily connected, parts of g^1 falling in, respectively, $g_{i_1}, \ldots, g_{i_{n_1}}$, i.e., $g_1^1 \subset g_{i_1}, \ldots, g_{n_1}^1 \subset g_{i_{n_1}}$, where $n_1 \leq m_1$, $\{g_{i_1}, \ldots, g_{i_{n_1}}\}$ is a subset of $\{g_1, \ldots, g_n\}$ such that $g^1 = \bigcup_{i=1}^{m_1} \tilde{g}_i^1 = \bigcup_{i=1}^{n_1} g_i^1$, and each g_i^1 is a union of one or more of the $\tilde{g}_j^1$. By using the induction hypothesis once more, we may rewrite $(-T_{\tilde{g}_1^1} A_{g_1^1}) \cdots (-T_{\tilde{g}_{m_1}^1} A_{g_{m_1}^1}) I_g$ in (5.4.5) as $(-T_{g_1^1} A_{g_1^1}) \cdots (-T_{g_{n_1}^1} A_{g_{n_1}^1}) I_g$. It should also be noted that since $g^1 \nsubseteq g$, it follows that the term $[(-T_{g_1} A_{g_1}) \cdots (-T_{g_n} A_{g_n})] I_g$, where $g_1, \ldots, g_n$ are the connected parts of g, does not occur in the sum in

(5.4.5). Accordingly, we may add this term to and subtract it from the sum on the right-hand side of (5.4.5) to write equivalently

$$(-T_g A_g)I_g = (-T_g)\left\{ \sum_{\emptyset \subset g^1 \subset g} (-T_{g_1^1} A_{g_1^1}) \cdots (-T_{g_{h_1}^1} A_{g_{h_1}^1}) \right.$$
$$\left. - \left[(-T_{g_1} A_{g_1}) \cdots (-T_{g_n} A_{g_n}) \right] \right\} I_g. \tag{5.4.6}$$

For convenience of notation, we now trivially complete the collection $(g_1^1, \ldots, g_{n_1}^1)$ of n_1 elements into a collection of n elements $(g_1^1, \ldots, g_{n_1}^1, \emptyset, \ldots, \emptyset) \equiv (g_1^1, \ldots, g_{n_1}^1, \ldots, g_n^1)$ by adjoining $n - n_1$ empty subdiagrams to the former collection. In an obvious notation, we may then write (5.4.6) as

$$(-T_g A_g)I_g = -T_g\left\{ \left[\sum_{\emptyset \subset g_1^1 \subset g_1} (-T_{g_1^1} A_{g_1^1}) \right] \cdots \left[\sum_{\emptyset \subset g_n^1 \subset g_n} (-T_{g_n^1} A_{g_n^1}) \right] \right.$$
$$\left. - \left[(-T_{g_1} A_{g_1}) \cdots (-T_{g_n} A_{g_n}) \right] \right\} I_g, \tag{5.4.7}$$

by further relabeling the subdiagrams $\{g_1^1, \ldots, g_{n_1}^1\}$. Definition (5.4.1) implies

$$\sum_{\emptyset \subset g_i^1 \subset g_i} (-T_{g_i^1} A_{g_i^1})I_{g_i} = (1 - T_{g_i})A_{g_i}I_{g_i}. \tag{5.4.8}$$

Using (5.4.8), we may rewrite (5.4.7) in the following simple form:

$$(-T_g A_g)I_g = (-T_g)[(1 - T_{g_1})A_{g_1} \cdots (1 - T_{g_n})A_{g_n}$$
$$- (-T_{g_1} A_{g_1}) \cdots (-T_{g_n} A_{g_n})]I_g. \tag{5.4.9}$$

We recall that $g = \bigcup_{j=1}^n g_j$ and $I_g = \prod_{j=1}^n I_{g_j}$. We scale the external variables of g, and hence of $g_1, \ldots, g_n$, by a parameter λ. If $d(g_j) \geq 0$ for all $j = 1, \ldots, n$, then the first expression in the square brackets in (5.4.9) applied to I_g is of the form

$$\prod_{j=1}^n \left\{ f_j(\lambda q^{g_j}) - \sum_{i=0}^{d(g_j)} \frac{(\lambda)^i}{i!} \left[\left(\frac{\partial}{\partial \eta} \right)^i f_j(\eta q^{g_j}) \right]_{\eta=0} \right\}, \tag{5.4.10}$$

where only the dependence on the external momenta of the g_j are shown in (5.4.10) and f_j are some explicit functions of q^{g_j} associated with the g_j. By using the fact that $d(g) = \sum_{j=1}^n d(g_j)$, we see that when $(-T_g)$ is applied to the expression (5.4.10), it reduces it to zero, by definition of T_g. That is, $(-T_g)$ annihilates the first expression in the square brackets in (5.4.9). If for some g_j, say, $g_1, \ldots, g_s$, we have $d(g_j) < 0$, i.e., $T_{g_j} = 0$ for $j = 1, \ldots, s < n$, then the product in (5.4.10) is restricted only over $j = s + 1, \ldots, n$ and the resulting expression is simply multiplied by $\prod_{j=1}^s f_j(\lambda q^{g_j})$. The operation $(-T_g)$ then still annihilates the first expression in the square bracket in (5.4.9), since in this case $d(g) < \sum_{j=s+1}^{n} d(g_j)$. On the other hand, $T_g T_{g_1} \cdots T_{g_n} = T_{g_1} \cdots T_{g_n}$. Accordingly, we finally obtain

$$- T_g A_g I_g = (- T_{g_1} A_{g_1} I_{g_1}) \cdots (- T_{g_n} A_{g_n} I_{gn}), \qquad (5.4.11)$$

which is the statement of the lemma.

Corollary 5.4.1: *In the definition of the renormalized Feynman integrand R in (5.2.1) one may, in obtaining the final expression for R, restrict the summation over renormalization sets D such that the dimensionality of each of the connected parts g_i of a $g \in D$ is nonnegative, as the other D sets will not contribute to the sum in (5.2.1).*

This useful result follows from (5.2.21), (5.2.22), and (5.4.2), which, in particular, imply that

$$R = (1 - T_G)A_G I_G, \qquad (5.4.12)$$

with $A_G I_G$ defined recursively by

$$A_{G'} I_{G'} = \sum_{\varnothing \subset G'' \nsubseteq G'} (- T_{G''})A_{G''} I_{G'}, \qquad (5.4.13)$$

and from the particular statement of Lemma 5.4.1, which says that $- T_{G''} A_{G''} I_{G''}$ is zero if the dimensionality of any one of the connected parts of G'' is strictly negative.

Lemma 5.4.1, together with (5.4.12) and (5.4.13), gives the following theorem.

Theorem 5.4.1   (The unifying theorem of renormalization): *The renormalized Feynman integrand R for a proper and connected graph G may be equivalently written*

$$R = (1 - T_G)A_G I_G, \qquad (5.4.14)$$

*where $A_G I_G$ is defined **recursively** by*

$$A_{G'} I_{G'} = \sum_{\varnothing \subset G'' \nsubseteq G'} (- T_{G_1'} A_{G_1'}) \cdots (- T_{G_m''} A_{G_m''})I_{G'}, \qquad (5.4.15)$$

*for each proper **and connected** subdiagram G'. The sum is over all proper subdiagrams G'' (for $G'' \neq \varnothing$) with proper and connected subdiagrams $G_1'', \ldots, G_m''$.*

The expression (5.4.14), with (5.4.15), provides the definition of the so-called Bogoliubov–Parasiuk–Hepp–Zimmermann subtraction scheme in the Zimmermann form. Since the latter eventually grew out of the approach of Bogoliubov, and our subtraction scheme in (5.2.1) eventually grew out of the approach of Salam, Theorem 5.4.1 leads essentially to the equivalence of the paths taken in the ingenious approaches of Salam and Bogoliubov (in momentum space). Theorem 5.4.1 allows us to make a transition from one form of R to the other whenever it is convenient.

For a proper and connected subdiagram g, we define by D'_g a D set such that its largest subdiagram is g and any subdiagram in it is a proper *and* connected subdiagram of g. If $\{D'_g\}$ is the set of all such D'_g sets *such that* for any $D'_{g,1}$ and $D'_{g,2}$ in $\{D'_g\}$, and for any $g_1 \in D'_{g,1}, g_2 \in D'_{g,2}$, we have $g_1 \cap g_2 = \varnothing, g_1 \subset g_2 \subset g$, or $g_2 \subset g_1 \subset g$, then we denote the union $\bigcup D'_g$ of the sets in $\{D'_g\}$ by $\tilde{D}(g)$. It is not difficult to see from the definition of the recursion relation (5.4.15) that R may be written as (5.4.14) with

$$
A_G I_G = \sum_{\varnothing \subset G' \nsubseteq G} \left[\sum_{\tilde{D}(G'_1)} \prod_{g \in \tilde{D}(G'_1)} (-T_g) \cdots \sum_{\tilde{D}(G'_n)} \prod_{g \in \tilde{D}(G'_n)} (-T_g) \right] I_G,
$$

(5.4.16)

where $G'_1, \ldots, G'_n$ are the connected points of G'. The sum $\sum_{\tilde{D}(G'_i)}$ is over all those sets having their largest subdiagram G'_i, and if $g_1, g_2 \in \tilde{D}(G'_i)$, then $g_1 \cap g_2 = \varnothing, g_1 \subset g_2$, or $g_2 \subset g_1$. Finally, (5.4.16) may be rewritten

$$
A_G I_G = \sum_{\varnothing \subset G' \nsubseteq G} \left[\sum_{S_{1, \ldots, n}(G')} \prod_{g \in S_{1, \ldots, n}(G')} (-T_g) \right] I_G,
$$

(5.4.17)

where $S_{1, \ldots, n}(G') = \tilde{D}(G'_1) \cup \cdots \cup \tilde{D}(G'_n)$, and the product is over all proper *and* connected subdiagrams g in $S_{1, \ldots, n}(G')$. Note that if $g_1, g_2 \in S_{1, \ldots, n}(G')$, then *either* $g_1 \in G'_i, g_2 \in G'_j$, for some $i \neq j$, *or* $g_1, g_2 \in G'_i$ for some i, with $g_1 \cap g_2 = \varnothing, g_1 \subset g_2, $ or $g_2 \subset g_1$. That is, in particular, if $g_1, g_2 \in S_{1, \ldots, n}(G')$, then $g_1 \cap g_2 = \varnothing$, or $g_1 \subset g_2$, or $g_2 \subset g_1$. Finally, recall that $\sum_{\varnothing \subset G' \nsubseteq G}$ in (5.4.16) and (5.4.17) is over all proper subdiagrams G' (for $G' \neq \varnothing$) with $G' \nsubseteq G$ and with $G'_1, \ldots, G'_n$ their proper and connected parts. For $G' = \varnothing$, we write as usual $(-T_\varnothing) = 1$.

NOTES

The subtraction scheme was introduced by Manoukian (1976), and we were guided by the classic work of Salam (1951b) in its development. It is important to note that the scheme is carried in *momentum space* with subtractions, over subdiagrams, applied directly to the Feynman integrand, and hence no questions of ultraviolet cutoffs arise. Canonical momenta were introduced cleverly by Zimmermann (1969), and we followed his definition of the canonical decomposition of the Q_{ijl} quite closely. The convergence proof (Section 5.3) was given in Manoukian (1982a), see also 1977. The "unifying theorem of renormalization" is due to Manoukian (1976). For some other studies of renormalization, apart from the Bogoliubov–Parasiuk (1957)–Hepp (1966)–Zimmermann (1969) one, see Steinmann (1966), Caianiello *et al.* (1969), Kuo and Yennie (1969), Speer (1969), Anikin *et al.* (1973), Bergère and Zuber (1974), Anikin and Polivanov (1974). See also Velo and Wightman (1976).

Chapter 6 / ASYMPTOTIC BEHAVIOR IN QUANTUM FIELD THEORY

The purpose of this chapter is to study the asymptotic behavior of subtracted-out Feynman amplitudes $\mathscr{A}_E(P, \mu)$, in Euclidean space as defined in (5.3.2), when some of the elements in P and μ "take on" asymptotic values. Throughout this chapter we write $\mathscr{A}(P, \mu)$ for $\mathscr{A}_E(P, \mu)$ to simplify the notation, and no confusion should arise.

Before plunging into a general asymptotic analysis of $\mathscr{A}(P, \mu)$ we derive in Section 6.1 some elementary asymptotic estimates for $\mathscr{A}$ directly from the general definition in (2.2.84) (with $\mathscr{I}_E$ in the latter identified with R_E) which require no detailed knowledge of the structure of R_E. We show in particular that if the dimensionality $d(G)$ of the graph G, with which $\mathscr{A}$ is associated, is nonnegative, then if the external momenta are led to approach zero, it follows that $\mathscr{A}$ vanishes (Theorem 6.1.1). This is consistent with (and is expected from) the overall subtraction at the origin one carries out in defining R. We also show, whether $d(G) \geq 0$ or $d(G) < 0$, if all the masses appearing in $\mathscr{A}$ are led to approach infinity, then $\mathscr{A}$ again vanishes (Theorem 6.1.2). This is a particular case of the so-called decoupling theorem of field theory, which essentially says that Feynman amplitudes involving "very heavy" masses may be "neglected."

In the remaining sections of this chapter we carry out a general asymptotic analysis for $\mathscr{A}$. In Section 6.2 a general dimensional analysis for R_E is given on which the remaining sections are based. In Section 6.3 a study of the high-energy behavior of $\mathscr{A}$ is given when all the masses are fixed and nonzero. The study in Section 6.4 generalizes the results in Section 6.3 to deal with the situation when not only some (or all) of the external momenta of G become large but some (or all) of its masses as well become large. A study of the zero mass behavior of $\mathscr{A}$ is given in Section 6.5. The latter is then applied in Section 6.6 to study the low-energy behavior of $\mathscr{A}$ when some of the masses in the theory are led to approach zero. Section 6.7 deals with a very general study of $\mathscr{A}$ when some of the external momenta become large, some become small,[1] and some of the masses become small. In general we let these various asymptotic components "approach" their asymptotic values at *different* rates. In Section 6.8 a generalized decoupling theorem is proved which establishes the vanishing property of $\mathscr{A}$ when any subset of the masses in the theory become large and, in general, at different rates, and gives sufficiency conditions for the validity of the decoupling theorem when any subset of the remaining nonasymptotic masses are scaled to zero as well. The breakdown of this chapter into the above mentioned sections, as just outlined, will, it is hoped, facilitate the reading of this rather difficult chapter. One thing worth noting in this chapter is that one is able to obtain the asymptotic behavior of $\mathscr{A}$ without the need of carrying out explicitly the rather complicated integrals defining $\mathscr{A}$, as usually no closed expression for $\mathscr{A}$ may be obtained when one is involved with complicated Feynman graphs.

6.1 PRELIMINARY ASYMPTOTIC ESTIMATES

We write[2]

$$\mathscr{A}(P, \mu) = \int_{\mathbb{R}^{4n}} dK \, R_{\mathrm{E}}(P, K, \mu), \qquad (6.1.1)$$

$$R_{\mathrm{E}}(P, K, \mu) = \mathscr{P}(P, K, \mu)/\prod_l (Q_l^2 + \mu_l^2), \qquad (6.1.2)$$

$$Q_l = \sum_{j=1}^{n} a_{lj} k_j + \sum_{j=1}^{m} b_{lj} p_j \equiv k(l) + p(l), \qquad (6.1.3)$$

and

$$\mathscr{P}(P, K, \mu) = \sum_a p^a \mathscr{P}_a(K, \mu), \qquad (6.1.4)$$

in a notation similar to the one in (2.2.18), where

$$a = (a_{01}, \ldots, a_{3m}), \qquad |a| = a_{01} + \cdots + a_{3m},$$
$$d_{\mu j} \geq a_{\mu j} \geq 0, \qquad j = 1, \ldots, m, \qquad \mu = 0, 1, 2, 3. \qquad (6.1.5)$$

---

[1] Of course, some of the external momenta may remain nonasymptotic.
[2] Since in this chapter we consider only a Euclidean metric, we omit the E subscript in $Q_{l\mathrm{E}}^2$.

Let $p_{j0}^{\mu}, \ldots, p_{jd_{\mu j}}^{\mu}$ denote $d_{\mu j} + 1$ distinct values for p_j^{μ}. Let $P_t^* = (p_{1t_{01}}^0, \ldots, p_{mt_{3m}}^3)$, where $0 \le t_{\mu j} \le d_{\mu j}$. Then the generalized Lagrange interpolating formula (4.1.11) states that we may find a constant $C_a(P_t^*)$ depending on a and P_t^* such that

$$\mathscr{P}_a(K, \mu) = \sum_t C_a(P_t^*)\mathscr{P}(P_t^*, K, \mu), \tag{6.1.6}$$

where the sum is over all $0 \le t_{\mu j} \le d_{\mu j}$, with $j = 1, \ldots, m$, $\mu = 0, 1, 2, 3$.

We note that with $Q = k + p$, we may write

$$(Q^2 + \mu^2) = (k^2 + \mu^2)\left[1 + \frac{2kp}{k^2 + \mu^2} + \frac{p^2}{k^2 + \mu^2}\right], \tag{6.1.7}$$

$$(k^2 + \mu^2) = (Q^2 + \mu^2)\left[1 - \frac{2Qp}{Q^2 + \mu^2} + \frac{p^2}{Q^2 + \mu^2}\right], \tag{6.1.8}$$

and upon using the elementary inequalities

$$\begin{aligned} +2kp \le 2|k||p|, \qquad -2Qp \le 2|Q||p|, \\ 2|Q|\mu \le Q^2 + \mu^2, \qquad 2|k|\mu \le k^2 + \mu^2, \end{aligned} \tag{6.1.9}$$

we obtain

$$(Q^2 + \mu^2) \le (k^2 + \mu^2)A, \tag{6.1.10}$$

$$(k^2 + \mu^2) \le (Q^2 + \mu^2)A, \tag{6.1.11}$$

or

$$(Q^2 + \mu^2)A^{-1} \le (k^2 + \mu^2) \le (Q^2 + \mu^2)A, \tag{6.1.12}$$

where

$$A = 1 + \frac{|p|}{\mu} + \frac{p^2}{\mu^2}. \tag{6.1.13}$$

Let

$$D(P, K, \mu) = \prod_l (Q_l^2 + \mu_l^2). \tag{6.1.14}$$

Then from (6.1.6), the inequalities in (6.1.12) and the definition (6.1.14) we may find a strictly positive constant G depending on P and P_t^*, respectively, such that[3]

$$G^{-1}(P)\frac{|\mathscr{P}_a(K, \mu)|}{D(P, K, \mu)} \le \frac{|\mathscr{P}_a(K, \mu)|}{D(0, K, \mu)}$$

$$\le \sum_t |C_a(P_t^*)|G(P_t^*)\frac{|\mathscr{P}(P_t^*, K, \mu)|}{D(P_t^*, K, \mu)}. \tag{6.1.15}$$

---

[3] $D(0, K, \mu)$ means that the elements in P are set equal to zero in $D(P, K, \mu)$.

The inequalities in (6.1.15) imply, in particular, that the absolute convergence of $\int_{\mathbb{R}^{4n}} dK \, \mathscr{P}(P_t^*, K, \mu) D^{-1}(P_t^*, K, \mu)$ imply the convergence of

$$\int_{\mathbb{R}^{4n}} dK \, \mathscr{P}_a(K, \mu) D^{-1}(0, K, \mu)$$

and of

$$\int_{\mathbb{R}^{4n}} dK \, \mathscr{P}_a(K, \mu) D^{-1}(P, K, \mu).$$

Let $\lambda P = (\lambda p_1^0, \ldots, \lambda p_m^3)$, $\lambda > 0$; then

$$\mathscr{A}(\lambda P, \mu) = \sum_a \lambda^{|a|} p^a \int_{\mathbb{R}^{4n}} dK \, \mathscr{P}_a(K, \mu) D^{-1}(\lambda P, K, \mu), \qquad (6.1.16)$$

where $|a| \geq d(G) + 1$ for $d(G) \geq 0$. Suppose $\lambda \leq 1$; then the left-hand side of the inequality in (6.1.15) implies that

$$\int_{\mathbb{R}^{4n}} dK \, |\mathscr{P}_a(K, \mu)| D^{-1}(\lambda P, K, \mu) \leq G(P) \int_{\mathbb{R}^{4n}} dK \, |\mathscr{P}_a(K, \mu)| D^{-1}(0, K, \mu), \tag{6.1.17}$$

and its right-hand side is independent of λ.[4]

Now we consider the limit $\lambda \to 0$ and apply the Lebesgue dominated convergence theorem [Theorem 1.2.2(ii)] and conclude from (6.1.17) that we may take the limit $\lambda \to 0$ inside the integral in (6.1.16), and for $d(G) \geq 0$ we have

$$\lim_{\lambda \to 0} \mathscr{A}(\lambda P, \mu) = \sum_a \lim_{\lambda \to 0} \lambda^{|a|} p^a \int_{\mathbb{R}^{4n}} dK \, \mathscr{P}_a(K, \mu) \lim_{\lambda \to 0} D^{-1}(\lambda P, K, \mu)$$

$$= \sum_a \lim_{\lambda \to 0} \lambda^{|a|} p^a \int_{\mathbb{R}^{4n}} dK \, \mathscr{P}_a(K, \mu) D^{-1}(0, K, \mu) = 0. \tag{6.1.18}$$

Accordingly, we may state the following theorem.

Theorem 6.1.1: *For* $d(G) \geq 0$,

$$\lim_{\lambda \to 0} \mathscr{A}(\lambda P, \mu) = 0, \tag{6.1.19}$$

and from the left-hand side inequality in (6.1.15) *we have*

$$|\mathscr{A}(\lambda P, \mu)| \leq C_0 \lambda^{d(G)+1}, \qquad 0 \leq \lambda \leq 1, \tag{6.1.20}$$

----

[4] Note that $G(\lambda P) \leq G(P)$ for $\lambda \leq 1$ [see (6.1.13)].

where

$$C_0 = \sum_a |p^a| \prod_l \left(1 + \frac{|p(l)|}{\mu_l} + \frac{p(l)^2}{\mu_l^2}\right) \int_{\mathbb{R}^{4n}} dK \frac{|\mathscr{P}_a(K, \mu)|}{D(0, K, \mu)}, \quad (6.1.21)$$

and, as we have seen above, the integral in (6.1.21) exists.

Finally we write $\eta\mu = (\eta\mu^1, \ldots, \eta\mu^\rho)$ to obtain

$$\mathscr{A}(P, \eta\mu) = (\eta)^{d_0(G)} \int_{\mathbb{R}^{4n}} dK \, R_E\left(\frac{P}{\eta}, K, \mu\right), \quad (6.1.22)$$

where

$$d_0(G) = d(G) - \sum_{i=1}^{\rho} \sigma_i \le d(G).^5 \quad (6.1.23)$$

Accordingly upon identifying η^{-1} in the integral in (6.1.22) with λ in (6.1.18) we obtain

$$\lim_{\eta \to \infty} \mathscr{A}(P, \eta, \mu)$$
$$= \sum_a \lim_{\eta \to \infty} \eta^{d_0(G) - |a|} p^a \int_{\mathbb{R}^{4n}} dK \, \mathscr{P}_a(K, \mu) \lim_{\eta \to \infty} D^{-1}\left(\frac{P}{\eta}, K, \mu\right) = 0$$
$$(6.1.24)$$

for all $d(G)$, since if $d(G) < 0$, then $|a| \ge 0$, and if $d(G) \ge 0$, then $|a| \ge d(G)$ + 1. Hence we may state the following theorem.

Theorem 6.1.2 (The decoupling theorem): *For all $d(G)$, i.e., for $d(G)$ < 0 or $d(G) \ge 0$,*

$$\lim_{\eta \to \infty} \mathscr{A}(P, \eta\mu) = 0 \quad (6.1.25)$$

and

$$|\mathscr{A}(P, \eta\mu)| \le \eta^{-1} C_0, \quad \eta \ge 1, \quad (6.1.26)$$

where C_0 is defined in (6.1.21).

6.2 GENERAL DIMENSIONAL ANALYSIS OF SUBTRACTED-OUT FEYNMAN INTEGRANDS

In this section we carry out a general dimensional analysis of subtracted-out Feynman integrands. This analysis will then be applied in the remaining part of this chapter to investigate the asymptotic behavior of subtracted-out Feynman amplitudes.

---

[5] See the definition of the positive integers σ_i in (2.2.17). We assume (without loss of generality) that $\deg r_{P,K} \mathscr{P} = \deg r_{P,K,\mu} \tilde{\mathscr{P}}$ in (2.2.17) [see also (6.5.4)].

In (5.4.14) and (5.4.17) we have reduced the expression for the renormalized Feynman integrand, associated with a proper and connected graph G, to the form

$$R = (1 - T_G)A_G I_G, \tag{6.2.1}$$

where

$$A_G I_G = \sum_{\varnothing \subset G' \nsubseteq G} \left[\sum_{S_{1,\dots,n}(G')} \prod_{g \in S_{1,\dots,n}(G')} (-T_g) \right] I_G, \tag{6.2.2}$$

with

$$S_{1,\dots,n}(G') = \tilde{D}(G'_1) \cup \cdots \cup \tilde{D}(G'_n). \tag{6.2.3}$$

The sum $\sum_{\varnothing \subset G' \nsubseteq G}$ in (6.2.2), for $G' \neq \varnothing$, is over all proper subdiagrams $G' \nsubseteq G$, with $G'_1, \dots, G'_n$ denoting the connected components of G' and $d(G'_1) \geq 0, \dots, d(G'_n) \geq 0$. $\tilde{D}(G'_i)$ is a set with G'_i the largest subdiagram in it; if $g \in \tilde{D}(G'_i)$, then $g \subset G'_i$ and g is proper and connected with nonnegative dimensionality; if $g_1, g_2 \in \tilde{D}(G'_i)$, then either $g_1 \cap g_2 = \varnothing$ or $g_1 \nsubseteq g_2$ or $g_2 \nsubseteq g_1$. We define $S_{1,\dots,n}(\varnothing) \equiv \varnothing$ and $(-T_\varnothing) = 1$. We note that if g_1, $g_2 \in S_{1,\dots,n}(G')$, then g_1, g_2 are proper and connected and if $g_1 \subset G'_i$, $g_2 \subset G'_j$, with $i \neq j$ in $[1, \dots, n]$, then $g_1 \cap g_2 = \varnothing$. If $g_1, g_2 \subset G'_i$, for some i in $[1, \dots, n]$, then either $g_1 \cap g_2 = \varnothing$ or $g_1 \subset g_2$ or $g_2 \subset g_1$. The proper and connected subdiagrams $G'_1, \dots, G'_n$ are called the maximal elements in $S_{1,\dots,n}(G')$ because for any $g \in S_{1,\dots,n}(G')$, there is an $i \in [1, \dots, n]$ such that $g \subset G'_i$. Let $g \in S_{1,\dots,n}(G')$ and suppose $g_1, \dots, g_m$ are the maximal elements in $S_{1,\dots,n}(G')$, with $g_1, \dots, g_m \nsubseteq g$. Again the latter means that if $g' \in S_{1,\dots,n}(G')$, with $g' \nsubseteq g$, then $g' \subset g_i$ for some $i \in [1, \dots, m]$. We denote $g/g_1 \cup \cdots \cup g_m$ by $\bar{g}$. By definition, $g_1, \dots, g_m$ are proper and connected and pairwise *disjoint*. As usual, for convenience, we shall remove the restriction on the g in (6.2.2) with $d(g) \geq 0$ by allowing the proper and connected subdiagrams g in (6.2.2) with $d(g) < 0$ as well by simply setting $(-T_g) = 0$ in the latter case.

We may rewrite (6.2.1), (6.2.2) in the form

$$R = \sum_{\varnothing \subset G' \subset G} \left[\sum_{S_{1,\dots,n}(G')} \prod_{g \in S_{1,\dots,n}(G')} (-T_g) \right] I_G, \tag{6.2.4}$$

where $n = 1$ for $G' = G$. We note that a set $S_1(G) \equiv S(G)$ is of the form $S(G) = \{G\} \cup S_{1,\dots,n}(G')$, where $G'_1, \dots, G'_n$ (the connected components of a proper subdiagram G') are the maximal elements in $S(G)$ contained in $G: G'_i \nsubseteq G$ for $i = 1, \dots, n$.

Let I be a $4n$-dimensional subspace of $\mathbb{R}^{4n+4m+\rho}$ associated with the $4n$ integration variables. Let E be a complement of I in $\mathbb{R}^{4n+4m+\rho}: \mathbb{R}^{4n+4m+\rho} = I \oplus E$ (see Section 1.3 and Chapter 3). In turn let E_1 be a $4m$-dimensional subspace of E associated with the components of the independent external momenta. Let E_2 be a complement of E_1 in E; then we may write $\mathbb{R}^{4n+4m+\rho}$ $= I' \oplus E_1 = I'' \oplus E_2$, and we denote by $\Lambda(E)$, $\Lambda(I')$, $\Lambda(I'')$, $\Lambda(I)$ the projection operations on the subspaces I, E_1, E_2, E along the subspaces E, I', I'', I, respectively. We choose E to be the orthogonal complement of I in $\mathbb{R}^{4n+4m+\rho}$ and E_2 to be the orthogonal complement of E_1 in E. The subspace E_2 will be associated with the relevant masses in the theory.

Let $\mathbf{P}'$ be a vector in $\mathbb{R}^{4n+4m+\rho}$ such that the elements in K, P, and μ may be written as some linear combinations of the components of $\mathbf{P}'$.

Suppose $\mathbf{P}'$ is of the form

$$\mathbf{P}' = \mathbf{L}'_1 \eta_1 \eta_2 \cdots \eta_k + \cdots + \mathbf{L}'_k \eta_k + \mathbf{C}', \tag{6.2.5}$$

where $1 \le k \le 4n + 4m + \rho$, and $\mathbf{L}'_1, \ldots, \mathbf{L}'_k$ are k independent vectors in $\mathbb{R}^{4n+4m+\rho}$, $\mathbf{C}'$ is a vector confined to a finite region in $\mathbb{R}^{4n+4m+\rho}$, such that $\mu^i \ne 0$ for all $i = 1 \ldots, \rho$.

Let $S'_r \equiv \{\mathbf{L}'_1, \ldots, \mathbf{L}'_r\}$, where r is a fixed integer in $1 \le r \le k$. Suppose that $\Lambda(E)S'_r \ne \{0\}$, i.e., $\Lambda(E)S'_r$ is not the zero subspace. We carry out an analysis similar to the one in Section 5.3.2 for grouping the Taylor operations in (6.2.4) with respect to the parameter η_r.

Consider a set $\{G\} \cup S_{1,\ldots,n}(G') = S(G)$. Let g be a (proper and connected) subdiagram in $S(G)$. Let $c(\bar{g})$ be the set of all lines in $\bar{g}$ which have their k^g_{ijl} independent of η_r.[6] We may, symbolically, use the convenient notation: $\varnothing \subset c(\bar{g}) \subset \bar{g}$.

(i) Suppose $\varnothing \nsubseteq c(g) \nsubseteq \bar{g}$, and introduce the subdiagram $g_0 = g - c(\bar{g})$ as the subdiagram consisting of all the lines $\mathcal{L}^g - c(\bar{g})$ and of the relevant vertices as the end points of these lines. Let $g'_1, \ldots, g'_m$ be the maximal elements in $S(G)$ contained in g: $g'_i \nsubseteq g$. Then, by construction, all the k^g_{ijl} in $\bar{g}_0 = g_0/g'_1 \cup \cdots \cup g'_m$ are dependent on η_r.[7] A similar analysis to that in Section 5.3.2 shows that g_0 is proper and all the $k^{g_0}_{ijl}$ in $\bar{g}_0$ are dependent on η_r. Let $g_{01}, \ldots, g_{0N}$ be the connected components of g_0. For any subdiagram $g \in S(G)$ such that $\varnothing \nsubseteq c(\bar{g}) \nsubseteq \bar{g}$ we introduce the set $\mathcal{O}(g) = \{g_{01}, \ldots, g_{0N}\}$. We then define the set $\mathcal{N} \supset S(G)$ by

$$\mathcal{N} = S(G) \cup \left(\bigcup_g{}' \mathcal{O}(g) \right), \tag{6.2.6}$$

where $\bigcup_g{}'$ denotes the union over all sets $\mathcal{O}(g)$ with $g \in S(G)$ such that $\varnothing \nsubseteq c(\bar{g}) \nsubseteq \bar{g}$.

---

[6] As usual, we say that a four-vector k^g_{ijl} depends on η_r if at least one of its four components depends on η_r. Otherwise we say that the four-vector k^g_{ijl} is independent of η_r.

[7] If g is a minimal element in $S(G)$, i.e., $g'_1 = \cdots = g'_m = \varnothing$, then $\bar{g} = g$ and $\bar{g}_0 = g_0$.

(ii)   Let $g \in S(G)$ with $g \nsubseteq G$ and consider the case where $c(\bar{g}) = \varnothing$. Suppose g is a maximal element in $S(G)$ contained in a subdiagram g': $g \nsubseteq g'$. If all the k^q_{ijl} in $\bar{g}'$ are independent of η_r, then we introduce the set $\beta(g) = \{g\}$, and otherwise we write $\beta(g) = \varnothing$.

(iii)   Let $g \in S(G)$ and consider the case where $c(\bar{g}) = \bar{g}$; then we write $\beta(g) = \varnothing$.

We introduce sets $\mathscr{D}_{\mathscr{N}}$ and $\mathscr{D}_0$ as obtained from the set $S(G)$ defined by

$$\mathscr{D}_{\mathscr{N}} = \mathscr{N} - \left(\bigcup_g {}' \mathcal{O}(g) \right) \cup \left(\bigcup_g {}'' \beta(g) \right) \cup \{G\}, \qquad (6.2.7)$$

$$\mathscr{D}_0 = \mathscr{N} - \mathscr{D}_{\mathscr{N}}, \qquad (6.2.8)$$

where $\bigcup_g {}'' \beta(g)$ is the union over all sets $\beta(g)$, with g in category (ii) or (iii), i.e., with $c(\bar{g}) = \varnothing$ or $c(\bar{g}) = \bar{g}$.

By construction, for any $g \in \mathscr{N}$ all the k^q_{ijl} in $\bar{g}$ are either all dependent on η_r or are all independent of η_r. From the set $\mathscr{D}_{\mathscr{N}}$ in (6.2.7) we may generate all possible sets which contain in addition to the elements in $\mathscr{D}_{\mathscr{N}}$ any *one*, or any *two*, or ... or finally *all* the elements in $\mathscr{D}_0$. The collection of all these sets together with the set $\mathscr{D}_{\mathscr{N}}$ will be denoted $\mathscr{K} \equiv \{\mathscr{D}_{\mathscr{N}}, \ldots, \mathscr{D}_{\mathscr{N}} \cup \mathscr{D}_0\}$. The corresponding Taylor operations in (6.2.4) for the subdiagrams in the set $\mathscr{N}$ may then be readily combined, and the corresponding *sum* over all the sets in $\mathscr{K}$ is then reduced to the elementary expression

$$F_{\mathscr{N}}(G) = \prod_{g \in \mathscr{N}} (\delta^{\mathscr{N}}_g - T_g) I_G, \qquad (6.2.9)$$

where

$$\delta^{\mathscr{N}}_g = \begin{cases} 1 & \text{if} \quad g \in \mathscr{D}_0 \\ 0 & \text{if} \quad g \in \mathscr{D}_{\mathscr{N}}. \end{cases} \qquad (6.2.10)$$

By summing over all distinct sets $\mathscr{N}$ we then obtain an equivalent expression for R in (6.2.4) given by

$$R = \sum_{\mathscr{N}} \prod_{g \in \mathscr{N}} (\delta^{\mathscr{N}}_g - T_g) I_G \qquad (6.2.11)$$

with $\delta^{\mathscr{N}}_G = 1$.[8]

We introduce the following notation:

$$\mathscr{N} = \mathscr{H}_1(\mathscr{N}) \cup \mathscr{H}_2(\mathscr{N}), \qquad (6.2.12)$$

where $g \in \mathscr{H}_1(\mathscr{N})$ if all the k^q_{ijl} in $\bar{g}$ are dependent on η_r, and $g \in \mathscr{H}_2(\mathscr{N})$ if all the k^q_{ijl} in $\bar{g}$ are independent of η_r. We also write

$$\mathscr{H}_1(\mathscr{N}) = \mathscr{F}_1(\mathscr{N}) \cup \mathscr{F}_2(\mathscr{N}), \qquad (6.2.13)$$

---

[8] Zimmermann (1969) was the first to reduce his Bogoliubov-type subtraction scheme to a form as in (6.2.11) and obtain the estimate of the form in (6.2.44) and (6.2.51).

with $\mathcal{F}_2(\mathcal{N}) = \mathcal{D}_0$ if $G \in \mathcal{H}_1(\mathcal{N})$ and $\mathcal{F}_2(\mathcal{N}) = \mathcal{D}_0 - \{G\}$ if $G \in \mathcal{H}_2(\mathcal{N})$. We note that for $g \nsubseteq G$, $\delta_g^{\mathcal{N}} = 1$ if $g \in \mathcal{F}_2(\mathcal{N})$, and $\delta_g^{\mathcal{N}} = 0$ if $g \in \mathcal{F}_1(\mathcal{N}) \cup \mathcal{H}_2(\mathcal{N})$. We note, in particular, that for $g \nsubseteq G$, with $g \in \mathcal{F}_2(\mathcal{N})$, if g is a maximal element in $\mathcal{N}$ contained in a subdiagram $g' \in \mathcal{N} : g \nsubseteq g'$, then all the $k_{ijl}^{g'}$ in $\bar{g}'$ are independent of η_r.

We use the following convenient recursion relation as obtained from (6.2.11):

$$F_g(\mathcal{N}) = (\delta_g^{\mathcal{N}} - T_g)I_{\bar{g}} \prod_i F_{g_i}(\mathcal{N}). \tag{6.2.14}$$

where

$$R = \sum_{\mathcal{N}} F_G(\mathcal{N}), \tag{6.2.15}$$

and $\{g_i\}_i$ in (6.2.14) denotes the set of the maximal elements in $\mathcal{N}$ contained in $g : g_i \nsubseteq g$.[9] We also write

$$\sigma(g) = 4 \sum_{\substack{g' \in \mathcal{H}_1(\mathcal{N}) \\ g' \subset g}} L(\bar{g}'), \tag{6.2.16}$$

and set $L(\varnothing) \equiv 0$.

A line ℓ_i in G joining a vertex v_i to a vertex v_j is, in Euclidean space, of the form (see (5.1.1))

$$D_{ijl}^+(Q_{ijl}, \mu_{ijl}) = P_{ijl}(Q_{ijl}, \mu_{ijl})/[Q_{ijl}^2 + \mu_{ijl}^2]. \tag{6.2.17}$$

We assume that

$$\operatorname*{degr}_{\mu_{ijl}} D_{ijl}^+ \leq -1,^{10} \tag{6.2.18}$$

and

$$\operatorname*{degr}_{Q_{ijl}, \mu_{ijl}} P_{ijl} \leq \operatorname*{degr}_{Q_{ijl}} P_{ijl},$$
$$\operatorname*{degr}_{Q_{ijl}, \mu_{ijl}} D_{ijl}^+ \leq \operatorname*{degr}_{Q_{ijl}} D_{ijl}^+. \tag{6.2.19}$$

---

[9] The reason for using the expression in (6.2.14), (6.2.15) rather than the one in (5.3.48), (5.3.54), (5.3.55) for R in our asympotic analysis is that it turns out that the latter leads, in general, to an overestimate for the degr R over the former one. The expression (5.3.48), (5.3.54), (5.3.55) for R is, however, by far much simpler in structure that the one in (6.2.14), (6.2.15).

[10] For example, for spins 0, 1, 2: $D_{ijl}^+ = 0(\mu_{ijl}^{-2})$, and for spins $\frac{1}{2}, \frac{3}{2}$: $D_{ijl}^+ = 0(\mu_{ijl}^{-1})$, and (6.2.19) is true as well. Equations (6.2.18), (6.2.19) will be used when dealing with the asymptotic behavior of $\mathcal{A}$ when some of the underlying masses "take on" some asymptotic values, and they will not be needed when all the masses have fixed nonzero finite values.

Throughout this chapter, since we are interested in the asymptotic behavior of $\mathscr{A}$ with respect to the external momenta and/or to the masses of G, we assume that $\Lambda(I)S'_r \neq \{0\}$. Now we prove the following very important lemma for subdiagrams $g \in \mathscr{N} - \{G\}$. The case for the graph G will be treated separately.

Lemma 6.2.1: *Let r be a fixed integer in $1 \leq r \leq k$. Suppose that $\Lambda(E)S'_r \neq \{0\}$. Let $F_g(\mathscr{N})$ be as defined in (6.2.14) and suppose that $g \in \mathscr{N} - \{G\}$. For $g \in \mathscr{F}_1(\mathscr{N}) \cup \mathscr{H}_2(\mathscr{N})$, if there is*

(i)   *a subdiagram $g' \nsubseteq g$ with $g' \in \mathscr{F}_2(\mathscr{N})$, and/or*
(ii) *a subdiagram $g'' \subset g$ with $g'' \in \mathscr{H}_2(\mathscr{N})$, such that at least one mass μ_{ijl} in $\bar{g}''$ depends on η_r,*

then

$$\underset{\eta_r}{\operatorname{degr}} F_g(\mathscr{N}) < d(g) - \sigma(g). \tag{6.2.20}$$

Otherwise, i.e., if (i) *and* (ii) *are not true for g, then we may replace the sign $<$ in* (6.2.20) *by $\leq$. The latter means that in such a case an equality in* (6.2.20) *may hold.*

The proof is by induction. We suppose that the lemma is true for all those maximal elements $g_i \nsubseteq g$ with $g_i \in \mathscr{F}_1(\mathscr{N}) \cup \mathscr{H}_2(\mathscr{N})$. In addition to this hypothesis, we suppose, as part of the induction hypotheses, that if we scale the $k^{g'}_{ijl}$ in $\bar{g}'$ for all $g' \in \mathscr{H}_1(\mathscr{N})$, $g' \subset g_i$, for all i, and scale as well the masses μ_{ijl} in the g_i, depending on η_r, by a parameter λ, then

$$\underset{\lambda}{\operatorname{degr}} F_{g_i}(\mathscr{N}) \begin{cases} \leq -1 - \sigma(g_i) \\ \text{or} \\ = 0 \end{cases} \tag{6.2.21}$$

for a $g_i \in \mathscr{H}_2(\mathscr{N})$, and

$$\underset{\lambda}{\operatorname{degr}} F_{g_i}(\mathscr{N}) \begin{cases} < \\ \text{or} \\ \leq \end{cases} \min[d(g_i), -1] - \sigma(g_i) \tag{6.2.22}$$

for a $g_i \in \mathscr{F}_2(\mathscr{N})$.[11] The condition $=0$ in (6.2.21) and an equality in (6.22) may hold if both conditions (i) and (ii) in the lemma are not true for the corresponding g_i with g in these conditions (i), (ii) formally replaced by g_i. In addition, suppose that for a $g_i \in \mathscr{F}_1(\mathscr{N}) \cup \mathscr{H}_2(\mathscr{N})$, an $F_{g_i}(\mathscr{N})$ has a structure as in

$$F_{g_i}(\mathscr{N}) = P_{g_i}(k^{g_i}, q^{g_i}, \mu) f^1_{g_i}(k^{g_i}, \mu) f^2_{g_i}(k^{g_i}, \mu), \tag{6.2.23}$$

[11] The notation in (6.2.21) and (6.2.22) will be used whenever convenient.

where P_{g_i} is a polynomial in the elements in the sets k^{g_i}, q^{g_i}, and μ, and, in general, in the $(\mu^i)^{-1}$ as well. Here we denote by $f_g^n(k^g, q^g, \mu)$ *any* function of the form

$$f_g^n(k^g, q^g, \mu) = \prod_{\substack{g' \in \mathcal{H}_n(\mathcal{N}) \\ g' \subset g}} \prod_{\substack{i jl \\ i < j}}^{\bar{g}'} [(\Theta_{ijl}^{g'})^2 + \mu_{ijl}^2]^{-\sigma_{ijl}^n}, \qquad (6.2.24)$$

for $n = 1, 2$, where $\Theta_{ijl}^{g'} \equiv Q_{ijl}^{g'}$, and for $g' \not\subseteq g$, $g' \in \mathcal{H}_n(\mathcal{N})$, $\Theta_{ijl}^{g'} \equiv k_{ijl}^{g'}$. The σ_{ijl}^n are strictly positive integers. We also denote by $f^n(k^g, q^g, \mu)$ *any* function of the form

$$f^n(k^g, q^g, \mu) = \prod_{\substack{ijl \\ i < j}}^{\bar{g}} [Q_{ijl}^2 + \mu_{ijl}^2]^{-\sigma_{ijl}^n}, \qquad (6.2.25)$$

for $n = 1, 2$. If the q_{ijl}^g appearing in (6.2.24) and (6.2.25) are set equal to zero, then we denote the corresponding functions $f_g^n(k^g, \mu)$ and $f^n(k^g, \mu)$, respectively. Finally, if a $g_i \in \mathcal{F}_2(\mathcal{N})$, then we suppose that the corresponding $F_{g_i}(\mathcal{N})$ has a structure as in

$$F_{g_i}(\mathcal{N}) = P_{g_i}(k^{g_i}, q^{g_i}, \mu) f_{g_i}^1(k^{g_i}, q^{g_i}, \mu) f_{g_i}^2(k^{g_i}, \mu). \qquad (6.2.26)$$

We prove the above lemma as well as the results in (6.2.21)–(6.2.26) for the subdiagram g as well.

[A]   Suppose $g \in \mathcal{H}_2(\mathcal{N})$. Then the $g_i \in \mathcal{H}_2(\mathcal{N}) \cup \mathcal{F}_2(\mathcal{N})$. We write $k^{g_i} = k^{g_i}(k^g)$, $q^{g_i} = q^{g_i}(k^g, q^g)$. Let $g_1, \ldots, g_{k_1} \in \mathcal{H}_2(\mathcal{N})$ and $g_{k_1+1}, \ldots, g_{k_1+k_2} \in \mathcal{F}_2(\mathcal{N})$. From the induction hypotheses we may then write in a compact notation

$$F_g(\mathcal{N}) = -\sum_{\substack{A, B \\ a, b, c}} (q^g)^{A+B} P_{\bar{g}}^A(k^g, 0, \mu) f_B^2(k^g, \mu) \prod_{i=1}^{k_1+k_2} f_{g_i}^2(k^{g_i}(k^g), \mu)$$

$$\times \prod_{i=1}^{k_1} \prod_{j=k_1+1}^{k_1+k_2} (q^g)^{a_i + b_j + c_j} P_{g_i}^{a_i}(k^{g_i}(k^g), q^{g_i}(k^g, 0), \mu) f_{g_i}^1(k^{g_i}(k^g), \mu)$$

$$\times P_{g_j}^{b_j}(k^{g_j}(k^g), q^{g_j}(k^g, 0), \mu) f_{g_j, c_j}^1(k^{g_j}(k^g), q^{g_j}(k^g, 0), \mu), \qquad (6.2.27)$$

where

$$|A| + |B| + \sum_{i=1}^{k_1} |a_i| + \sum_{j=k_1+1}^{k_1+k_2} (|b_j| + |c_j|) \le d(g), \qquad (6.2.28)$$

$$a = (a_1, \ldots, a_{k_1}), \qquad b = (b_{k_1+1}, \ldots, b_{k_1+k_2}), \qquad c = (c_{k_1+1}, \ldots, c_{k_1+k_2}),$$

$$\operatorname*{degr}_\lambda P_{g_i}^{a_i} + \operatorname*{degr}_\lambda f_{g_i}^1 + \operatorname*{degr}_\lambda f_{g_i}^2 \begin{cases} \le -1 - \sigma(g_i) \\ \text{or} \\ = 0 \end{cases}, \qquad (6.2.29)$$

$$\operatorname*{degr}_\lambda P_{g_j}^{b_j} + \operatorname*{degr}_\lambda f_{g_j, c_j}^1 + \operatorname*{degr}_\lambda f_{g_j}^2 \begin{cases} < \\ \text{or} \\ \le \end{cases} \min[d(g_j), -1] - \sigma(g_j). \qquad (6.2.30)$$

Accordingly if $\mathrm{degr}_\lambda f_B^2 \leq -1$ and/or at least one of the conditions $=0$ (for some i) in (6.2.29) or an equality in (6.2.30) (for some j) does not hold, then we obtain

$$\mathrm{degr}_\lambda F_g(\mathcal{N}) \leq -1 - \sigma(g).^{12} \tag{6.2.31}$$

If $k_2 = 0$, (i) and (ii) in the lemma are not true for the g_i, $i = 1, \ldots, k_1$, i.e., the conditions $=0$ in (6.2.29) hold for the g_i, and $\mathrm{degr}_\lambda f_B^2 = 0$, then

$$\mathrm{degr}_\lambda F_g(\mathcal{N}) = 0, \tag{6.2.32}$$

and $\sigma(g) = 0$. Finally, from (6.2.27)–(6.2.31) we have

$$\mathrm{degr}_{\eta_r} F_g(\mathcal{N}) \begin{Bmatrix} < \\ \text{or} \\ \leq \end{Bmatrix} d(g) - \sigma(g), \tag{6.2.33}$$

where an equality in (6.2.33) may hold if the conditions (i) and (ii) in the lemma are not true for g. We also note from (6.2.27) that $F_g(\mathcal{N})$ has a structure as in (6.2.23).

[B]   Suppose $g \in \mathscr{F}_1(\mathcal{N})$. Then the $g_i \in \mathscr{F}_1(\mathcal{N}) \cup \mathscr{H}_2(\mathcal{N})$. According to the induction hypotheses we may write

$$F_g(\mathcal{N}) = - \sum_{A,B} (q^g)^{A+B} P_{\bar{g}}^A(k^g, 0, \mu) f_B^1(k^g, \mu)$$

$$\times \prod_i (q^g)^{a_i} P_{g_i}^{a_i}(k^{g_i}(k^g), q^{g_i}(k^g, 0), \mu) f_{g_i}^1(k^{g_i}(k^g), \mu)$$

$$\times f_{g_i}^2(k^{g_i}(k^g), \mu), \tag{6.2.34}$$

where

$$|A| + |B| + \mathrm{degr}_\lambda P_{\bar{g}}^A + \mathrm{degr}_\lambda f_B^1 \leq d(\bar{g}) - 4L(\bar{g}), \tag{6.2.35}$$

$$|a_i| + \mathrm{degr}_\lambda P_{g_i}^{a_i} + \mathrm{degr}_\lambda f_{g_i}^1 + \mathrm{degr}_\lambda f_{g_i}^2 \begin{Bmatrix} < \\ \text{or} \\ \leq \end{Bmatrix} d(g_i) - \sigma(g_i). \tag{6.2.36}$$

Accordingly we have

$$\mathrm{degr}_{\eta_r} F_g(\mathcal{N}) \begin{Bmatrix} < \\ \text{or} \\ \leq \end{Bmatrix} d(g) - \sigma(g), \tag{6.2.37}$$

---

[12] For $\mathrm{degr}_\lambda F_g(\mathcal{N}) \neq 0$, the expression on the right-hand side of the inequality in (6.2.31) gives an upper bound value for $\mathrm{degr}_\lambda F_g(\mathcal{N})$.

where, again, an equality in (6.2.37) may hold if both (i) and (ii) in the lemma are not true for g. From (6.2.34) we also note that $F_g(\mathcal{N})$ has a structure as in (6.2.23).

[C]   Finally, suppose that $g \in \mathscr{F}_2(\mathcal{N})$. Then the $g_i \in \mathscr{H}_2(\mathcal{N}) \cup \mathscr{F}_1(\mathcal{N})$, and by the induction hypotheses we may write

$$F_g(\mathcal{N}) = \sum_a \prod_i (q^g)^{a_i} P_{g_i}^{a_i}(k^{g_i}(k^g), q^{g_i}(k^g, 0), \mu) f_{g_i}^{1}(k^{g_i}(k^g), \mu)$$

$$\times f_{g_i}^{2}(k^{g_i}(k^g), \mu) [1 - T_g^{d(g) - |a|}] I_g. \tag{6.2.38}$$

For $d(g) \geq |a|$, Eq. (5.3.26) in Lemma 5.3.3 implies that

$$\operatorname*{degr}_{\lambda} [1 - T_g^{d(g) - |a|}] I_{\bar{g}} \leq d(\bar{g}) - 4L(\bar{g}) - d(g) + |a| - 1, \tag{6.2.39}$$

where we also have

$$\operatorname*{degr}_{\lambda} I_{\bar{g}} \leq d(\bar{g}) - 4L(\bar{g}). \tag{6.2.40}$$

From (6.2.38), (6.2.39), and (6.2.36) we then obtain for $d(g) \geq 0$

$$\operatorname*{degr}_{\lambda} F_g(\mathcal{N}) \begin{Bmatrix} < \\ \text{or} \\ \leq \end{Bmatrix} - 1 - \sigma(g). \tag{6.2.41}$$

If $d(g) < 0$, then (6.2.38), (6.2.40), and (6.2.36) imply that

$$\operatorname*{degr}_{\lambda} F_g(\mathcal{N}) \begin{Bmatrix} < \\ \text{or} \\ \leq \end{Bmatrix} d(g) - \sigma(g). \tag{6.2.42}$$

Equations (6.2.41) and (6.2.42) may be combined to yield

$$\operatorname*{degr}_{\lambda} F_g(\mathcal{N}) \begin{Bmatrix} < \\ \text{or} \\ \leq \end{Bmatrix} \min[d(g), -1] - \sigma(g), \tag{6.2.43}$$

in the notation in (6.2.22). On the other hand, (6.2.38) shows that $F_g(\mathcal{N})$ has a structure as given in (6.2.26). This completes the proof of the lemma together with the results in (6.2.21)–(6.2.26) for the subdiagram g itself.

Now we apply the above lemma to the graph G in question.

[I]   Suppose $G \in \mathscr{F}_2(\mathcal{N})$. Then (6.2.43) implies that

$$\operatorname*{degr}_{\lambda} F_G(\mathcal{N}) < -\sigma(G), \tag{6.2.44}$$

in the notation of (6.2.22).

Let $G_1, \ldots, G_m$ be the maximal elements in $\mathscr{N}$ contained in G: $G_i \nsubseteq G$. We may then directly apply the estimate in (6.2.37) to obtain for $d(G) \geq 0$

$$\operatorname*{degr}_{\eta_r}(-T_G)I_{\bar{G}} \prod_{i=1}^{m} F_{G_i}(\mathscr{N}) \left\{ \begin{matrix} < \\ \text{or} \\ \leq \end{matrix} \right\} d(G) - \sigma(G). \qquad (6.2.45)$$

Also, (6.2.20) and Lemma 6.2.1 imply that

$$\operatorname*{degr}_{\eta_r} I_{\bar{G}} \prod_{i=1}^{m} F_{G_i}(\mathscr{N}) \left\{ \begin{matrix} < \\ \text{or} \\ \leq \end{matrix} \right\} \operatorname*{degr}_{\eta_r} I_{\bar{G}} + \sum_{i=1}^{m} [d(G_i) - \sigma(G_i)]. \qquad (6.2.46)$$

To simplify the notation, let G' be that subdiagram of G with $\bar{G}' = G'/\bigcup_{i=1}^{m} G_i$ corresponding to all those lines in $\bar{G}$ depending on η_r. Then from (6.2.45) and (6.2.46) we may write

$$\operatorname*{degr}_{\eta_r} F_G(\mathscr{N}) \left\{ \begin{matrix} < \\ \text{or} \\ \leq \end{matrix} \right\} \max\left[d(G) - 4L(\bar{G}) - \sum_{i=1}^{m} \sigma(G_i), d(G') \right.$$

$$\left. - 4L(\bar{G}') - \sum_{i=1}^{m} \sigma(G_i) \right] \qquad (6.2.47)$$

for $d(G) \geq 0$, and (6.2.46) implies that

$$\operatorname*{degr}_{\eta_r} F_G(\mathscr{N}) \left\{ \begin{matrix} < \\ \text{or} \\ \leq \end{matrix} \right\} d(G') - 4L(\bar{G}') - \sum_{i=1}^{m} \sigma(G_i) \qquad (6.2.48)$$

for $d(G) < 0$. An equality in (6.2.47), (6.2.48) may hold if there is no subdiagram $g' \nsubseteq G$ such that $g' \in \mathscr{F}_2(\mathscr{N})$, and there is no subdiagram $g'' \subset G$ in $\mathscr{H}_2(\mathscr{N})$ such that at least one of the masses μ_{ijl} in $\bar{g}''$ depends on η_r.

[II]   Suppose that $G \in \mathscr{H}_2(\mathscr{N})$. Then (6.2.31) implies that for $d(G) \geq 0$

$$\operatorname*{degr}_{\lambda}(-T_G)I_{\bar{G}} \prod_{i} F_{G_i}(\mathscr{N}) < -\sigma(G), \qquad (6.2.49)$$

where the G_i denote the maximal elements in $\mathscr{N}$ contained in G: $G_i \nsubseteq G$. Since $\operatorname{degr}_\lambda I_{\bar{G}} = 0$, (6.2.31) and (6.2.43) imply, when applied to the G_i, that

$$\operatorname*{degr}_{\lambda} I_{\bar{G}} \prod_{i} F_{G_i}(\mathscr{N}) < -\sigma(G), \qquad (6.2.50)$$

where $\sigma(G) = \sum_i \sigma(G_i')$. If $d(G) < 0$, $F_G(\mathscr{N})$ coincides with $I_{\bar{G}} \prod_i F_{G_i}(\mathscr{N})$; accordingly from (6.2.49) and (6.2.50) we always have, i.e., for $d(G) < 0$ or for $d(G) \geq 0$,

$$\operatorname*{degr}_{\lambda} F_G(\mathscr{N}) < -\sigma(G). \qquad (6.2.51)$$

Suppose that the G_i are such that $G_1, \ldots, G_{k_1}, G_{k_1+1}, \ldots, G_{k_1+k_2} \in \mathscr{F}_2(\mathscr{N})$ with $d(G_i) \geq 0$ for $i = 1, \ldots, k_1$, and $d(G_i) < 0$ for $i = k_1 + 1, \ldots, k_1 + k_2$; and $G_{k_1+k_2+1}, \ldots, G_m \in \mathscr{H}_2(\mathscr{N})$. Let $\{G_{ij}\}$ be the set of maximal elements in $\mathscr{N}$ contained in G_i: $G_{ij} \nsubseteq G_i$. Let G_i' be that subdiagram of G_i with $\bar{G}_i = G_i' / \bigcup_j G_{ij}$ corresponding to all those lines in G_i depending on η_r. Let G_{ij}' be that subdiagram of G_{ij} with $\bar{G}_{ij}'$ corresponding to all those lines in $\bar{G}_{ij}$ depending on η_r. We may use (6.2.33), (6.2.47), and (6.2.48) to write

$$\operatorname*{degr}_{\eta_r} F_G(\mathscr{N}) \begin{Bmatrix} < \\ \text{or} \\ \leq \end{Bmatrix} d(\bar{G}') - 4L(\bar{G}') + \sum_{i=1}^{k_1} \left\{ \max[d(G_i) - 4L(\bar{G}_i), d(G_i') \right.$$

$$- 4L(\bar{G}_i')] - \sum_j \sigma(G_{ij}) \Big\} + \sum_{i=k_1+1}^{k_1+k_2} \left[d(G_i') - 4L(\bar{G}_i') \right.$$

$$- \sum_j \sigma(G_{ij}) \Big] + \sum_{i=k_1+k_2+1}^{m} [d(G_i) - \sigma(G_i)] \qquad (6.2.52)$$

for $d(G) < 0$, and

$$\operatorname*{degr}_{\eta_r} F_G(\mathscr{N}) \begin{Bmatrix} < \\ \text{or} \\ \leq \end{Bmatrix} \max[d(\bar{G}), d(\bar{G}') - 4L(\bar{G}')]$$

$$+ \sum_{i=1}^{k_1+k_2} \left\{ \max[d(G_i) - 4L(\bar{G}_i), d(G_i') - 4L(\bar{G}_i')] - \sum_j \sigma(G_{ij}) \right\}$$

$$+ \sum_{i=k_1+k_2+1}^{m} [d(G_i) - \sigma(G_i)] \qquad (6.2.53)$$

for $d(G) \geq 0$. An equality in (6.2.52) may hold if the conditions (i) and (ii) in Lemma 6.2.1 are not true for all the G_i. An equality in (6.2.53) may hold if (i) and (ii) in Lemma 6.2.1 are not true for all the G_i and, in addition to these constraints, we have

$$\operatorname*{degr}_{\eta_r} I_{\bar{G}} \prod_i F_{G_i}(\mathscr{N}) \geq \operatorname*{degr}_{\eta_r} (-T_G) I_{\bar{G}} \prod_i F_{G_i}(\mathscr{N}).$$

Now consider the situation when $\Lambda(E)S_r' = \{0\}$. In this case the integration variables are independent of η_r. Then we may simply replace $\delta_g^{\mathscr{N}}$ by 0 for $g \nsubseteq G$ in (6.2.14) and, as usual, replace $\delta_G^{\mathscr{N}}$ by 1, and $\mathscr{N}$ "becomes" simply $\mathscr{H}_2(\mathscr{N}) \equiv S(G)$. We may then use directly the estimates in (6.2.52) and (6.2.53) by replacing k_1 and k_2 in them by zero, as well as setting $\sigma(G_i) \equiv 0$, and obtain

$$\operatorname{degr} F_G(\mathcal{N}) \underset{\eta_r}{\left\{ \begin{matrix} < \\ \text{or} \\ \leq \end{matrix} \right\}} d(\bar{G}') - 4L(\bar{G}') + \sum_{i=1}^{m} d(G_i) \qquad (6.2.54)$$

for $d(G) < 0$, and

$$\operatorname{degr} F_G(\mathcal{N}) \underset{\eta_r}{\left\{ \begin{matrix} < \\ \text{or} \\ \leq \end{matrix} \right\}} \max[d(\bar{G}'), d(\bar{G}') - 4L(\bar{G}')] + \sum_{i=1}^{m} d(G_i) \qquad (6.2.55)$$

for $d(G) \geq 0$. An equality in (6.2.54) and (6.2.55) may hold if the G_i have no masses depending on η_r.

This completes our dimensional analysis of subtracted-out Feynman integrands R and will be applied in the remaining sections of this chapter.

We note in particular that we may write

$$\sigma(G) + 4 \sum_{\substack{g' \in \mathcal{H}_2(\mathcal{N}) \\ g' \subset G}} L(\bar{g}') = 4L(G), \qquad (6.2.56)$$

and quite generally we have

$$\dim \Lambda(E)S_r' \leq \sigma(G). \qquad (6.2.57)$$

In particular, to have an equality in (6.2.57) the subspaces S_r' must be such that $\dim \Lambda(E)S_r'$ is a multiple of 4, and all the four components of the k_{ijl}^q in $\bar{g}$ with $g \in \mathcal{H}_1(\mathcal{N})$, depend on η_r.

Let $\mathbf{P}$ be a vector in E, and $\mathbf{L}_1, \ldots, \mathbf{L}_k$ be k independent vectors in E such that

$$\mathbf{P} = \mathbf{L}_1 \eta_1 \cdots \eta_k + \cdots + \mathbf{L}_r \eta_k + \cdots + \mathbf{L}_k \eta_k + \mathbf{C}, \qquad (6.2.58)$$

where $\mathbf{C}$ is confined to a finite region in E, with the $\mu^i \neq 0$ for $i = 1, \ldots, \rho$, and the external momenta and the masses of the graph G in question may be written as some linear combinations of the components of $\mathbf{P}$. We may then write

$$\mathcal{A}(P, \mu) \equiv \mathcal{A}(\mathbf{L}_1 \eta_1 \cdots \eta_k + \cdots + \mathbf{L}_k \eta_k + \mathbf{C}). \qquad (6.2.59)$$

According to Theorem 3.1.1, the power asymptotic coefficient $\alpha_I(S_r)$ of $\mathcal{A}(P, \mu)$ associated with a subspace $S_r \equiv \{\mathbf{L}_1, \ldots, \mathbf{L}_r\}$ is given by

$$\alpha_I(S_r) = \max_{\Lambda(I)S = S_r} [\alpha(S) + \dim S - \dim S_r] \qquad (6.2.60)$$

with $S \subset \mathbb{R}^{4n+4m+\rho}$, where $\alpha(S)$ is the power asymptotic coefficient of the integrand R, and, according to the analysis in Chapter 2, may be identified with the degree of R. In the subsequent sections we shall construct the class of the maximizing subspaces $\mathcal{M}$ (see Chapter 3) for the various situations at hand directly from the dimensional analysis carried out in this section. This will lead to both the power and logarithmic behavior of $\mathcal{A}$. We recall that if $S \in \mathcal{M}$, then

$$\alpha_I(S_r) = \alpha(S) + \dim S - \dim S_r. \tag{6.2.61}$$

Note that for any $S \subset \mathbb{R}^{4n+4m+\rho}$, $\dim S \leq \dim \Lambda(I)S + \dim \Lambda(E)S$ or $\dim S - \dim \Lambda(I)S \leq \dim \Lambda(E)S$.

6.3   HIGH-ENERGY BEHAVIOR

In this section we are interested in the high-energy behavior of renormalized Feynman amplitudes with all the masses involved in the graph G in question fixed and nonzero. Technically we are interested in the behavior of

$$\mathcal{A}(\mathbf{L}_1\eta_1 \cdots \eta_k + \cdots + \mathbf{L}_r\eta_r \cdots \eta_k + \cdots + \mathbf{L}_k\eta_k + \mathbf{C}), \tag{6.3.1}$$

for $\eta_1, \eta_2, \ldots, \eta_k \to \infty$ independently, where $\mathbf{L}_1, \ldots, \mathbf{L}_k$ are k independent vectors in E_1 and $1 \leq k \leq 4m$. $\mathbf{C}$ is a vector confined to a finite region in E such that $\mu^i \neq 0$ for all $i = 1, \ldots, \rho$. As usual we write $S_r \equiv \{\mathbf{L}_1, \ldots, \mathbf{L}_r\}$ with $1 \leq r \leq k$. (Recall that $\mathbb{R}^{4n+4m+\rho} = I \oplus E$, $E = E_1 \oplus E_2$, where E_1 is associated with the external momenta.)

We may specialize Lemma 6.2.1 to the problem at hand through the following corollary [see also (6.2.5)].

Corollary 6.3.1: *Let r be a fixed integer in $1 \leq r \leq k$. Suppose that $\Lambda(E)S'_r \neq \{0\}$. Let $F_g(\mathcal{N})$ be as defined in (6.2.14) and suppose that $g \in \mathcal{N} - \{G\}$. For $g \in \mathcal{F}_1(\mathcal{N}) \cup \mathcal{H}_2(\mathcal{N})$, if there is a subdiagram $g' \nsubseteq g$ with $g' \in \mathcal{F}_2(\mathcal{N})$, then*

$$\operatorname*{degr}_{\eta_r} F_g(\mathcal{N}) < d(g) - \sigma(g).$$

Otherwise, if there is no subdiagram $g' \nsubseteq g$ with $g' \in \mathcal{F}_2(\mathcal{N})$, then the $<$ sign may be replaced by $\leq$, which means that an equality may hold for the latter case.

By treating the situation for the graph G in the light of Corollary 6.3.1 we arrive at the estimates for $\operatorname{degr}_{\eta_r} F_G(\mathcal{N})$ in (6.2.47) and (6.2.48) for $G \in \mathcal{F}_2(\mathcal{N})$, and to the estimates (6.2.52) and (6.2.53) for $G \in \mathcal{H}_2(\mathcal{N})$ by completely *deleting*, in the process, condition (ii) in Lemma 6.2.1, as all the μ^i are fixed (nonzero), i.e., are independent of η_r. Similarly, for $\Lambda(E)S'_r = \{0\}$

we have the estimate in (6.2.54) and (6.2.55) with a possible equality holding, as, again, all the masses are kept fixed.

For each line ℓ_1 joining a vertex v_i to a vertex v_j of the graph G, we introduce vectors $\mathbf{V}_0(ijl), \ldots, \mathbf{V}_3(ijl)$ in $\mathbb{R}^{4n+4m+\rho}$ such that, with $\mathbf{P}'$ as given in (6.2.5),

$$\mathbf{V}_0(ijl) \cdot \mathbf{P}' = Q_{ijl}^0,$$
$$\vdots \tag{6.3.2}$$
$$\mathbf{V}_3(ijl) \cdot \mathbf{P}' = Q_{ijl}^3.$$

We denote by $S^0(ijl)$ the subspace generated by the vectors $\mathbf{V}_0(ijl), \ldots, \mathbf{V}_3(ijl)$. We also introduce vectors $\mathbf{V}_0'(ijl), \ldots, \mathbf{V}_3'(ijl)$ such that

$$\mathbf{V}_0'(ijl) \cdot \mathbf{P}' = k_{ijl}^0,$$
$$\vdots \tag{6.3.3}$$
$$\mathbf{V}_3'(ijl) \cdot \mathbf{P}' = k_{ijl}^3.$$

A close examination of Corollary 6.3.1, together with the estimates (6.2.47), (6.2.48), (6.2.52)–(6.2.55) for the present situation with fixed masses, as discussed above, after summing over $\mathcal{N}$ in (6.2.15), suggests defining the following class $\mathcal{M}_0$ of subspaces, which turns out to form, in general, a subset of the *maximizing* subspaces for the I integration of R relative to $S_r \equiv \{\mathbf{L}_1, \ldots, \mathbf{L}_r\}$, with $\mathbf{L}_1, \ldots, \mathbf{L}_r$ as given in (6.3.1).

Definition of class $\mathcal{M}_0$: We define a class $\mathcal{M}_0 = \{S', \ldots\}$ of subspaces $S' \subset \mathbb{R}^{4n+4m+\rho}$ and a set of subdiagrams $\tau_0 = \{G', \ldots\}$ in such a way that the following are consistent:

(i) $\Lambda(I)S' = S_r$.

(ii)   Let G' be the subdiagram of G, corresponding to all those lines in G (and, of course, corresponding to the vertices as the end points of these lines), such that all the subspaces $S^0(ijl)$ of G in G' are not orthogonal to S'.[13]

(iii)   The proper part G_0' of G' corresponds to all those lines in G with their $\mathbf{V}_0(ijl), \ldots, \mathbf{V}_3(ijl)$ not orthogonal to S'.[14]

(iv)   If $S'' \subset \mathbb{R}^{4n+4m+\rho}$ is such that (i)–(iii) above are consistently true, with G'', in particular, corresponding to all the lines in G with the subspaces $S^0(ijl)$ of G in G'' not orthogonal to S'', then $d(G'') \leq d(G')$. If $d(G'') = d(G')$, then $S'' \in \mathcal{M}_0$ and $G'' \in \tau_0$.

In light of this definition, we say that the subdiagram G' is *associated with* the subspace S'.

---

[13] The subspaces $S^0(ijl)$ are defined following Eq. (6.3.2). By all subspaces $S^0(ijl)$ of G in G' we mean the subspaces $S^0(ijl)$ for all $i, j,$ and l pertaining to the lines and vertices of G in G'.

[14] The condition that all the $\mathbf{V}_0(ijl), \ldots, \mathbf{V}_3(ijl)$ in the lines of G appearing in G' be not orthogonal to S' will be necessary in order to have $\dim \Lambda(E)S' = 4L(G_0')$.

We note that G'/G'_0 (if not empty), with $G' \in \tau_0$ associated with a subspace $S' \in \mathcal{M}_0$ and G'_0 being the proper part of G', corresponds to the external and improper (if any) lines of G'. An analysis very similar to the one given in Section 5.3.2 in reference to Fig. 5.11 shows that all the vectors $\mathbf{V}'_0(ijl), \ldots,$ $\mathbf{V}'_3(ijl)$ of the external and improper (if any) lines of G' in G are orthogonal to S'. This is consistent with condition (iii) in the definition. Note, however, that condition (iii) requires that the *whole* proper part of G', not just a proper subdiagram of G', have the $\mathbf{V}'_0(ijl), \ldots, \mathbf{V}'_3(ijl)$ of G in its lines not orthogonal to S'. We also note that for any proper subdiagram $g \not\supseteq G'_0$, we may write $k^0_{ijl} - (q^g_{ijl})^0(k^G, 0) = (k^g_{ijl})^0$, or conveniently as $\mathbf{V}'_0(ijl) - (q^g_{ijl})^0(\mathbf{V}'_0(ijl), \ldots,$ $\mathbf{V}'_3(ijl)) \equiv \mathbf{V}^g_0(ijl)'$ and $\mathbf{V}^g_0(ijl)' \cdot \mathbf{P}' = (k^g_{ijl})^0$. Here $(q^g_{ijl})^0$ is a linear combination of the $\mathbf{V}'_0(ijl), \ldots, \mathbf{V}'_3(ijl)$ with $i, j,$ and l pertaining to the lines and vertices in G/g [see (5.1.117)]. Since these latter vectors are orthogonal to S', and $\mathbf{V}'_0(ijl)$, with $i, j,$ and l pertaining to G/G'_0, is orthogonal to S' as well, it follows that the $\mathbf{V}^g_0(ijl)'$ in g/G'_0 are orthogonal to S'. Repeating the same analysis for the $\mathbf{V}^g_1(ijl)', \ldots, \mathbf{V}^g_3(ijl)'$, similarly defined, as well, we arrive to the conclusion that for any proper $g \not\supseteq G'_0$, all the $\mathbf{V}^g_0(ijl)', \ldots, \mathbf{V}^g_3(ijl)'$ in g/G'_0 are orthogonal to S' since the $\mathbf{V}'_0(ijl), \ldots, \mathbf{V}'_3(ijl)$ in G/G'_0 are orthogonal to S'.

If the whole graph $G \in \tau_0$ and is associated with a subspace $S \in \mathcal{M}_0$, then (6.2.47), (6.2.48), applied to the problem at hand, imply for the set $\mathcal{N} = \{G\}$ that we may take for an estimated degree of $F_G(\mathcal{N})$, with respect to η_r, the expression $d(G) - 4L(G)$, i.e.,

$$\operatorname*{degr}_{\eta_r} F_G(\mathcal{N}) = d(G) - 4L(G), \tag{6.3.4}$$

with $4L(G) = \dim \Lambda(E)S$.

More generally, if $G' \in \tau_0$, with $G' \not\subseteq G$, and is associated with a subspace $S' \in \mathcal{M}_0$, then (6.2.52), (6.2.53), applied to the problem at hand, imply for the set $\mathcal{N} = \{G, G'_{01}, \ldots, G'_{0n}\}$, where $G'_{01}, \ldots, G'_{0n}$ are the connected parts of the proper part G'_0 of G', that we may take

$$\operatorname*{degr}_{\eta_r} F_G(\mathcal{N}) = d(G') - \dim \Lambda(E)S', \tag{6.3.5}$$

where $\dim \Lambda(E)S' = 4\sum_{i=1}^n L(G'_{0i}) = 4L(G'_0)$. On the other hand, if $G'_{01}, \ldots,$ G'_{0n} are contained (as maximal elements) in some maximal elements $\tilde{G}_1, \ldots, \tilde{G}_m, m \leq n$, with the latter contained in $G: \tilde{G}_i \not\subseteq G$, in some set $\mathcal{N}$, then according to Corollary 6.3.1 and the definition of $(\mathcal{M}_0, \tau_0)$ [in particular, condition (iv)], $\operatorname{degr}_{\eta_r} F_G(\mathcal{N})$ for such a set $\mathcal{N}$ will not be greater than the one in (6.3.5).

A moment's reflection then shows that the power asymptotic coefficient of R itself [see (6.2.15)] for a subspace $S' \in \mathcal{M}_0$ may be taken to be

$$\alpha(S') = d(G') - \dim \Lambda(E)S', \tag{6.3.6}$$

and that all the subspaces $S' \in \mathcal{M}_0$ are maximizing subspaces for the I integration relation to $S_r \equiv \{\mathbf{L}_1, \ldots, \mathbf{L}_r\}$. The latter, in particular, follows from conditions (i)–(iv) in the definition of $\mathcal{M}_0$, the power counting conditions (6.2.44), (6.2.51) [see criterion [A], (3.1.3) in Theorem 3.1.1], and (3.1.4), which imply that the power asymptotic coefficient $\alpha_I(S_r)$ of $\mathcal{A}$ in reference to the parameter η_r in (6.3.1) may be taken for the bound of $|\mathcal{A}|$ [see (6.2.60), (6.2.61), (6.3.6)]:

$$\alpha_I(S_r) = d(G'). \tag{6.3.7}$$

Accordingly, we may state the following theorem.

Theorem 6.3.1: *The power asymptotic coefficient $\alpha_I(S_r)$ of the renormalized amplitude*

$$\mathcal{A}(\mathbf{L}_1 \eta_1 \cdots \eta_k + \cdots + \mathbf{L}_r \eta_r \cdots \eta_k + \cdots + \mathbf{L}_k \eta_k + \mathbf{C})$$

is simply given by

$$\alpha_I(S_R) = d(G'), \quad S_r \subset E_1, \tag{6.3.8}$$

where G' is any subdiagram in τ_0, respectively in r, with $1 \leq r \leq k \leq 4m$.

We note that when some (or all) of the external independent momenta of the graph G becomes large, specified by a parameter, say, $\eta_r \to \infty$ in (6.3.1), then in reference to this parameter a subdiagram $G' \in \tau_0$ cannot have an extral vertex at which all the momenta carried by all the external lines *to* G, at this vertex, are nonasymptotic.[15] As a matter of fact, if G' has an extral vertex, then the total external momentum at that vertex must be asymptotic. This follows from the fact that an extral vertex v_j of G' is necessarily an external vertex of G and an external line ℓ_l of G' attached to this vertex has its $k_{ijl}^{G'}$ independent of η_r, and the corresponding $q_{ijl}^{G'}$ is dependent on η_r, by definition of G'. Momentum conservation then requires that q_j^G must depend on η_r ($\eta_r \to \infty$). Conversely, G' contains all those vertices of G at which the total momentum carried by the external lines *to* G, at each of these vertices, is asymptotic. Note also (for $G' \not\subseteq G$), G' cannot have a subdiagram, say, G_i, as one of its connected components with *all* the external momenta of G_i being nonasymptotic. Because, whether $d(G_i) < 0$ or $d(G_i) \geq 0$, this can only "decrease the value" of $\alpha(S_r)$ below $d(G')$, since then $\deg r_{\eta_r} F_{G_i}(\mathcal{N}) < -\sigma(G_i)$ [see (6.2.43)]. Therefore the determination of the subdiagrams $G' \in \tau_0$ is not difficult. Finally, we recall that if any other subdiagram G'' similarly defined as G' is such that $d(G'') \leq d(G')$, then $G'' \in \tau_0$ only if $d(G'') = d(G')$.

---

[15] Such a point was also emphasized by Weinberg (1960, p. 847).

The subspaces in $\mathcal{M}_0$ do not, in general, constitute all of the maximizing subspaces for the I integration relative to $S_r = \{\mathbf{L}_1, \ldots, \mathbf{L}_r\}$ due to the simple fact that if there is a proper and connected subdiagram g' in a set $\mathcal{N}$, with $d(g') \geq 0$ and $g' \not\subseteq g$, $g' \in \mathcal{H}_2(\mathcal{N})$, in Corollary 6.3.1, then $\mathrm{degr}_{n_r} F_g(\mathcal{N})$ may still coincide with $d(g) - \sigma(g)$. Accordingly we may readily extend the definition of the class $\mathcal{M}_0$ as follows.

Consider a subdiagram G' in τ_0 associated with a subspace S' in $\mathcal{M}_0$. Let $J_{G'} = \{g'_1, g'_2, \ldots, g'_{N'}\}$ be the set of all proper, but not necessarily connected, subdiagrams of the proper part G'_0 of G' *such that* the connected part of each of the subdiagrams in $J_{G'}$ has a nonnegative dimensionality. In particular, we note that if each of the connected parts G'_0 has a nonnegative dimensionality, then $G'_0 \in J_{G'}$, by definition of $J_{G'}$.

Let $g'_1 \in J_{G'}$. We define a *generalized subdiagram* $(G'|g'_1)$ obtained from G' by shrinking g'_1 in it to a point and replacing the analytical expression $I_{g'_1}$, in the unrenormalized integrand $I_{G'}$ for G', by a polynomial in the *external variables* of g'_1 of degree $\leq d(g'_1)$. Therefore $(G'|g'_1)$ is nothing but the subdiagram G' with g'_1 in it replaced by a vertex, which we call a *generalized vertex*, with the corresponding analytical expression for the latter as a polynomial in the external varibles of g'_1. In this respect, we also note from (5.1.117), that the external variables of g'_1 may be expressed as linear combinations of the Q_{ijl} in G'/g'_1, with the v_i being vertices in G'/g'_1, but not in g'_1, and the v_j being external vertices of g'_1, and also, in general, as linear combinations of external variables of G'. Accordingly we may formally define an integrand $I_{(G'|gi)}$ in the same way as we defined $I_{G'}$. By considering all the elements in $J_{G'}$, we may generate the following generalized subdiagrams: $(G'|g'_1), \ldots, (G'|g'_{N'})$. Finally, we repeat the above construction by considering *all* of the remaining subdiagrams in τ_0 and generate:

$$(G''|g''_1), \ldots, (G''|g''_{N''}), \ldots; G'', \ldots \in \tau_0.$$

We extend the definition of the class $\mathcal{M}_0$ to the class $\mathcal{M}$ by simultaneously enlarging the set τ_0 to include as well all the generalized subdiagrams $(G'|g'_1), \ldots, (G'|g'_{N'}); (G''|g''_1), \ldots, (G''|g''_{N''}); \ldots$ for all $G', G'', \ldots \in \tau_0$, as follows. To this end we define $(G'|\varnothing) \equiv G'$.

Definition of class $\mathcal{M}$:   We define the class

$$\mathcal{M} = \{S', S'_1, \ldots, S'_{N'}; S'', S''_1, \ldots, S''_{N''}; \ldots\},$$

where $\mathcal{M}_0 = \{S', S'', \ldots\}$, and the set of subdiagrams $\tau = \{G', (G'|g'_1), \ldots, (G'|g'_N), G'', (G''|g''_1), \ldots, (G''|g''_{N''}); \ldots\}$ such that the following are consistent: For any $\tilde{S} \in \mathcal{M}$

(i) $\Lambda(I)\tilde{S} = S_r$.

(ii)   Let $\hat{G}$ be the subdiagram of G, corresponding to all the lines in G, such that all the subspaces $S^0(ijl)$ of G in $\hat{G}$ are not orthogonal to $\tilde{S}$.

(iii)   The set of all the lines of G in $\hat{G}$ having all their $V'_0(ijl), \ldots, V'_3(ijl)$ not orthogonal to $\tilde{S}$ coincide with the set of all the lines in the proper part of the generalized subdiagram $(\hat{G}|\hat{g}) \equiv \tilde{G}$, where $\hat{g}$ is either empty, or otherwise $\hat{g}$ is a proper subdiagram of $\hat{G}$ such that each of the connected components of $\hat{g}$ has nonnegative dimensionality.

(iv) $\hat{G}$ belongs to τ_0 (and obviously to τ) and $\tilde{G}$ belongs to τ.[16]

As before, we also say that the generalized subdiagrams $(G'|g'_1), \ldots,$ $(G'|g'_{N'}); (G''|g''_1), \ldots, (G''|g''_{N''}); \ldots$ are associated with the subspaces $S'_1, \ldots, S'_{N'}; S''_1, \ldots, S''_{N''}; \ldots$, respectively.[17]

The basic difference between a subspace S' and S'_1, say, is that

$$\alpha(S') = d(G') - \dim \Lambda(E)S' \qquad (6.3.9)$$

and

$$\alpha(S'_1) = d(G') - \dim \Lambda(E)S'_1 \qquad (6.3.10)$$

where $4L(G'_0) = \dim \Lambda(E)S'$, $4L(G'_0/g'_1) = \dim \Lambda(E)S'_1$,   and   hence $\dim \Lambda(E)S'_1 < \dim \Lambda(E)S'$. Both subspaces S' abd S'_1 are obviously, however, *maximizing* subspaces for the I integration relative to S_r, and from the definition of the set τ_0 together with (6.2.60), (6.2.61), we may take

$$\alpha_I(S_r) = d(G') \qquad (6.3.11)$$

when considering both subspaces S' and S'_1.

Now with the class $\mathcal{M}$ of the maximizing subspaces for the I integration of R relative to the subspace $S_r \equiv \{L_1, \ldots, L_r\}$, with $L_1, \ldots, L_r$ as given in (6.3.1), we determine the logarithmic asymptotic coefficients $\beta_I(S_r)$ of $\mathcal{A}$.

We decompose I as a direct sum of $4n$ one-dimensional subspaces: $I = I_1 \oplus I_2 \oplus \cdots \oplus I_{4n}$.

Lemma 6.3.1:   *All the maximizing subspaces for I_1 integration relative to S_r, after performing the $I_2 \oplus \cdots \oplus I_{4n}$, are given in the set*

$$\mathcal{M}^1 = \{\Lambda(I_2 \oplus \cdots \oplus I_{4n})S : \text{all } S \in \mathcal{M}\}. \qquad (6.3.12)$$

All the maximizing subspaces for the I_2 integration relative to any one of the subspaces in $\mathcal{M}^1$, say, $S^1 \in \mathcal{M}^1$, after performing the $I_3 \oplus \cdots \oplus I_{4n}$, are given in the set

$$\mathcal{M}^2 = \{\Lambda(I_3 \oplus \cdots \oplus I_{4n})S : S \in \mathcal{M} \text{ and } \Lambda(I_2 \oplus \cdots \oplus I_{4n})S = S^1\}. \qquad (6.3.13)$$

---

[16] Of course the classes $\mathcal{M}_0$ and $\mathcal{M}$ depend on the integer r.

[17] Note that with the connected components $g'_{1i} \in \mathcal{H}_2(\mathcal{N})$ of a g'_1, all the $Fg'_{1i}(\mathcal{N})$ are some polynomials of degree $\leq d(g'_{1i})$ in η_r. This follows from the fact that we may write $k_{ijl} + q_{ijl} = k_{ijl}^{g'_{1i}} + q_{ijl}^{g'_{1i}}$, and by noting that the $k_{ijl}, k_{ijl}^{g'_{1i}}$ in g'_{1i} are independent of η_r. Any dependence of $k_{ijl}^{g'_{1i}}$ on η_r must necessarily come from q_{ijl}.

More generally we have recursively that all the maximizing subspaces for the I_i integration relative to any one of the maximizing subspaces in $\mathcal{M}^{i-1}$, say, $S^{i-1} \in \mathcal{M}^{i-1}$, after performing the $I_{i+1} \oplus \cdots \oplus I_{4n}$ integration are given in the set

$$\mathcal{M}^i = \{\Lambda(I_{i+1} \oplus \cdots \oplus I_{4n})S : S \in \mathcal{M} \text{ and } \Lambda(I_i \oplus \cdots \oplus I_{4n})S = S^{i-1}\},$$
(6.3.14)

for $i < 4n$. We set $\mathcal{M}^0 \equiv \{S_r\}$ and we define for some $S^{4n-1} \in \mathcal{M}^{4n-1}$

$$\mathcal{M}^{4n} = \{S : S \in \mathcal{M} \text{ and } \Lambda(I_{4n})S = S^{4n-1}\}.$$
(6.3.15)

Suppose that the lemma is true for all i in $1 \le i \le k < 4n$. Let S^k be any given subspace in $\mathcal{M}^k$. The latter means, in particular, that there is some subspace $S' \in \mathcal{M}$ such that

$$\Lambda(I_{k+1} \oplus \cdots \oplus I_{4n})S' \equiv S^k.$$
(6.3.16)

We now show that for any subspace $S \in \mathcal{M}$ such that

$$\Lambda(I_{k+1} \oplus \cdots \oplus I_{4n})S = S^k$$
(6.3.17)

it follows that $\Lambda(I_{k+2} \oplus \cdots \oplus I_{4n})S$ is a maximizing subspace for the I_{k+1} integration relative to S^k after performing the $I_{k+2} \oplus \cdots \oplus I_{4n}$ integration. This follows from the following chain of inequalities:

$$\alpha_{I_{k+1} \oplus \cdots \oplus I_{4n}}(S^k)$$

$$= \max_{\Lambda(I_{k+1})\tilde{S} = S^k} [\alpha_{I_{k+2} \oplus \cdots \oplus I_{4n}}(\tilde{S}) + \dim \tilde{S} - \dim S^k]$$

$$\ge \alpha_{I_{k+2} \oplus \cdots \oplus I_{4n}}(\Lambda(I_{k+2} \oplus \cdots \oplus I_{4n})S)$$
$$+ \dim \Lambda(I_{k+2} \oplus \cdots \oplus I_{4n})S - \dim S^k$$

$$= \max_{\Lambda(I_{k+2} \oplus \cdots \oplus I_{4n})S'' = \Lambda(I_{k+2} \oplus \cdots \oplus I_{4n})S} [\alpha(S'') + \dim S'' - \dim S^k]$$

$$\ge \alpha(S) + \dim S - \dim S^k$$

$$= \alpha_I(S_r) + \dim S_r - \dim S^k$$

$$= \max_{\Lambda(I_1 \oplus \cdots \oplus I_k)\tilde{S} = S_r} [\alpha_{I_{k+1} \oplus \cdots \oplus I_{4n}}(\tilde{S}) + \dim \tilde{S} - \dim S^k]$$

$$\ge \alpha_{I_{k+1} \oplus \cdots \oplus I_{4n}}(S^k) + \dim S^k - \dim S^k$$

$$= \alpha_{I_{k+1} \oplus \cdots \oplus I_{4n}}(S^k),$$
(6.3.18)

where we have used the facts that S' and S in (6.3.16) and (6.3.17) are in $\mathcal{M}$ and the fact that

$$\Lambda(I_{k+1})\Lambda(I_{k+2} \oplus \cdots \oplus I_{4n})S = S^k,$$
$$\Lambda(I_1 \oplus \cdots \oplus I_k)S^k = S_r.$$
(6.3.19)

Since the extreme left-hand side and the extreme right-hand side of the inequalities in (6.3.18) are identical, we may replace all the inequality signs in it by equalities. Therefore we obtain, in particular, that

$$\alpha_{I_{k+1} \oplus \, \cdots \, \oplus I_{4n}}(S^k) = \alpha_{I_{k+2} \oplus \, \cdots \, \oplus I_{4n}}(\Lambda(I_{k+2} \oplus \cdots \oplus I_{4n})S)$$
$$+ \dim \Lambda(I_{k+2} \oplus \cdots \oplus I_{4n})S - \dim S^k;$$

$$(6.3.20)$$

i.e., $\Lambda(I_{k+2} \oplus \cdots \oplus I_{4n})S$ is a maximizing subspace for the I_{k+1} integration relative to S^k after performing the $I_{k+2} \oplus \cdots \oplus I_{4n}$ integration. Now we prove that any maximizing subspace for the I_{k+1} integration relative to S^k after performing the $I_{k+2} \oplus \cdots \oplus I_{4n}$ integration is in the set

$$\mathcal{M}^{k+1} = \{\Lambda(I_{k+2} \oplus \cdots \oplus I_{4n})S : S \in \mathcal{M} \text{ and } \Lambda(I_{k+1} \oplus \cdots \oplus I_{4n})S = S^k\},$$

$$(6.3.21)$$

thus completing the proof of the lemma by induction.[18]

Suppose that S_0 is a maximizing subspace for the I_{k+1} integration relative to S^k, after performing the $I_{k+2} \oplus \cdots \oplus I_{4n}$ integration, and that $S_0 \notin \mathcal{M}^{k+1}$. We shall then reach a contradiction. By hypothesis

$$\alpha_{I_{k+1} \oplus \, \cdots \, \oplus I_{4n}}(S^k) = \alpha_{I_{k+2} \oplus \, \cdots \, \oplus I_{4n}}(S_0) + \dim S_0 - \dim S^k,$$

$$\Lambda(I_{k+1})S_0 = S^k. \tag{6.3.22}$$

Let $\tilde{S}$ be a maximizing subspace for the $I_{k+2} \oplus \cdots \oplus I_{4n}$ relative to S_0, i.e.,

$$\alpha_{I_{k+2} \oplus \, \cdots \, \oplus I_{4n}}(S_0) = \alpha(\tilde{S}) + \dim \tilde{S} - \dim S_0,$$

$$\Lambda(I_{k+2} \oplus \cdots \oplus I_{4n})\tilde{S} = S_0. \tag{6.3.23}$$

From (6.3.22) and (6.3.23) we have

$$\alpha_{I_{k+1} \oplus \, \cdots \, \oplus I_{4n}}(S^k) = \alpha(\tilde{S}) + \dim \tilde{S} - \dim S^k,$$

$$\Lambda(I_{k+1} \oplus \cdots \oplus I_{4n})\tilde{S} = S^k. \tag{6.3.24}$$

Equation (6.3.18), however, implies that

$$\alpha_{I_{k+1} \oplus \, \cdots \, \oplus I_{4n}}(S^k) = \alpha_I(S_r) + \dim S_r - \dim S^k, \tag{6.3.25}$$

which upon comparison with (6.3.24) shows that $\tilde{S} \in \mathcal{M}$ and hence, by definition of $\mathcal{M}^{k+1}$, $\Lambda(I_{k+2} \oplus \cdots \oplus I_{4n})\tilde{S} = S_0 \in \mathcal{M}^{k+1}$, thus leading to a contradiction of the initial hypothesis that $S_0 \notin \mathcal{M}^{k+1}$. Note that the proof does not depend on the fact that the I_i are one dimensional, and the same

---

[18] We note that the set $\mathcal{M}^{k+1}$ is not empty since $\Lambda(I_{k+2} \oplus \cdots \oplus I_{4n})S'$, with S' introduced in (6.3.16), belongs to $\mathcal{M}^{k+1}$.

analysis as above shows that all the maximizing subspaces for the I_1 integration relative to S_r after performing the $I_2 \oplus \cdots \oplus I_{4n}$ integration are in $\mathcal{M}^1$. This completes the proof of the lemma by induction.

Since the dimension numbers $p_j, j = 1, \ldots, 4n$, in the expression for the logarithmic asymptotic coefficients $\beta_I(S_r)$ of $\mathcal{M}$ in (3.1.6) may be computed relative to any one of the maximizing subspaces for the I_{j-1} integration, after performing the $I_j \oplus \cdots \oplus I_{4n}$ integration, we may use the results in Lemma 6.3.1 and (3.1.6) to state

Theorem 6.3.2: *The logarithmic asymptotic coefficients $\beta_I(S_r)$ of the renormalized amplitude $\mathcal{A}$ are given by*

$$\beta_I(S_r) = \sum_{j=1}^{4n} p_j, \tag{6.3.26}$$

where $p_j, j = 1, \ldots, 4n$, is equal to zero if all the subspaces in $\mathcal{M}^j$ have the same dimension, and $p_j = 1$ otherwise.

We shall simplify Theorem 6.3.2 further to a form more suitable for applications. Choose some subspace in $\mathcal{M}_0$; call it S^0. Let $\mathcal{M}^2$ in (6.3.13) be the set of all the maximizing subspaces for the I_2 integration *relative to* $\Lambda(I_2 \oplus \cdots \oplus I_{4n})S^0$ after performing the $I_3 \oplus \cdots \oplus I_{4n}$ integration. By definition, we observe that $\Lambda(I_3 \oplus \cdots \oplus I_{4n})S^0$ is necessarily in $\mathcal{M}^2$ [see (6.3.13)]. Recursively, then, let $\mathcal{M}^j$ be the set of all the maximizing subspaces for the I_j integration relative to $\Lambda(I_j \oplus \cdots \oplus I_{4n})S^0$ after performing the $I_{j+1} \oplus \cdots \oplus I_{4n}$ integration. Obviously $\Lambda(I_{j+1} \oplus \cdots \oplus I_{4n})S^0$ belongs to $\mathcal{M}^j$ since $\Lambda(I_j)\Lambda(I_{j+1} \oplus \cdots \oplus I_{4n})S^0 = \Lambda(I_j \oplus \cdots \oplus I_{4n})S^0$. Accordingly we may replace Theorem 6.3.2 by the following simpler version:

Theorem 6.3.3: *Choose S^0 to be any subspace in $\mathcal{M}_0$. Let $\{S', S'', \ldots\}$ be the set of all those subspaces in $\mathcal{M}$ such that* [19]

$$\Lambda(I_j \oplus \cdots \oplus I_{4n})S' = \Lambda(I_j \oplus \cdots \oplus I_{4n})S'' = \cdots = \Lambda(I_j \oplus \cdots \oplus I_{4n})S^0, \tag{6.3.27}$$

then

$$\beta_I(S_r) = \sum_{j=1}^{4n} p_j, \tag{6.3.28}$$

where $p_j = 0$ if all the elements in

$$\{\dim \Lambda(I_{j+1} \oplus \cdots \oplus I_{4n})S' - \dim S_r, \dim \Lambda(I_{j+1} \oplus \cdots \oplus I_{4n})S'' - \dim S_r,$$
$$\ldots, \dim \Lambda(I_{j+1} \oplus \cdots \oplus I_{4n})S^0 - \dim S_r\} \tag{6.3.29}$$

---

[19] Obviously this set is not empty since it contains the subspace S^0 itself.

are equal, and $p_j = 1$ otherwise. The subspaces S', S'', ... are given by (6.3.27). The dimensions of the $\Lambda(I_{j+1} \oplus \cdots \oplus I_{4n})S'$, ... have been measured relative to dim S_r. *The chosen subspace S^0 in $\mathcal{M}_0$ will be called a reference subspace.*

In many applications, $\mathcal{M}_0$ consists of only one element $\mathcal{M}_0 = \{S^0\}$ and Theorem 6.3.3 may be readily applied.

From (2.1.1), the work in Chapter 2, Theorem 6.3.1, and Theorem 6.3.3 we may then state

Theorem 6.3.4

$$\mathcal{A}(\mathbf{L}_1 \eta_1 \cdots \eta_k + \cdots + \mathbf{L}_r \eta_r \cdots \eta_r + \cdots + \mathbf{L}_k \eta_k + \mathbf{C})$$

$$= 0\{\eta_1^{\alpha_I(\{\mathbf{L}_1\})} \cdots \eta_k^{\alpha_I(\{\mathbf{L}_1, \ldots, \mathbf{L}_k\})} \sum_{\gamma_1, \ldots, \gamma_k} (\ln \eta_{\pi_1})^{\gamma_1} \cdots (\ln \eta_{\pi_k})^{\gamma_k}\}, \quad (6.3.30)$$

where $\{\mathbf{L}_1, \ldots, \mathbf{L}_r\} \equiv S_r \subset E_1$, $1 \le r \le k \le 4m$, $\mathbf{C}$ *is confined to a finite region in E, with the masses* $\mu^i \ne 0$ *for all* $i = 1, \ldots, \rho$. *The sum in (6.3.30) is over all nonnegative integers* $\gamma_1, \ldots, \gamma_k$ *such that*

$$\sum_{i=1}^{t} \gamma_i \le \beta(\{\mathbf{L}_1, \ldots, \mathbf{L}_{\pi_t}\}), \quad (6.3.31)$$

for all $1 \le t \le k$, *and the logarithmic coefficients* β *have been arranged in increasing order*

$$\beta(\{\mathbf{L}_1, \ldots, \mathbf{L}_{\pi_1}\}) \le \cdots \le \beta(\{\mathbf{L}_1, \ldots, \mathbf{L}_{\pi_k}\}), \quad (6.3.32)$$

where $\{\pi_1, \ldots, \pi_k\}$ *is a permutation of the integers in* $\{1, \ldots, k\}$. *The power* $\alpha_I(S_r)$ *and the logarithmic* $\beta_I(S_r)$ *asymptotic coefficients are, respectively, given in Theorem 6.3.1 and Theorem 6.3.3.*

Example 6.1: Consider the self-energy graph G of a fermion shown in Fig. 6.1. Suppose we let the external momentum q become large. Consider the amplitude $\mathcal{A}(\eta q, m, \mu)$ and let $\eta \to \infty$. Here m denotes the mass of the fermion and μ denotes the mass of the boson, with both masses assumed to be nonzero.[20]

Let $\hat{I}_1 = I_1 \oplus \cdots \oplus I_4$ be associated with integration variables k_1, and $\hat{I}_2 = I_5 \oplus \cdots \oplus I_8$ be associated with integration variables k_2. $I_1, \ldots, I_8$ denote one-dimensional subspaces. We note that in Fig. 6.1, $\mathrm{degr}_Q D_{13}^+ = \mathrm{degr}_Q D_{24}^+ = -2$ for the spin 0 propagation, and $\mathrm{degr}_Q D_{12}^+ = \mathrm{degr}_Q D_{23}^+ = \mathrm{degr}_Q D_{34}^+ = -1$ for the spin $\frac{1}{2}$ particle. Also the dimensionalities of G, g_1, and g_2 are as follows: $d(G) = 1$, $d(g_1) = d(g_2) = 0$. Obviously $\tau_0 = \{G\}$. As a matter of fact any other subdiagram $g \nsubseteq G$ has $d(g) < 1$. Canonical decompositions of the Q_{ijl} for G in Fig. 6.1 are given in (5.1.40), and the corresponding expressions for $k_{ijl}^{g_1}, q_{ijl}^{g_1}, k_{ijl}^{g_2}, q_{ijl}^{g_2}$ are given in (5.1.81)–(5.1.84).

---

[20] Further generalizations to this will be given in subsequent sections.

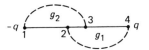

Fig. 6.1   A self-energy graph G of a fermion with a $\bar{\psi}\psi\phi$ coupling contributing to it with the dashed lines representing a spin 0 and the fermion is of spin $\frac{1}{2}$.

The graph $G \in \tau_0$ is associated with the subspace $S^0 = S(q) \oplus \hat{I}_1 \oplus \hat{I}_2$, where $S(q)$ is a subspace associated with the momentum q. We readily infer from Theorem 6.3.1 with no further work that

$$\alpha_I(S(q)) = d(G) = 1. \tag{6.3.33}$$

From (5.1.82) we note that $\hat{I}_1$ is associated with $k_{iji}^{q_1}$ and $\hat{I}_2$ is associated with $k_{iji}^{q_2}$. From the definition of $(\mathcal{M}, \tau)$ we may write

$$\mathcal{M} = \{S', S'_1, S'_2, S'_3\} \tag{6.3.34}$$

$$\tau = \{G, (G|g_1), (G|g_2), (G|G)\}, \tag{6.3.35}$$

where

$$\begin{aligned} S' &\equiv S^0 = S(q) \oplus \hat{I}_1 \oplus \hat{I}_2, \\ S'_1 &= S(q) \oplus \hat{I}_2, \\ S'_2 &= S(q) \oplus \hat{I}_1, \\ S'_3 &= S(q). \end{aligned} \tag{6.3.36}$$

The subdiagrams $G, (G|g_1), (G|g_2), (G|G)$ are associated, respectively, with the subspaces S', S'_1, S'_2, S'_3. Hence in the notation of Theorem 6.3.2 and Lemma 6.3.1 we have

$$\mathcal{M}^1 = \{\Lambda(I_2 \oplus \cdots \oplus I_8)S : S = S^0, S'_1, S'_2, S'_3\}, \tag{6.3.37}$$

$$\mathcal{M}^j = \{\Lambda(I_{j+1} \oplus \cdots \oplus I_8)S^0\}, \qquad j = 2, 3, 4, 6, 7, 8, \tag{6.3.38}$$

$$\mathcal{M}^5 = \{\Lambda(I_6 \oplus I_7 \oplus I_8)S : S = S^0, S'_2\}. \tag{6.3.39}$$

Accordingly by the application of Theorem 6.3.3, using

$$\{\dim \Lambda(I_2 \oplus \cdots \oplus I_8)S - \dim S(q) : S = S^0, S'_1, S'_2, S'_3\} = \{1, 0, 1, 0\}, \tag{6.3.40}$$

we obtain $p_1 = 1$, and from (6.3.38) we obtain $p_j = 0$ for $j = 2, 3, 4, 6, 7, 8$ since the $\mathcal{M}^j$, for such j, constitute only one element. Finally, from

$$\{\dim \Lambda(I_6 \oplus I_7 \oplus I_8)S - \dim S(q) : S = S^0, S'_2\} = \{5, 4\}, \tag{6.3.41}$$

we have $p_5 = 1$. From (6.3.28) we then obtain

$$\beta_I(S(q)) = \sum_{j=1}^{8} p_j = 2. \tag{6.3.42}$$

Therefore we may write

$$\mathscr{A}(\eta q, m, \mu) = 0 \left\{ \eta \sum_{\gamma=0}^{2} (\ln \eta)^\gamma \right\}, \qquad \eta \to \infty. \tag{6.3.43}$$

6.4 GENERAL ASYMPTOTIC BEHAVIOR I

In this section we generalize the results given in Section 6.3 and consider the asymptotic behavior of $\mathscr{A}$ when not only some of the momenta become large but also some of the masses in the theory become large as well. The analysis here is similar to the one carried out in Section 6.3, and we shall be brief. Applications of this analysis will be given in the remaining part of this chapter.

As in (6.3.2), we introduce for each line ℓ_i joining a vertex v_i to a vertex v_j of G, vectors $\mathbf{V}_0(ijl), \ldots, \mathbf{V}_3(ijl)$ *and* now an additional vector $\mathbf{V}_4(ijl)$ such that, with $\mathbf{P}'$ as given in (6.2.5),

$$\mathbf{V}_0(ijl) \cdot \mathbf{P}' = Q^0_{ijl},$$
$$\vdots \tag{6.4.1}$$
$$\mathbf{V}_3(ijl) \cdot \mathbf{P}' = Q^3_{ijl},$$
$$\mathbf{V}_4(ijl) \cdot \mathbf{P}' = \mu_{ijl}.$$

We denote by $S(ijl)$ the subspace generated by the vectors $\mathbf{V}_0(ijl), \ldots,$ $\mathbf{V}_3(ijl), \mathbf{V}_4(ijl)$. We also introduce vectors $\mathbf{V}'_0(ijl), \ldots, \mathbf{V}'_3(ijl)$ as in Section 6.3 satisfying (6.3.3).

Technically we are interested in the behavior of

$$\mathscr{A}(\mathbf{L}_1 \eta_1 \cdots \eta_k + \cdots + \mathbf{L}_r \eta_r \cdots \eta_k + \cdots + \mathbf{L}_k \eta_k + \mathbf{C}) \tag{6.4.2}$$

for $\eta_1, \eta_2, \ldots, \eta_k \to \infty$, independently, where $\mathbf{L}_1, \ldots, \mathbf{L}_k$ are k independent vectors in $E = E_1 \oplus E_2$, $1 \le k \le 4m + \rho$, $\mathbf{C}$ is a vector confined to a finite region in E, such that $\mu^i \ne 0$ for all $i = 1, \ldots, \rho$.

Definition of class $\mathscr{M}_0$:   We define a class $\mathscr{M}_0 = \{S', \ldots\}$ of subspaces $S' \subset \mathbb{R}^{4n+4m+\rho}$ and a set of subdiagrams $\tau_0 = \{G', \ldots\}$ in such a way that the following are consistent:

   (i) $\Lambda(l)S' = S_r$.

   (ii)   Let G' be the subdiagram of G, corresponding to all the lines in G (and, of course, corresponding to the vertices as their end points) such that all the subspaces $S(ijl)$ of G in G' are not orthogonal to S'.

   (iii)   The proper part G'_0 of G' corresponds to all those lines in G with all their $\mathbf{V}'_0(ijl), \ldots, \mathbf{V}'_3(ijl)$ [see (6.3.3)] not orthogonal to S'.

(iv)   If $S'' \subset \mathbb{R}^{4n+4m+\rho}$, with which a subdiagram G'' is associated, is such that (i)–(iii) are true, then $d(G'') \leq d(G')$. If $d(G'') = d(G')$, then $S'' \in \mathcal{M}_0$ and $G'' \in \tau_0$.

By the same analysis leading to Theorem 6.3.1 we have

Theorem 6.4.1: *The power asymptotic coefficients $\alpha_I(S_r)$ of the renormalized amplitude*

$$\mathscr{A}(\mathbf{L}_1\eta_1 \cdots \eta_k + \cdots + \mathbf{L}_r\eta_r \cdots \eta_k^* + \cdots + \mathbf{L}_k\eta_k + \mathbf{C}), \qquad (6.4.3)$$

with $1 \leq r \leq k \leq 4m + \rho$ are given, respectively in r, by

$$\alpha_I(S_r) = d(G'), \qquad (6.4.4)$$

where G' is any subdiagram in τ_0.

The subdiagrams $G' \in \tau_0$ are determined as in the case for the high-energy behavior given in Section 6.3. There are, however, some differences in the nature of the diagrams in this case. For example, a subdiagram may contain an extral vertex at which the total external momentum carried by the external line to G' at that vertex is nonasymptotic as long as the external line of G' attached to this vertex carries an asymptotic mass.[21]

Definition of class $\mathcal{M}$:   We define the class

$$\mathcal{M} = \{S', S'_1, \ldots, S'_{N'}; S'', S''_1, \ldots, S''_{N''}; \ldots\},$$

where $\mathcal{M}_0 = \{S', S''\ldots\}$, and the set of subdiagrams

$$\tau = \{G', (G'|g'_1), \ldots, (G'|g'_{N'}); G'', (G''|g''_1), \ldots, (G''|g''_{N''}); \ldots\}$$

with $\tau_0 = \{G', G'', \ldots\}$, such that the following are consistent: For any $\tilde{S} \in \mathcal{M}$

(i) $\Lambda(I)\tilde{S} = S_r$.

(ii)   Let $\hat{G}$ be the subdiagram of G, corresponding to all the lines in G, such that all the subspaces $S(ijl)$ of G in $\hat{G}$ are not orthogonal to $\tilde{S}$.

(iii)   The set of all the lines of G in $\hat{G}$ having all their $\mathbf{V}_0(ijl), \ldots, \mathbf{V}_3(ijl)$ not orthogonal to $\tilde{S}$ coincide with the set of all the lines in the proper part of the generalized subdiagram $(\hat{G}|\hat{g}) \equiv \tilde{G}$, where $\hat{g}$ is either empty or otherwise $\hat{g}$ is a proper subdiagram of $\hat{G}$ such that each of the connected components of $\hat{g}$ has a nonnegative dimensionality, and all the masses in the lines in $\hat{g}$ are *independent* of η_r. In the notation in (6.4.1) the latter means that the $\mathbf{V}_4(ijl)$ of the lines in $\hat{g}$ are orthogonal to $\tilde{S}$.

(iv) $\hat{G} \in \tau_0 \subset \tau$ and $\tilde{G} \in \tau$ for all nonempty $\hat{g}$ as defined above.

---

[21] In such a case, of course, the momentum-dependent part of this external line is nonasymptotic by momentum conservation.

Theorem 6.4.2

$$\mathscr{A}(\mathbf{L}_1\eta_1 \cdots \eta_k + \cdots + \mathbf{L}_r\eta_r \cdots \eta_k + \cdots + \mathbf{L}_k\eta_k + \mathbf{C})$$

$$= 0\{\eta_1^{\alpha_I(\{\mathbf{L}_1\})} \cdots \eta_k^{\alpha_I(\{\mathbf{L}_1, \ldots, \mathbf{L}_k\})} \sum_{\gamma_1, \ldots, \gamma_k} (\ln \eta_{\pi_1})^{\gamma_1} \cdots (\ln \eta_{\pi_k})^{\gamma_k}\}, \quad (6.4.5)$$

where $\{\mathbf{L}_1, \ldots, \mathbf{L}_r\} \equiv S_r \subset E$, $1 \le r \le k \le 4m + \rho$, $\mathbf{C}$ *is confined to a finite region in* E, *with the masses* $\mu^i \ne 0$ *for all* $i = 1, \ldots, \rho$. *The sum in* (6.4.5) *is over all nonnegative integers* $\gamma_1, \ldots, \gamma_k$ *such that*

$$\sum_{i=1}^{t} \gamma_i \le \beta(\{\mathbf{L}_1, \ldots, \mathbf{L}_{\pi_t}\}), \quad 1 \le t \le k, \quad (6.4.6)$$

and the logarithmic coefficients β *have been ordered in increasing order*

$$\beta(\{\mathbf{L}_1, \ldots, \mathbf{L}_{\pi_1}\}) \le \cdots \le \beta(\{\mathbf{L}_1, \ldots, \mathbf{L}_{\pi_k}\}), \quad (6.4.7)$$

where $\{\pi_1, \ldots, \pi_k\}$ *is a permutation of the integers in* $\{1, \ldots, k\}$. $\alpha_I(S_r)$ *is given in Theorem* 6.4.1, *and the* $\beta_I(S_r)$ *may be obtained from Theorem* 6.3.2 *or Theorem* 6.3.3 *from the classes* $\mathscr{M}_0$ *and* $\mathscr{M}$ *defined above.*[22]

6.5 ZERO-MASS BEHAVIOR

The purpose of this section is to study the zero-mass behavior of renormalized Feynman amplitudes (Section 6.5.1) and finally give sufficient conditions to guarantee the existence of the zero-mass limit of renormalized Feynman amplitudes (Section 6.5.2). The study is general enough to deal with the most general cases when some (not necessarily all) of the masses of a Feynman graph G become small and, in general, at different rates. That we have to consider (i) the most general cases when some of the masses as well and not necessarily all the masses become small and (ii) the approach of such masses to zero at different rates as well are clearly physical requirements. In quantum electrodynamics, for example, one would be interested in the behavior of a renormalized Feynman amplitude for $\mu \to 0$, $m \to 0$, and $(\mu/m) \to 0$, where μ is a photon "mass" and m is the mass of the electron. For a propagator D_{ijl}^+, carrying a mass μ_{ijl} that we wish to scale to zero, we write (in Euclidean space)

$$D_{ijl}^+ = \tilde{P}_{ijl}(Q_{ijl}, \mu_{ijl})/(Q_{ijl}^2 + \mu_{ijl}^2), \quad (6.5.1)$$

where $\tilde{P}_{ijl}(Q_{ijl}, \mu_{ijl})$ is a polynomial in Q_{ijl} and μ_{ijl} but not in $(1/\mu_{ijl})$ such that

$$D_{ijl}^+(Q_{ijl}, 0) = \tilde{P}_{ijl}(Q_{ijl}, 0)/Q_{ijl}^2 \quad (6.5.2)$$

---

[22] Of course, Theorems 6.3.2 and 6.3.3, as they stand, apply to the present situation as well.

denotes the mass $\mu_{ijl} = 0$ propagator.[23] With the propagator in (6.5.1) we then carry out the subtractions of renormalization as usual to obtain the final expression for R. All those propagators carrying masses which we do not wish to approach zero will be written as in (5.1.1). We note that in general we may rewrite (5.1.1) as (in Euclidean space with $\varepsilon = 0$)

$$\frac{P_{ijl}(Q_{ijl}, \mu_{ijl})}{(Q_{ijl}^2 + \mu_{ijl}^2)} = (\mu_{ijl})^{-\delta_{ijl}} \frac{\tilde{P}_{ijl}(Q_{ijl}, \mu_{ijl})}{(Q_{ijl}^2 + \mu_{ijl}^2)}, \tag{6.5.3}$$

where δ_{ijl} is some nonnegative integer, and $\tilde{P}_{ijl}$ is a polynomial in Q_{ijl} and μ_{ijl}, but not in $1/\mu_{ijl}$. Also quite generally we suppose that

$$\begin{aligned} \operatorname*{degr}_{Q_{ijl}} P_{ijl}(Q_{ijl}, \mu_{ijl}) &= \operatorname*{degr}_{Q_{ijl}, \mu_{ijl}} \tilde{P}_{ijl}(Q_{ijl}, \mu_{ijl}) \\ &= \operatorname*{degr}_{Q_{ijl}} \tilde{P}_{ijl}(Q_{ijl}, \mu_{ijl}), \end{aligned} \tag{6.5.4}$$

and with $D_{ijl}^+ = (\mu_{ijl})^{-\delta_{ijl}} \tilde{D}_{ijl}^+$, we have

$$\operatorname*{degr}_{Q_{ijl}, \mu_{ijl}} \tilde{D}_{ijl}^+ = \operatorname*{degr}_{Q_{ijl}} \tilde{D}_{ijl}^+. \tag{6.5.5}$$

The expressions (6.5.4) and (6.5.5) will be assumed explicitly. By working with the propagators D_{ijl}^+ in (6.5.1) and $\tilde{D}_{ijl}^+$ in (6.5.5) we may introduce an unrenormalized integrand $\bar{I}_G$ given by

$$I_G = \prod_{j=1}^{\rho} (\mu^j)^{-\sigma_j} \tilde{I}_G \tag{6.5.6}$$

in a form as in (2.2.17), where the σ_j are some positive integers.

6.5.1   Zero-Mass Behavior of $\mathscr{A}$

The structure of a Feynman amplitude associated with a proper and connected graph G is, from (2.2.3), of the form

$$\mathscr{A}(p_1^0, \dots, p_m^3; \mu^1, \dots, \mu^\rho) = \int_{\mathbb{R}^{4n}} dK\, R(p_1^0, \dots, p_m^3; k_1^0, \dots, k_n^3; \mu^1, \dots, \mu^\rho), \tag{6.5.7}$$

---

[23] For example, in quantum electrodynamics we write for the spin $\frac{1}{2}$ propagator $(\gamma p - m)/(p^2 + m^2)$ and for the photon propagator (in the Feynman gauge) $g_{\mu\nu}/(Q^2 + \mu^2)$. Quite generally, we may also allow higher powers of the denominator in (6.5.1): $(Q_{ijl}^2 + \mu_{ijl}^2)^{-n}$, with integers $n \geq 1$. The latter does not necessarily mean that one is allowing a double, triple, $\dots$, etc., pole term in the propagator, as this depends very much on the structure of the polynomial P_{ijl}. In any case we always have to take the correct dimensionality $\operatorname{degr}_{Q_{ijl}} D_{ijl}^+(Q_{ijl}, \mu_{ijl})$ of $D_{ijl}^+(Q_{ijl}, \mu_{ijl})$ when carrying out the subtractions of renormalization.

where [see (2.2.14), (2.2.17), (2.2.18)–(2.2.20), and (6.5.6)]

$$R = \prod_{j=1}^{\rho} (\mu^j)^{-\sigma_j} \tilde{R}, \tag{6.5.8}$$

$$\tilde{R}(p_1^0, \ldots, p_m^3; k_1^0, \ldots, k_n^3; \mu^1, \ldots, \mu^\rho) = \sum_i A_{s_i t_i u_i}^i k^{s_i} p^{t_i} \mu^{u_i} / \prod_{l=1}^{L} [Q_l^2 + \mu_l^2], \tag{6.5.9}$$

$$k^{s_i} \equiv (k_1^0)^{s_{01}^i} \cdots (k_n^3)^{s_{3n}^i}, \qquad s_{\mu j}^i \geq 0,$$

$$p^{t_i} \equiv (p_1^0)^{t_{01}^i} \cdots (p_m^3)^{t_{3m}^i}, \qquad t_{\mu j}^i \geq 0, \tag{6.5.10}$$

$$\mu^{u_i} \equiv (\mu^1)^{u_1^i} \cdots (\mu^\rho)^{u_\rho^i}, \qquad u_j^i \geq 0,$$

with $Q_l = \sum_i a_l^i k_i + \sum_j b_l^j p_j$. The $A_{s_i t_i u_i}^i$ are some suitable coefficients.

Suppose that $\{\mu^1, \ldots, \mu^s\}$, with $s \leq \rho$, denotes the subset of the masses that we wish to approach zero. We scale the masses in the set $\{\mu^1, \ldots, \mu^s\}$ as follows:

$$\mu^1 \to \lambda_1 \mu^1,$$

$$\mu^2 \to \lambda_1 \lambda_2 \mu^2, \tag{6.5.11}$$

$$\vdots$$

$$\mu^s \to \lambda_1 \lambda_2 \cdots \lambda_s \mu^s.$$

The masses in the subset $\{\mu^1, \ldots, \mu^s\}$ have been arbitrarily labeled from 1 to s for convenience of notation. Without loss of generality, we assume that all those masses that we wish to approach zero at the same rate have been identified with μ^1, or μ^2, or $\ldots$, or μ^s depending on the rate we wish them to approach zero, etc. By the definition in (6.5.2), the factor $\prod_{j=1}^{\rho} (\mu^j)^{-\sigma_j}$ in (6.5.8) is invariant under the scaling in (6.5.11); in particular, $\sigma_1 = \sigma_2 = \cdots = \sigma_s = 0$. Quite generally we have under the scaling (6.5.11).

$$k^{s_i} p^{t_i} \mu^{u_i} \to (\lambda_1 \cdots \lambda_s)^{d(N)} (k')^{s_i} (p')^{t_i} (\mu')^{u_i}, \tag{6.5.12}$$

where

$$k_j'^\mu = k_j^\mu / \lambda_1 \cdots \lambda_s, \qquad p_j'^\mu = p_j^\mu / \lambda_1 \cdots \lambda_s,$$

$$\mu'^1 = \mu^1 / \lambda_2 \cdots \lambda_s,$$

$$\mu'^2 = \mu^2 / \lambda_3 \cdots \lambda_s,$$

$$\vdots \tag{6.5.13}$$

$$\mu'^{s-1} = \mu^{s-1} / \lambda_s,$$

$$\mu'^s = \mu^s,$$

$$\mu'^j = \mu^j / \lambda_1 \cdots \lambda_s, \qquad s < j \leq \rho,$$

and $d(N)$ is the dimensionality of the expression on the left-hand side of (6.5.12), i.e.,

$$d(N) = (s_{01}^i + \cdots + s_{3n}^i) + (t_{01}^i + \cdots + t_{3m}^i) + (u_1^i + \cdots + u_\rho^i), \quad (6.5.14)$$

and is a fixed number of all i, for which $A_{s_i t_i u_i}^i \neq 0$, and coincides with the dimensionality of the numerator in (6.5.9). Similarly the denominator in (6.5.9) is transformed as

$$\prod_{l=1}^{L} [Q_l^2 + \mu_l^2] \to (\lambda_1 \cdots \lambda_s)^{d(D)} \prod_{l=1}^{L} [Q_l'^2 + \mu_l'^2], \quad (6.5.15)$$

where $Q_l' = Q_l/\lambda_1 \cdots \lambda_s$. $d(D)$ is the dimensionality of the denominator on the left-hand side of (6.5.15), i.e.,

$$d(D) = 2L. \quad (6.5.16)$$

Accordingly $\tilde{R}$ in (6.5.9) is transformed to

$$(\lambda_1 \cdots \lambda_s)^{d(\tilde{R})} \tilde{R}(p_1'^0, \ldots, p_m'^0; k_1'^0, \ldots, k_n'^3; \mu'^1, \ldots, \mu'^\rho), \quad (6.5.17)$$

where

$$d(\tilde{R}) = d(N) - d(D). \quad (6.5.18)$$

Hence finally the renormalized amplitude $\mathscr{A} \equiv \prod_{j=1}^{\rho} (\mu_j)^{-\sigma_j} \tilde{\mathscr{A}}$ is transformed to

$$\prod_{j=1}^{\rho} (\mu^j)^{-\sigma_j} (\lambda_1 \cdots \lambda_s)^{d(G)} \tilde{\mathscr{A}} \left(\frac{P}{\lambda_1 \cdots \lambda_s}, \frac{\mu^1}{\lambda_2 \cdots \lambda_s}, \right.$$
$$\left. \ldots, \frac{\mu^{s-1}}{\lambda_s}, \mu^s, \frac{\mu^{s+1}}{\lambda_1 \cdots \lambda_s}, \ldots, \frac{\mu^\rho}{\lambda_1 \cdots \lambda_s} \right), \quad (6.5.19)$$

where

$$d(G) = d(N) - 2L + 4n, \quad (6.5.20)$$

and we infer from (6.5.5) that $d(G)$ coincides with the dimensionality of the graph G in question. We choose the external (independent) momenta $p_1, \ldots, p_m$ such that no partial sums of these momenta vanish; i.e., $p_{i_1} + \cdots + p_{i_t} \neq 0$ for all subsets $\{i_1, \ldots, i_t\} \subset \{1, \ldots, m\}$.[24] This, in particular,

---

[24] Such external Euclidean momenta have been called nonexceptional momenta (cf. Symanzik, 1971). We recall that these external momenta are momenta carried by the external lines to the graph G taken, by convention, in a direction away from the external vertices. In general, at each external vertex, the total external momentum carried away from that vertex may be written in the form $\pm (p_{i_1} + \cdots + p_{i_t})$.

means that the total external momentum carried away from each exernal vertex is nonzero. As the elements in μ are also chosen to be fixed and non-zero, the factor $\prod_{j=1}^{\rho} (\mu^j)^{-\sigma_j}$ is independent of the parameters $\lambda_1, \ldots, \lambda_s$.

The amplitude $\tilde{\mathscr{A}}$ in (6.5.19) is of a particular form of the amplitude $\mathscr{A}$ analyzed in Section 6.4, where in the former the propagators are of the form in (6.5.1) and $\tilde{D}_{ijl}^+$ introduced below Eq. (6.5.4). We decompose E_2 as $E_2 = E_2^1 \oplus E_2^2$, where E_2^1 is an $(s-1)$-dimensional subspace and E_2^2 is its orthogonal complement in E_2. We introduce orthogonal vectors $\mathbf{L}_2, \ldots, \mathbf{L}_s$ in E_2^1 with nonvanishing components $\mu^1, \ldots, \mu^{s-1}$, respectively. We may also introduce a vector $\mathbf{L}_1$ in $E_1 \oplus E_2^2$ with nonvanishing components $p_1^0, \ldots,$ $p_m^3, \mu^{s+1}, \ldots, \mu^\rho$ and a vector $\mathbf{C} \in E_2^2$ orthogonal to $\mathbf{L}_1$ with nonvanishing component μ^s. Finally, we may write $\tilde{\mathscr{A}}$ in (6.5.19) in the form $\tilde{\mathscr{A}}(\mathbf{P})$, where

$$\mathbf{P} = (\lambda_1 \cdots \lambda_s)^{-1} \mathbf{L}_1 + (\lambda_2 \cdots \lambda_s)^{-1} \mathbf{L}_2 + \cdots + \lambda_s^{-1} \mathbf{L}_s + \mathbf{C}. \quad (6.5.21)$$

The conditions $\lambda_1, \ldots, \lambda_s \to 0$ in (6.5.21) mean, in particular, that all the external momenta become asymptotic and the total external momenta at *all* the external vertices are asymptotic. Obviously the vectors $\mathbf{L}_1, \ldots, \mathbf{L}_s$ are s independent vectors in $\mathbb{R}^{4n+4m+\rho}$.

We may now apply Theorem 6.4.2 (see Section 6.4 for details) and infer from (6.5.19), together with (6.5.21), that

$$\mathscr{A} = 0\left\{\lambda_1^{\delta_1} \cdots \lambda_s^{\delta_s} \sum_{\gamma_1, \ldots, \gamma_s} \left(\ln \frac{1}{\lambda_{\pi_1}}\right)^{\gamma_1} \cdots \left(\ln \frac{1}{\lambda_{\pi_s}}\right)^{\gamma_s}\right\} \quad (6.5.22)$$

for $\lambda_1, \lambda_2, \ldots, \lambda_s \to 0$, and where

$$\delta_i = d(G) - \alpha_I(S_i) \quad (6.5.23)$$

for $i = 1, \ldots, s$, $S_i \equiv \{\mathbf{L}_1, \ldots, \mathbf{L}_i\}$, with $\alpha_I(S_i)$ being the power asymptotic coefficients of $\tilde{\mathscr{A}}(\mathbf{P})$ with respect to the parameters $1/\lambda_i$. The sum in (6.5.22) is over all nonnegative integers $\gamma_1, \ldots, \gamma_s$ such that

$$\sum_{i=1}^{t} \gamma_i \leq \beta(\{\mathbf{L}_1, \ldots, \mathbf{L}_{\pi_t}\}) \quad (6.5.24)$$

for all t in $1 \leq t \leq s$, and the β have been arranged in an increasing order:

$$\beta(\{\mathbf{L}_1, \ldots, \mathbf{L}_{\pi_1}\}) \leq \cdots \leq \beta(\{\mathbf{L}_1, \ldots, \mathbf{L}_{\pi_s}\}). \quad (6.5.25)$$

From the definition of the classes $\mathscr{M}_0$ and $\mathscr{M}$ in Section 6.4 and (6.5.22) we may then readily give sufficient conditions for the existence, with non-exceptional external momenta, of the zero-mass limit of renormalized Feynman amplitudes $\mathscr{A}$ for $\lambda_1, \ldots, \lambda_s \to 0$.

6.5.2 Rules (Sufficiency Conditions) for the Existence of lim $\mathscr{A}$

Consider the amplitude $\tilde{\mathscr{A}}$, as given in (6.5.19), associated with a proper and connected graph G. Let i be fixed in $1 \leq i \leq s$. Let T_i be the set of all the subdiagrams of G such that the following are true. If $G_i \in T_i$, then $G_i (\subset G)$ contains all of the external vertices of G but not necessarily all of its lines. The lines in G/G_i (if not empty) do not carry any external momenta, and all their masses are from the set $\{\mu^i, \mu^{i+1}, \ldots, \mu^s\}$.[25] Any external line of G_i depends on the elements from the set P and/or $\{\mu^1, \ldots, \mu^{i-1}, \mu^{s+1}, \ldots, \mu^p\}$.[26] We repeat the definition of T_i for all $i = 1, \ldots, s$, thus generating the sets $T_1, \ldots, T_s$.

If the following are true, for all $i = 1, \ldots, s$, then $\lim_{\gamma_1, \ldots, \lambda_s \to 0} \mathscr{A}$ exists.[27] (i) for any $G_i \in T_i$, $d(G_i) \leq d(G)$. (ii) If $d(G_i) = d(G)$, then G_i does not have a proper subdiagram $g_i \subset G_i$ in it such that the masses in the lines in g_i are from the set $\{\mu^i, \mu^{i+1}, \ldots, \mu^s\}$ (if not empty), and the dimensionality of each of the connected components of g_i is nonnegative.

The above rules are very simple to apply and one may, in general, infer the existence of $\lim_{\lambda_1, \ldots, \lambda_s \to 0} \mathscr{A}$ by a mere examination of the graph G from these rules with almost no extra work. In particular, the rules state as one of the sufficiency conditions for the existence of the latter limit that the graph G itself is not to contain a proper subdiagram g such that all the masses in g are from the set $\{\mu^1, \ldots, \mu^s\}$, and that the dimensionality of each of the connected components of g is to be nonnegative.

Before giving some examples applying these rules we wish to note the following. This will save us time in applications. The above conditions if satisfied imply that $\delta_i \geq 0$ for all $i = 1, \ldots, s$ and for the corresponding i for which $\delta_i = 0$ we have $\beta_I(S_i) = 0$. If $\beta_I(S_j) = 0$ for some j, then no $\ln(1/\lambda_j)$ terms will appear in (6.5.22). The reason is that with the ordering as in (6.5.25) with $\pi_1 = j$, i.e., $0 = \beta(S_j) \leq \cdots \leq \beta(\{\mathbf{L}_1, \ldots, \mathbf{L}_{\pi_s}\})$, then from (6.5.24), we have $0 \leq \gamma_1 \leq \beta(S_j) = 0$, and hence we obtain $\gamma_1 = 0$.

Example 6.2: Consider the graph in Fig. 6.2 representing the lowest-order contribution to the self-energy of the electron in quantum electrodynamics. We write the photon propagator (in the Feynman gauge) as $D_{\mu\nu}(Q) = g_{\mu\nu}/(Q^2 + \mu^2)$. We consider the behavior of the amplitude $\mathscr{A}(q, m, \lambda\mu)$ for $\lambda \to 0$, where m is the electron mass and $q \neq 0$. Any subdiagram $G' \subset G$ is such that $d(G') \leq d(G)$. We also have $d(G') = d(G)$ only if G' is the graph G

---

[25] This simply means that the lines in G/G_i, in reference to the amplitude $\tilde{\mathscr{A}}$, are independent of λ_i^{-1}.

[26] This simply means that the external lines of G_i depend on λ_i^{-1}.

[27] More precisely, we should say that $\mathscr{A}$ remains bounded in this limit.

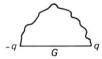

Fig. 6.2 Lowest-order electron self-energy graph in quantum electrodynamics. The wavy line represents a photon, and the straight line represents the spin $\frac{1}{2}$ particle.

itself. Also G does *not* contain a proper subdiagram having all its masses from the set $\{\mu\}$. Accordingly the $\lim_{\lambda \to 0} \mathscr{A}(q, m, \lambda\mu)$ exists. This example demonstrates the simplicity of the application of the rules for the existence of $\lim \mathscr{A}$ given before.

For the convenience of the reader we give the explicit expression for the renormalized amplitude corresponding to Fig. 6.2 with subtractions performed at the origin and with $\mu = 0$:

$$\mathscr{A}(q, m, 0) = -\frac{\alpha}{2\pi} \gamma q \int_0^1 x \, dx \ln\left(1 + \frac{q^2}{m^2} x\right)$$

$$-\frac{\alpha}{\pi} m \int_0^1 dx \ln\left(1 + \frac{q^2}{m^2} x\right), \qquad (6.5.26)$$

where α is the fine-structure constant $\alpha = e^2/4\pi$. The expression $\mathscr{A}(q, m, 0)$ in (6.5.26) obviously exists for $q^2 > 0$ (also for $q = 0$) and $m \neq 0$.

Example 6.3: Consider the photon self-energy graphs in quantum electrodynamics. Some of these graphs are shown in Fig. 6.3. For the photon propagator we write $D_{\mu\nu} = g_{\mu\nu}/(Q^2 + \mu^2)$. Let G be any graph in Fig. 6.3. We note that for any $g \subset G$, we have $d(g) \leq d(G)$. Also, G does not contain any proper subdiagram having all its masses equal to μ. Accordingly the limit $\lambda \to 0$ of the renormalized amplitudes $\mathscr{A}(q, m, \lambda\mu)$, corresponding to all the graphs in Fig. 6.3. exist in the limit $\lambda \to 0$.

Finally, we give an example where the behavior of $\mathscr{A}$ may be studied as of one of the masses of the graph in question becomes small and another one goes to zero.

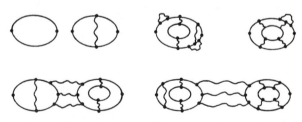

Fig. 6.3   Some low- and high-order photon self-energy graphs in quantum electrodynamics.

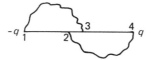

Fig. 6.4   A fourth-order electron self-energy graph in quantum electrodynamics.

Example 6.4: Consider the behavior of the renormalized amplitude $\mathscr{A}(q, \lambda_1 m, \lambda_1 \lambda_2 \mu)$ associated with a fourth-order self-energy graph of the electron in quantum electrodynamics, shown in Fig. 6.4., for $\lambda_1, \lambda_2 \to 0$. Consider the amplitude $\mathscr{A}(q/\lambda_1, m, \lambda_2 \mu)$. We refer to Example 6.1 to infer, in the notation in (6.3.33) and (6.3.42), that $\alpha_I(S(q)) = 1$, $\beta_I(S(q)) = 2$. On the other hand, repeating the same analysis as the one given in the previous two examples shows that the limit $\lambda_2 \to 0$ of $\mathscr{A}(q/\lambda_1, m, \lambda_2 \mu)$ exists. Accordingly we have for $\lambda_1, \lambda_2 \to 0$

$$\mathscr{A}(q, \lambda_1 m, \lambda_1 \lambda_2 \mu) = 0\left\{(\lambda_1)^{d(G)-1} \sum_{\gamma_2=0}^{2} \left(\ln \frac{1}{\lambda_1}\right)^{\gamma_2}\right\}$$

$$= 0\left\{\sum_{\gamma_2=0}^{2} \left(\ln \frac{1}{\lambda_1}\right)^{\gamma_2}\right\}, \qquad (6.5.27)$$

where we have used the fact that $d(G) = 1$.

6.6   LOW-ENERGY BEHAVIOR

In Theorem 6.1.1 we have seen that $\mathscr{A}(\lambda P, \mu) \to 0$ $(\lambda \to 0)$ for $d(G) \geq 0$, with fixed nonzero masses, as expected. The same analysis leading to this theorem shows that $\mathscr{A}(\lambda P, \mu)$ remains bounded for $\lambda \to 0$ when $d(G) < 0$ since $|a|$, in (6.1.18), is positive.[28]

In this section we generalize these results to the cases when some of the underlying masses approach zero as well, and some, not necessarily all, of the external momenta become small. On physical grounds we let these masses approach zero at a rate faster than the corresponding external momentum become small.[29] For generality we let these vanishing external momentum components and those vanishing masses become small at different rates.

We write

$$P = P' \cup P'', \qquad (6.6.1)$$

[28] Note also that $|a|$ is bounded above.

[29] Applications may be also given when some of the external momenta become small at a faster rate than some of the vanishing masses; however, these are of less interest than the general cases given here and will not be discussed.

where P'' is that subset of P containing those elements we wish to scale to zero, and P' constitutes the remaining elements in P. We decompose E_1 as

$$E_1 = E_1^1 \oplus E_1^2, \tag{6.6.2}$$

where E_1^2 is, say, a k-dimensional subspace of E_1 associated with the vanishing momenta, and E_1^1 is the orthogonal complement of E_1^2 in E_1. We also decompose E_2 as

$$E_2 = E_2^1 \oplus E_2^2, \tag{6.6.3}$$

where E_2^1 is, say, of s dimensions and will be associated with the masses that we wish to become small, and E_2^2 is the orthogonal complement of E_2^1 in E_2.

Let $\mathbf{P}_1$ be a vector in E_1^1 such that the elements in P' may be written as some linear combinations of the components of $\mathbf{P}_1$. Let $\mathbf{P}_2$ be a vector in E_1^2 of the form

$$\mathbf{P}_2 = \lambda_1 \mathbf{L}_2 + \cdots + \lambda_1 \cdots \lambda_k \mathbf{L}_{k+1}, \tag{6.6.4}$$

where $\mathbf{L}_2, \ldots, \mathbf{L}_{k+1}$ are k independent vectors in E_1^2, and suppose that the elements in P'' may be written as some linear combinations of the components of $\mathbf{P}_2$ such that every element $p_{i_1}^\mu$ in P'' may be conveniently written as $p_{i_1}^\mu = \lambda_1 \cdots \lambda_{j(i_1)} \tilde{p}_{i_1}^\mu$ for some $j(i_1) \le k$, and where $\tilde{p}_{i_1}^\mu$ is independent of $\lambda_1, \ldots, \lambda_k$.

Let $\mathbf{P}_3$ be a vector in E_2 of the form

$$\begin{aligned}\mathbf{P}_3 &= \lambda_1 \cdots \lambda_k (\lambda_{k+1} \mathbf{L}_{k+2} + \cdots + \lambda_{k+1} \cdots \lambda_{k+s} \mathbf{L}_{k+s+1}) + \mathbf{L} \\ &\equiv \mathbf{P}_3' + \mathbf{L}, \tag{6.6.5}\end{aligned}$$

with $\mathbf{P}_3' \in E_2^1$, $\mathbf{L} \in E_2^2$, where $\mathbf{L}_{k+2}, \ldots, \mathbf{L}_{k+s+1}$ are s independent vectors in E_2^1 such that the masses that we wish to become small may be written as some linear combinations of the components of $\mathbf{P}_3'$ in such a way that every mass μ^{i_1} we wish to approach zero may be written in the form $\mu^{i_1} = \lambda_1 \cdots \lambda_k \lambda_{k+1} \cdots \lambda_{k+j(i_1)} \tilde{\mu}^{i_1}$ for some $1 \le j(i_1) \le s$, and where $\tilde{\mu}^{i_1}$ is independent of $\lambda_1, \ldots, \lambda_{k+s}$, and the remaining nonasymptotic masses may be written as some linear combinations of the components of $\mathbf{L}$.

Accordingly the subtracted-out amplitude $\mathscr{A}(P, \mu)$ may be written as $\mathscr{A}(\mathbf{P})$, with

$$\mathbf{P} = \mathbf{P}_1 + \mathbf{P}_2 + \mathbf{P}_3. \tag{6.6.6}$$

Introducing the vector $\mathbf{P}'$ by

$$\mathbf{P} = (\lambda_1 \cdots \lambda_{k+s}) \mathbf{P}', \tag{6.6.7}$$

defining

$$C = L_{k+s+1},$$ (6.6.8)

$$L_1 = P_1 + L,$$ (6.6.9)

and hence writing

$$P' = \sum_{i=1}^{k+s} \left(\frac{1}{\lambda_i} \cdots \frac{1}{\lambda_{k+s}} \right) L_i + C,$$ (6.6.10)

we obtain the following expression for $\mathscr{A}(P)$:

$$\mathscr{A}(P) = \prod_{j=1}^{\rho} (\mu^j)^{-\sigma_j} (\lambda_1 \cdots \lambda_{k+s})^{d(G)} \tilde{\mathscr{A}}(P')$$ (6.6.11)

[see (6.5.19)], and for all those masses that we wish to become small the corresponding $\sigma_j \equiv 0$ (see Section 6.5.1). From (6.6.9) and (6.6.10) we note that *all* the external momenta in $\tilde{\mathscr{A}}(P')$ are proportional to $(\lambda_{k+1} \cdots \lambda_{k+s})^{-1}$. For $\lambda_1 = \cdots = \lambda_{k+s} = 1$, we choose the external momenta $p_1, \ldots, p_m$ such that no partial sums of these momenta vanish, i.e., $p_{i_1} + \cdots + p_{i_t} \neq 0$ for all $\{i_1, \ldots, i_t\} \subset \{1, \ldots, m\}$. For $\lambda_1, \ldots, \lambda_{k+s} \to 0$ independently, the total external momenta carried away from each external vertex is asymptotic.

We may then apply Theorem 6.4.2 to investigate the behavior of $\mathscr{A}(P)$ from (6.6.11). We may also apply the rules for the existence of $\lim \mathscr{A}$ for $\lambda_{k+1}, \ldots, \lambda_{k+s} \to 0$, given in Section 6.5.2, to infer the behavior of $\mathscr{A}$ at low energies, in the zero-mass limit.

Example 6.5: Consider the elementary graph G in Fig. 6.2. We are interested in the behavior of $\mathscr{A}(\lambda_1 q, m, \lambda_1 \lambda_2 \mu)$ for $\lambda_1, \lambda_2 \to 0$, where μ is the mass of the photon and m is the mass of the electron. We write for the photon propagator $D_{\mu\nu}(Q) = g_{\mu\nu}/(Q^2 + \mu^2)$. From Example 6.2 we know that $\lim \mathscr{A}$ for $\lambda_2 \to 0$ exists. We may write

$$\mathscr{A}(\lambda_1 q, m, \lambda_1 \lambda_2 \mu) = \lambda_1 \mathscr{A}(q, m/\lambda_1, \lambda_2 \mu) = \lambda_1 \lambda_2 \mathscr{A}(q/\lambda_2, m/\lambda_1 \lambda_2, \mu).$$

In reference to the parameter $1/\lambda_2$, it is easy to show that for the amplitude $\mathscr{A}(q/\lambda_2, m/\lambda_1 \lambda_2, \mu)$, we have $\tau_0 = \{G\} = \tau$, $\alpha_I(S(q, m)) = d(G)$, and $\beta_I(S(q, m)) = 0$. In reference to the parameter $1/\lambda_1$, we note that for $\mathscr{A}(q/\lambda_2, m/\lambda_1 \lambda_2, \mu)$, as in Example 6.2, $\tau_0 = \{G\}$ and hence $\alpha_I(S(m)) = d(G) = 1$. From the definition of class $\mathscr{M}$ in Section 6.4 we also note that G has no proper subdiagram g [with $d(g) \geq 0$] dependent only on the mass μ [i.e., independent of $1/\lambda_1$, in reference to the amplitude $\mathscr{A}(q/\lambda_2, m/\lambda_1 \lambda_2, \mu)$]. Accordingly $\beta_I(S(m)) = 0$. Hence we obtain from (6.6.11) and (6.4.5)

$$\lim_{\lambda_1, \lambda_2 \to 0} \mathscr{A}(\lambda_1 q, m, \lambda_1 \lambda_2 \mu) = 0 \left(\lambda_1 \lambda_2 \frac{1}{\lambda_1 \lambda_2} \right) = 0(1).$$ (6.6.12)

It is instructive to study the expression explicitly for $\mathcal{A}(q, m, \mu)$, corresponding to Fig. 6.2, with $\mu = 0$ and to determine its behavior for $q \to 0$. The expression for $\mathcal{A}(q, m, 0)$ is given in (6.5.26). We use the identity ($x \geq 0$)

$$\ln\left(1 + \frac{q^2}{m^2}x\right) = \frac{q^2}{m^2}x - \frac{q^4}{m^4}x^2 \int_0^1 \frac{y\,dy}{[1 + (q^2/m^2)xy]}, \qquad (6.6.13)$$

and hence the bound

$$\left|\ln\left(1 + \frac{q^2}{m^2}x\right)\right| \leq \frac{q^2}{m^2}x + \frac{1}{2}\frac{q^4}{m^4}x^2, \qquad (6.6.14)$$

to bound the expressions

$$a\left(\frac{q^2}{m^2}\right) = -\frac{\alpha}{2\pi}\int_0^1 x\,dx\,\ln\left(1 + \frac{q^2}{m^2}x\right), \qquad (6.6.15)$$

$$b\left(\frac{q^2}{m^2}\right) = -\frac{\alpha}{\pi}\int_0^1 dx\,\ln\left(1 + \frac{q^2}{m^2}x\right) \qquad (6.6.16)$$

as

$$\left|a\left(\frac{q^2}{m^2}\right)\right| \leq \frac{\alpha}{6\pi}\frac{q^2}{m^2} + \frac{\alpha}{16\pi}\frac{q^4}{m^4}, \qquad (6.6.17)$$

$$\left|b\left(\frac{q^2}{m^2}\right)\right| \leq \frac{\alpha}{2\pi}\frac{q^2}{m^2} + \frac{\alpha}{6\pi}\frac{q^4}{m^4}. \qquad (6.6.18)$$

Accordingly we have

$$\mathcal{A}(\lambda_1 q, m, 0) = \lambda_1\gamma q a\left(\frac{\lambda_1^2 q^2}{m^2}\right) + mb\left(\frac{\lambda_1^2 q^2}{m^2}\right), \qquad (6.6.19)$$

and for $\lambda_1 \to 0$, we have from (6.6.17)–(6.6.19) that $\mathcal{A}$ remains bounded (actually, it vanishes) consistent with the result in (6.6.12), as expected.

Example 6.6: Consider the behavior of $\mathcal{A}(\lambda_1 q, \lambda_1\lambda_2 m, \lambda_1\lambda_2\lambda_3\mu)$ for $\lambda_1, \lambda_2, \lambda_3 \to 0$ corresponding to the graph in Fig. 6.4. We may write $\mathcal{A}(\lambda_1 q, \lambda_1\lambda_2 m, \lambda_1\lambda_2\lambda_3\mu) = \lambda_1\lambda_2\lambda_3\mathcal{A}(q/\lambda_2\lambda_3, m/\lambda_3, \mu)$. In reference to the latter amplitude we have from Example 6.1, $\alpha_I(S(q)) = 1$, $\beta_I(S(q)) = 2$. From Example 6.4., we may infer that $\alpha_I(S(q, m)) = 1$, $\beta_I(S(q, m)) = 0$. Accordingly we have from (6.6.11) and (6.4.5)

$$\lim_{\lambda_1, \lambda_2, \lambda_3 \to 0} \mathcal{A}(\lambda_1 q, \lambda_1\lambda_2 m, \lambda_1\lambda_2\lambda_3\mu) = 0\left[\lambda_1\lambda_2\lambda_3\frac{1}{\lambda_2\lambda_3}\sum_{\gamma=0}^2 \ln^\gamma\left(\frac{1}{\lambda_2}\right)\right]$$

$$= 0\left[\lambda_1\sum_{\gamma=0}^2 \ln^\gamma\left(\frac{1}{\lambda_2}\right)\right]. \qquad (6.6.20)$$

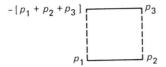

Fig. 6.5   A fermion–fermion (spin $\frac{1}{2}$) scattering graph with a $\bar{\psi}\psi\phi$ coupling.

Example 6.7:   Consider the process depicted in Fig. 6.5, with a $\bar{\psi}\psi\phi$ coupling, where the dashed lines denote scalar bosons, and for simplicity the fermion and the scalar bosons are assumed to have the same mass μ. Here $d(G) = -2$. We are interested in the behavior of $\mathscr{A}(\lambda_1 p_1, \lambda_1 p_2, p_3, \lambda_1 \lambda_2 \mu)$ for $\lambda_1, \lambda_2 \to 0$. As in Example 6.4, the limit $\lambda_2 \to 0$ may be taken. In reference to the parameter $1/\lambda_1$, we readily obtain for the amplitude

$$\mathscr{A}(p_1/\lambda_2, p_2/\lambda_2, p_3/\lambda_1\lambda_2, \mu)$$

that $\alpha_I(S(p_3)) = -1$. Also, G does not contain any proper subdiagram g [with $d(g) \geq 0$], and hence $\beta_I(S(p_3)) = 0$. Accordingly we have

$$\lim_{\lambda_1, \lambda_2 \to 0} \mathscr{A}(\lambda_1 p_1, \lambda_1 p_2, p_3, \lambda_1 \lambda_2 \mu) = 0\left((\lambda_1)^{-2}\left(\frac{1}{\lambda_1}\right)^{-1}\right) = 0\left(\frac{1}{\lambda_1}\right), \quad (6.6.21)$$

for nonexceptional momenta.

6.7   GENERAL ASYMPTOTIC BEHAVIOR II

In this section we generalize our applications to cases when, in general, some (or all) of the external momenta become large, some (or all) become small, and some of the masses are led to approach zero.

We repeat the definition of P in (6.6.1), with the subset P' now consisting of those elements in P becoming either large or remaining nonasymptotic. Let E_1 and E_2 be written as in (6.6.2) and (6.6.3), respectively.

Let $\mathbf{P}_1$ be a vector in E_1^1 of the form

$$\mathbf{P}_1 = \mathbf{L}_1 \eta_1 \cdots \eta_a + \cdots + \mathbf{L}_a \eta_a + \mathbf{L}'_{a+1}, \quad (6.7.1)$$

where $\mathbf{L}_1, \ldots, \mathbf{L}_a, \mathbf{L}'_{a+1}$, with $a + 1 \leq 4m - k$, are $a + 1$ independent vectors in E_1^1. As in (6.6.4) and in a notation similar to it, let $\mathbf{P}_2 \in E_1^2$ be such that

$$\mathbf{P}_2 = \lambda_1 \mathbf{L}_{a+2} + \cdots + \lambda_1 \cdots \lambda_k \mathbf{L}_{a+k+1}. \quad (6.7.2)$$

As in (6.6.5), let $\mathbf{P}_3 \in E_2$ be such that

$$\mathbf{P}_3 = \lambda_1 \cdots \lambda_k (\lambda_{k+1} \mathbf{L}_{a+k+2} + \cdots + \lambda_{k+1} \cdots \lambda_{k+s} \mathbf{L}_{a+k+1+s}) + \mathbf{L}''_{a+1},$$

$$\equiv \mathbf{P}'_3 + \mathbf{L}''_{a+1}. \quad (6.7.3)$$

We may then write the amplitude $\mathscr{A}(P, \mu)$ as $\mathscr{A}(\mathbf{P})$, with

$$\mathbf{P} = \mathbf{P}_1 + \mathbf{P}_2 + \mathbf{P}_3, \tag{6.7.4}$$

or

$$\mathbf{P} = (\lambda_1 \cdots \lambda_{k+s})\mathbf{P}', \tag{6.7.5}$$

with

$$\mathbf{P}' = \sum_{i=1}^{a+k+s} \eta_i \cdots \eta_{a+k+s} \mathbf{L}_i + \mathbf{C}, \tag{6.7.6}$$

$$\mathbf{C} = \mathbf{L}_{a+k+1+s},$$

$$\mathbf{L}_{a+1} = \mathbf{L}'_{a+1} + \mathbf{L}''_{a+1},$$

$$1/\lambda_i = \eta_{a+i}, \qquad i = 1, \ldots, k + s \tag{6.7.7}$$

The amplitude $\mathscr{A}(\mathbf{P})$ may be then written as in (6.6.11), and the limits $\eta_1, \ldots, \eta_{a+k+s} \to \infty$ may be then studied directly from (6.4.5) for the amplitude $\mathscr{A}(\mathbf{P}')$ with $\mathbf{P}'$ now defined in (6.7.6). We note that $\mathbf{L}_1, \ldots, \mathbf{L}_{a+k+s}$ are independent vectors. Again, with $p_{i_1} + \cdots + p_{i_t} \neq 0$ for all $\{i_1, \ldots, i_t\} \subset \{1, \ldots, m\}$, all the external momenta at the external vertices then become asymptotic for $\eta_1, \ldots, \eta_{a+k+s} \to \infty$ for the amplitude $\mathscr{A}(\mathbf{P}')$.

Example 6.8: Consider the renormalized amplitude $\mathscr{A}(q, p, m, \mu)$ associated with the graph G in Fig. 6.6, where a dashed line represents a scalar particle of mass μ, and a solid line represents a spin-$\frac{1}{2}$ particle of mass m. Let $(k_2^0, \ldots, k_2^3)$ be integration variables, with $k^g = k_2$, associated with the subdiagram g, and $(k_1^0, \ldots, k_2^3)$ be the integration variables associated with the graph G. We note that $d(g) = 2$, $d(G) = 0$. We write $R^{18} = I \oplus E$. We also write $I = \hat{I}_1 \oplus \hat{I}_2, \hat{I}_1 = I_1 \oplus \cdots \oplus I_4, \hat{I}_2 = I_5 \oplus \cdots \oplus I_8$(dim $I_j = 1, j = 1, \ldots, $ 8), and $E = E_1 \oplus E_2$. The subspace $\hat{I}_2$ is associated with the integration variable k_2. We make the further decomposition $E_1 = E_1^1 \oplus E_1^2$ and

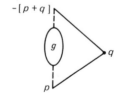

Fig. 6.6   A vertex correction with a $\bar{\psi}\psi\phi$ coupling, with the dashed line representing a scalar boson with mass μ, and a solid line representing a spin $\frac{1}{2}$ particle of mass m.

$E_2 = E_2^1 \oplus E_2^2$, with $E_1^1, E_1^2, E_2^1, E_2^2$ associated, respectively, with q, p, m, μ. We define the subspaces

$$S(q) = E_1^1,$$

$$S(q, \mu) = E_1^1 \oplus E_2^2,$$

$$S^0 = I \oplus E_1^1,$$

$$S' = \hat{I}_1 \oplus E_1^1, \qquad (6.7.8)$$

$$S_1 = I \oplus E_1^1 \oplus E_2^2,$$

$$S_2 = \hat{I}_1 \oplus E_1^1 \oplus E_2^2,$$

$$S(p, q, \mu) = E_2 \oplus E_2^2.$$

We are interested in the behavior of the amplitude

$$\mathscr{A}(\eta_1 q, \lambda_1 p, \lambda_1 \lambda_2 m, \mu) \quad \text{for} \quad \eta_1 \to \infty, \lambda_1 \to 0, \lambda_2 \to 0.$$

We consider the parameter η_1 first. In reference to this parameter, it is readily seen, as in Example 6.1, that $\alpha_I(S(q)) = d(G) = 0$, as $\tau_0 = \{G\}$. From the definition of $(\mathcal{M}, \tau)$ in Section 6.3, we have $\tau = \{G, (G|g), (G|G)\}$ with the latter subdiagrams associated with the subspaces in $\mathcal{M} = \{S^0, S', S'' \equiv S(q)\}$. By noting that

$$\Lambda(I)S^0 = \Lambda(I)S' = \Lambda(I)S'' = S(q),$$

$$\Lambda(I_j \oplus \cdots \oplus I_8)S^0 = \Lambda(I_j \oplus \cdots \oplus I_8)S' = I_1 \oplus \cdots \oplus I_{j-1} \oplus E_1^1,$$
$$j = 2, \ldots, 5, \quad (6.7.9)$$

we obtain from Theorem 6.3.2 $p_1 = p_5 = 1$, and $p_j = 0$ for $j \neq 1, 5$, and hence $\beta_I(S(q)) = 2$.

We note that

$$\mathscr{A}(\eta_1 q, \lambda_1 p, \lambda_1 \lambda_2 m, \mu) = \mathscr{A}(\eta_1 \lambda_1^{-1} \lambda_2^{-1} q, \lambda_2^{-1} p, m, \lambda_1^{-1} \lambda_2^{-1} \mu).$$

Consider the parameter λ_1^{-1}. We note that $\tau_0 = \{G\}$, i.e., $\alpha_I(S(q, \mu)) = 0$, and $\tau = \{G, (G|g)\}$. Note that $(G|G) \notin \tau$ since G (trivially) contains the mass μ. The subdiagrams in τ are associated with the subspaces in $\mathcal{M} = \{S_1, S_2\}$. By repeating an analysis similar to the one given before, we obtain $p_5 = 1$ and $p_i = 0$ for $i \neq 5$, thus leading to $\beta_I(S(q, \mu)) = 1$.

Finally, we note that the subdiagram g does not depend on the mass μ, and as before we obtain $\alpha_I(S(q, p, \mu)) = 0$ and $\beta_I(S(q, p, \mu)) = 1$ in reference to the parameter λ_2^{-1}.

Accordingly we have from Theorem 6.4.2, for nonexceptional momenta,

$$\mathscr{A}(\eta_1 q, \lambda_1 p, \lambda_1 \lambda_2 m, \mu) = 0 \left[\sum_{\gamma_1, \gamma_2, \gamma_3} \left(\ln \frac{1}{\lambda_1} \right)^{\gamma_1} \left(\ln \frac{1}{\lambda_2} \right)^{\gamma_2} (\ln \eta_1)^{\gamma_3} \right], \quad (6.7.10)$$

for $\eta_1 \to \infty$, $\lambda_1 \to 0$, $\lambda_2 \to 0$, and the sum is over all nonnegative integers $\gamma_1, \gamma_2, \gamma_3$ such that

$$\gamma_1 \leq 1,$$

$$\gamma_1 + \gamma_2 \leq 1, \quad (6.7.11)$$

$$\gamma_1 + \gamma_2 + \gamma_3 \leq 2.$$

Other examples may be also carried out where, for example, only some (or all) of the external momenta, of the graph in question, become large, and some (or all) of the masses become small.

6.8   GENERALIZED DECOUPLING THEOREM

In this section we generalize the decoupling theorem given in Section 6.1. In Section 6.8.1 we consider the behavior of $\mathscr{A}$ when any subset of the masses in $\mathscr{A}$ become large, and in general, at different rates. In section 6.8.2 this theorem is generalized further to cases when some of the remaining masses in the theory are led to become small as well, corresponding to theories which, on experimental grounds, may contain zero-mass particles.

6.8.1   Generalized Decoupling Theorem I

Consider the renormalized integrand R with argument $\mathbf{P}'$ as defined in (6.2.5). In this section all the external momenta are kept fixed and hence we require that $\Lambda(I \oplus E_2)S'_r = \{0\}$. For the problem at hand we suppose that $\Lambda(I \oplus E_1)S'_r \neq \{0\}$.

First suppose that $\Lambda(E)S'_r \neq \{0\}$. Then we may apply (6.2.21) and (6.2.22) to the graph G itself. If $G \in \mathscr{F}_2(\mathscr{N})$ [see (6.2.13)], then (6.2.22) applied to G implies that

$$\operatorname*{degr}_{\eta_r} F_G(\mathscr{N}) \begin{Bmatrix} < \\ \text{or} \\ \leq \end{Bmatrix} \min[d(G), -1] - \sigma(G).^{30} \quad (6.8.1)$$

---

[30] Since the external momenta of G are independent of η_r, we took the liberty of replacing $\operatorname{degr}_\lambda$ by $\operatorname{degr}_{\eta_r}$ in (6.8.1).

An equality in (6.8.1) may hold if there is no subdiagram $g \subset G$ in $\mathcal{H}_2(\mathcal{N})$ such that at least one of the masses in the lines in $\bar{g}$ is dependent on η_r, and there is no subdiagram $g' \not\subset G$ in $\mathcal{F}_2(\mathcal{N})$. Finally suppose that $G \in \mathcal{H}_2(\mathcal{N})$. Then we may apply (6.2.21), (6.2.22) [and (6.2.18), in general, applied to $I_{\bar{G}}$] to conclude that

$$\operatorname{degr}_{\eta_r} I_{\bar{G}} \prod_i F_{G_i}(\mathcal{N}) \le \sum_{i=1}^{k} \min[d(G_i), -1] - \sigma(G), \qquad (6.8.2)$$

where $G_1, \ldots, G_{k_1}$ are those maximal elements in $\mathcal{N}$ contained in $G: G_i \not\subset G$ such that $G_i \in \mathcal{F}_2(\mathcal{N})$, with $i = 1, \ldots, k_1$. We denote the remaining maximal elements in $\mathcal{N}$ contained in G by $G_{k_1+1}, \ldots, G_n$. We may also apply (6.2.21) directly with g_i in it simply replaced by G, for $d(G) \ge 0$, and use the estimate in (6.8.2) to conclude for $d(G) \ge 0$ or $d(G) < 0$ that

$$\operatorname{degr} F_G(\mathcal{N}) \le -1 - \sigma(G), \qquad (6.8.3)$$

since $k_1 \ge 1$, for $\Lambda(E)S'_r \ne \{0\}$.

From (6.8.1) or (6.8.3) and the fact that

$$\dim \Lambda(E)S'_r \le \sigma(G), \qquad (6.8.4)$$

we obtain upon summing over the sets $\mathcal{N}$ in (6.2.11) that

$$\operatorname{degr}_{\eta_r} R \le -1 - \dim \Lambda(E)S'_r. \qquad (6.8.5)$$

If $\Lambda(E)S'_r = \{0\}$, then directly from (6.2.4) and (6.2.18) we conclude that

$$\operatorname{degr}_{\eta_r} R \le -1. \qquad (6.8.6)$$

Now consider the renormalized amplitude

$$\mathscr{A}(P, \eta_1 \cdots \eta_s \mu^1, \eta_2 \cdots \eta_s \mu^2, \ldots, \eta_s \mu^s, \mu^{s+1}, \ldots, \mu^p), \qquad (6.8.7)$$

with $s \le p$. We conveniently decompose the subspace E_2 into s one-dimensional orthogonal subspaces E^i and introduce s vectors $\mathbf{L}_1 \in E_1, \ldots, \mathbf{L}_s \in E^s$ with nonvanishing components $\mu^1, \ldots, \mu^s$, respectively. We then rewrite (6.8.7) as

$$\mathscr{A}(\mathbf{L}_1 \eta_1 \cdots \eta_s + \cdots + \mathbf{L}_s \eta_s + \mathbf{C}), \qquad (6.8.8)$$

where $\mathbf{C}$ is confined to a finite region in E, with the $\mu^i \ne 0$. We are mainly interested in the vanishing property of $\mathscr{A}$ for $\eta_1, \ldots, \eta_s \to \infty$, and we shall not carry out an analysis regarding the logarithmic asymptotic coefficients $\beta_l(S_r)$, as the latter is quite cumbersome.

From (6.2.44), (6.2.51), (6.2.60), (6.8.5), and (6.8.6) we may then state

Theorem 6.8.1: *For* $\eta_1, \ldots, \eta_s \to \infty$, $\mathscr{A}$ *in* (6.8.8) *vanishes, as the power asymptotic coefficients* $\alpha_I(S_r)$ *are bounded above as*

$$\alpha_I(S_r) \leq -1. \tag{6.8.9}$$

In particular, since the $\beta_I(S_r)$ *are finite, we can always find positive integers* $N_1, \ldots, N_s$ *and a real constant* $C_0 > 0$ *such that*

$$|\mathscr{A}(\mathbf{L}_1 \eta_1 \cdots \eta_s + \cdots + \mathbf{L}_s \eta_s + \mathbf{C})| \leq \frac{C_0}{\eta_1 \cdots \eta_s} \prod_{i=1}^{s} (\ln \eta_i)^{N_i}. \tag{6.8.10}$$

6.8.2   Generalized Decoupling Theorem II

Here we are interested in generalizing Theorem 6.8.1 to the cases when some of the remaining masses $\mu^{s+1}, \ldots, \mu^\rho$ vanish; i.e., we are interested in the behavior of

$$\mathscr{A}(P, \eta_1 \cdots \eta_s \mu^1, \ldots, \eta_s \mu^s, \lambda_1 \mu^{s+1}, \ldots, \lambda_1 \cdots \lambda_k \mu^{s+k}, \mu^{s+k+1}, \ldots, \mu^\rho), \tag{6.8.11}$$

where $s + k \leq \rho$, for $\eta_1, \ldots, \eta_s \to \infty$ and $\lambda_1, \ldots, \lambda_k \to 0$. We have already established sufficiency conditions for taking the limits $\lambda_1, \ldots, \lambda_k \to 0$ of $\mathscr{A}$ in Section 6.5.2. Accordingly Theorem 6.8.1 remains true if the sufficiency conditions stated in Section 6.5.2 are satisfied with respect to the parameters $\lambda_1, \ldots, \lambda_k$.

The renormalized amplitudes $\mathscr{A}(q, m, \mu)$ corresponding to the graphs in Figs. 6.2 and 6.3, for example, all vanish for $m \to \infty, \mu = $ fixed and $m \to \infty$, $\mu \to 0$. We note, in particular, from the estimates in (6.6.17) and (6.6.18) that for the graph G in Fig. 6.2 we have for $\mu = 0$

$$\mathscr{A}(q, \eta m, 0) = \gamma q a\left(\frac{q^2}{\eta^2 m^2}\right) + \eta m b\left(\frac{q^2}{\eta^2 m^2}\right), \tag{6.8.12}$$

with

$$\left| a\left(\frac{q^2}{\eta^2 m^2}\right) \right| \leq \frac{\alpha}{6\pi} \frac{q^2}{\eta^2 m^2} + \frac{\alpha}{16\pi} \frac{q^4}{\eta^4 m^4} \tag{6.8.13}$$

and

$$\eta m \left| b\left(\frac{q^2}{\eta^2 m^2}\right) \right| \leq \frac{\alpha}{2\pi} \frac{q^2}{\eta m} + \frac{\alpha}{6\pi} \frac{q^4}{\eta^3 m^3}, \tag{6.8.14}$$

which show the vanishing property of $\mathscr{A}(q, \eta m, 0)$ for $\eta \to \infty$ — a result that is consistent with our conclusions.

NOTES

Section 6.1 is based on Manoukian (1981a) and on some analysis, in particular, estimate (6.1.12), due to Hahn and Zimmermann (1968). Section 6.2 is based on Manoukian (1980a, 1981b), Section 6.3 on Manoukian (1978, see also 1980c), Section 6.4 on Manoukian (1980c, 1981c), Section 6.5 on Manoukian (1980b, 1981c), Section 6.6 on Manoukian (1980d, see also 1979a), Section 6.7 on Manoukian (1981c), and Section 6.8 on Manoukian (1981d). Asymptotics were also studied by completely different methods (in the so-called α-parameter representation) by, e.g., Bergère *et al.* (1978). The decoupling theorem with only one mass scale becoming large and with no zero mass particles was proved by Ambjørn (1979) (again in the α-parameter representation). A proof of the general case with several mass scales becoming large and with zero-mass-particle limit was given in Manoukian (1981d). Interesting applications of the decoupling theorem have been carried out [see Appelquist and Carrazzonne (1975), Poggio *et al.* (1977), Collins *et al.* (1978), Toussaint (1978), Kazama and Yao (1979), Ovrut and Schnitzer (1981), and Hagiwara and Nakazawa (1981)], and many other papers on the subject are still appearing.

Appendix / SUBTRACTIONS VERSUS COUNTERTERMS

In this appendix we show by a direct and simple method that the subtractions introduced in Section 5.2, as carried out directly in momentum space, are equivalent to formally adding counterterms to the (unrenormalized) interaction Lagrangian density. This equivalence theorem had a very important role in the history of quantum field theory.

A.1 THE FORMAL UNRENORMALIZED THEORY

Let $K = \{\Phi_1, \Phi_2, \ldots\}$ be the set of all basic (free) fields and their adjoints of interest in the theory, added to which are all their derivatives of arbitrary orders, suppressing, for simplicity of notation, all tensor and spinor indices.

We may define the interaction Lagrangian density without counterterms (in the interaction picture, i.e., in terms of the elements in the set K) in the form

$$\mathscr{L}_I(x) = \sum_{G \in \mathscr{G}(1)} c_G \, {:}\Phi_{11}(x) \cdots \Phi_{1e(G)}(x){:}, \tag{A.1}$$

where $\mathscr{L}_I(x)$ is formally hermitian and the c_G are constant matrices (couplings) depending, in general, on tensor and spinor indices and summation over them, when the c_G are multiplied by the fields, is understood. The double dots $:\;:$ in (A.1) denote the so-called Wick ordering.[1] $\mathscr{G}(1)$

---

[1] Wick (1950), or see Schweber (1961).

corresponds to the set of all the terms in the interaction specified with the combinations $\{G, :\Phi_{11}(x)\cdots\Phi_{1e(G)}(x):\}$ and such an element is denoted by $\tilde{G}$. This notation will be indispensable when we come to counterterms. In momentum space variables we may write $\mathcal{L}_I(x)$ as

$$
\mathcal{L}_I(x) = \sum_{\tilde{G} \in \mathcal{G}(1)} c_G
$$

$$
\times \int dQ_G \exp\{i[Q_{11} + \cdots + Q_{1e(G)}]x\} :\Phi_{11}(Q_{11})\cdots\Phi_{1e(G)}(Q_{1e(G)}):,
\tag{A.2}
$$

using the same symbols for the Fourier transform of the fields, for simplicity of notation, and

$$
Q_G = (Q_{11}^0, \ldots, Q_{ie(G)}^3)
\tag{A.3}
$$

Consider the chronological product[2] of $n(n \geq 2)$ Wick-ordered terms each of the form in (A.2):

$$
c_{G_1}\cdots c_{G_n}\left(\prod_{i=1}^{n} : \prod_{j=1}^{e(G_i)} \Phi_{ij}(Q_{ij}):\right)_+^{c,p}
$$

$$
= \sum_{\tilde{G}_0} \sigma_{\tilde{G}_0}(c_{G_1}\cdots c_{G_n}\hat{I}_{G_0})(\prod \delta_{G_0(Q_{..}-Q_{..})}) :\Phi_{1G_0}^f(Q_{1G_0})\cdots\Phi_{kG_0}^f(Q_{kG_0}):,
\tag{A.4}
$$

where the superscripts c, p denote connected and proper parts; we shall come back to their precise meaning. We have introduced the superscripts f on the fields on the right-hand side of (A.4) to remind us that they resulted in the final expression after carrying out the (covariant) time ordering product. $\sigma_{\tilde{G}_0}$ is the so-called parity associated with the commutation of the fields within a chronological product, on the left-hand side of (A.4), so that a pair of fields which are to be contracted together are brought next to each other. The chronological pairing of two fields has the general structure

$$
\overline{\Phi_1(Q_1)\Phi_2(Q_2)} = \delta(Q_1 + Q_2)\Delta_{12}(Q_1),
\tag{A.5}
$$

and $\hat{I}_{G_0}$ in (A.4) is defined by

$$
\hat{I}_{G_0} = \prod_{\text{cont.}} \Delta(Q_i),
\tag{A.6}
$$

as a product of the contracted pairs of fields. We shall not go into the details in defining explicit forms of the $\Delta(Q_i)$, as we are working at a very general level. The factor $(\prod \delta_{G_0})$ in (A.4) occurs as the product of δ-functions appearing in pairings such as in (A.5). The sum over $\tilde{G}_0$ is over all possible contractions among the fields on the left-hand side of (A.4) not occurring

---

[2] We assume covariant chronological products. We shall not go into details of such definitions as these details are not necessary in understanding the content of the appendix. For such details cf. Nishijima (1969).

within the same Wick-ordering signs : : such that when we integrate over the momentum components $Q_{1G_0}, \ldots$, we obtain for (A.4), after having it multiplied by $\prod_{i=1}^{n}(\delta(\sum_{j=1}^{e(G_i)} Q_{ij}))$ only proper and connected subgraphs G_0 of order n, i.e., an expression of the form

$$\sum_{G_0} \int dQ_{G_0} \, dK_{G_0} \, c_{G_1} \cdots c_{G_n} I'_{G_0} \delta\left(\sum_{i=1}^{k} Q_{iG_0}\right) : \Phi^{f}_{1G_0}(Q_{1G_0}) \cdots \Phi^{f}_{kG_0}(Q_{kG_0}): , \quad \text{(A.7)}$$

where G_0 is a proper and connected subgraph with precisely n vertices. The expression

$$I_{G_0} = c_{G_1} \cdots c_{G_n} I'_{G_0} \quad \text{(A.8)}$$

defines the unrenormalized Feynman integrand associated with G_0. The expression $\hat{I}_{G_0}$ in (A.6), by definition, contains no vertices to be distinguished from I_{G_0} corresponding to a genuine subdiagram of order n. The introduction of the factor I'_{G_0} in (A.8) will be useful later on. For simplicity of notation we have absorbed the σ_{G_0} factor in I'_{G_0}. The one overall $\delta(\sum_{i=1}^{k} Q_{iG_0})$ occurs in (A.7) as the result of a restriction over connected parts only and momentum conservation. K_{G_0} denotes the internal momenta of G_0 and Q_{G_0} its external momenta.

Expression (A.7) prepares us to define the concept of a Wick polynomial of order n.

Definition of a Wick polynomial of order n: We denote the range of possible expressions $\tilde{G}_0$ in (A.7) by $\mathscr{G}(n)$ and we call the expression

$$I_{G_0} : \Phi^{f}_{1G_0} \cdots \Phi^{f}_{kG_0}: , \quad \text{(A.9)}$$

a Wick polynomial of order n, symbolized by $\tilde{G}_0$. The latter may be defined by the combination $\{G_0, :\Phi^{f}_{1G_0} \cdots \Phi^{f}_{kG_0}:\}$, where G_0 is a proper and connected graph of order n. A Wick polynomial of order n is specified by both G_0 and $:\Phi^{f}_{1G_0} \cdots \Phi^{f}_{kG_0}:$.

We may then summarize by saying that $\mathscr{G}(n)$ is the set of all possible Wick polynomials of order n as arising in the expansion (A.4) leading to (A.7). This definition will be quite useful later on.

Definition of counterterms: Let G_0 be a graph with a set of vertices $\mathscr{V}$ and a set of lines $\mathscr{L}$. Consider n *disjoint* subsets of $\mathscr{V} : \mathscr{V}'_1, \ldots, \mathscr{V}'_n$ associated with subgraphs $G'_1, \ldots, G'_n$ of G. By definition the G'_i contain all those lines in $\mathscr{L}$ joining the vertices in $\mathscr{V}'_i$. Let $\mathscr{L}'_i$ denote the set of lines of G'_i. Consider the sets $\mathscr{V}' = \mathscr{V}'_1 \cup \cdots \cup \mathscr{V}'_n$ and $\mathscr{L}' = \mathscr{L}'_1 \cup \cdots \cup \mathscr{L}'_n$. The latter obviously define a proper subdiagram G'. Note, however, that G' is not necessarily a subgraph of G_0 since the set $\mathscr{L}'$ does not necessarily contain all those lines in $\mathscr{L}$ which join the vertices in $\mathscr{V}'$. We call such a proper subdiagram G' as a union of vertex-disjoint *subgraphs*. We may then repeat the proof of Lemma 5.4.1 to prove *for such a* G' that

$$(-T_{G'})\tilde{A}_{G'}I_{G'} = (-T_{G'_1})\tilde{A}_{G'_1}\cdots(-T_{G'_n})\tilde{A}_{G'_n}I_{G'}, \qquad (A.10)$$

where the $\tilde{A}$ are defined recursively by

$$\tilde{A}_{G'}I_{G'} = \sum_{\varnothing \subset G'' \nsubseteq G'}' (-T_{G''})\tilde{A}_{G''}I_{G'}, \qquad (A.11)$$

where the notation $\sum'_{\varnothing \subset G'' \nsubseteq G'}$ corresponds to a sum over all proper subdiagrams $\nsubseteq G'$ (not necessarily connected) with the latter as unions of vertex-disjoint *subgraphs*.

We then define an interaction Lagrangian density including counterterms as

$$\mathcal{L}_c(x) = \sum_{m=1}^{\infty} (-i)^{m-1} \sum_{G \in \mathcal{G}(m)}$$

$$\times \int dQ_G\, dK_G \exp\{i[Q_{m1} + \cdots + Q_{me(G)}]x\}\mathcal{L}_c^{\tilde{G}}(Q_G, K_G), \quad (A.12)$$

where

$$\mathcal{L}_c^{\tilde{G}}(Q_G, K_G) = (-T_G)\tilde{A}_G I_G(Q_G, K_G):\Phi_{m1}(Q_{m1})\cdots\Phi_{me(G)}(Q_{me(G)}):, \quad (A.13)$$

with $\tilde{A}_G$ defined by (A.11).

For $m = 1$, $\mathcal{G}(1)$ corresponds to the terms of the interaction Lagrangian density $\mathcal{L}_I(x)$ of the unrenormalized theory as discussed above with $(-T_G)\tilde{A}_G I_G$, formally, replaced by c_G. For $m \neq 1$, $\mathcal{G}(m)$, as also defined above, is the set of all Wick polynomials of order m arising in expressions (A.4) and (A.7) with n replaced by m. Accordingly the Wick polynomial in the sum and in (A.12) is

$$I_G:\Phi_{m1}\cdots\Phi_{me(G)}:, \qquad (A.14)$$

where G is proper and connected graph of order m. By definition $(-T_G) = 0$ if $d(G) < 0$.

In the study of renormalized Feynman amplitudes $\int R$ we do *not* have to introduce any cutoff in actual computations, since we perform the subtractions in momentum space directly on I_G to define R. To make all the subsequent analysis rigorous when working with *counterterms*, however, we have to introduce cutoffs. Evidently then with such an assumption the $\mathcal{L}_c^{\tilde{G}}$ will be cutoff dependent. Throughout this appendix we shall assume that the cutoffs are strong enough, if necessary, so that all the subsequent integrations are meaningful. We shall not, however, go into the nature of such cutoffs.

The proper and connected parts of the formal Dyson–Feynman perturbation expansion of the theory with interaction Lagrangian density $\mathscr{L}_c(x)$ is[3]

$$1 + \sum_{n=1}^{\infty} \frac{(-i)^n}{n!} \int (dx_1) \cdots (dx_n)(\mathscr{L}_c(x_1) \cdots \mathscr{L}_c(x_n))_+^{c,\,p}$$

$$= 1 + \sum_{N=1}^{\infty} (-i)^N \sum_{\substack{n=1 \\ m_1+\cdots+m_n=N}}^{N} \frac{1}{n!} \left[\prod_{i=1}^{n} \sum_{\tilde{G}_i \in \mathscr{G}(m_i)} \int dQ_{G_i}\, dK_{G_i} \right.$$

$$\times \left. \delta\!\left(\sum_{j=1}^{e(G_i)} Q_{m_i j} \right) \right] (-T_{G_1})\tilde{A}_{G_1} \cdots (-T_{G_n})\tilde{A}_{G_n} I_{G_1,\ldots,n} \prod_{i=1}^{n}$$

$$\times \left(:\prod_{j=1}^{e(G_i)} \Phi_{m_i j}(Q_{m_i j}): \right)_+^{c,\,p}, \tag{A.15}$$

where we have $m_1 + \cdots + m_n = N$ and

$$I_{G_1,\ldots,n} = I_{G_1} \cdots I_{G_n}. \tag{A.16}$$

The $\delta(\sum_{j=1}^{e(G_i)} Q_{m_i j})$ in (A.15) arises upon integration over $x_1, \ldots, x_n$. Recall that $G_{1,\ldots,n}$ is a subdiagram as a union of vertex-disjoint (connected) subgraphs.

A.2 EQUIVALENCE

Using the expansion (A.4) now applied to (A.15) we obtain ($n \geq 2$)

$$\prod_{i=1}^{n} \left(:\prod_{j=1}^{e(G_i)} \Phi_{m_i j}(Q_{m_i j}): \right)_+^{c,\,p}$$

$$= \sum_{\tilde{G}_0} \hat{I}_{G_0}(\prod \delta_{G_0}(Q_{..} - Q_{..})):\Phi^f_{1G_0}(Q_{1G_0}) \ldots \Phi^f_{kG_0}(Q_{kG_0}): . \tag{A.17}$$

Upon replacing this expression in (A.15) and carrying out intermediate integrations to make use of the delta-functions in (A.17) we obtain for each fixed $N \geq 2, n \geq 2,[4]$ $N \geq n$.

$$\sum_{m_1+\ldots+m_n=N} \prod_{i=1}^{n} \sum_{\tilde{G}_i \in \mathscr{G}(m_i)} \sum_{\tilde{G}_0 \in \mathscr{G}(n)} \int dQ_G\, dK_G\, \delta\!\left(\sum_{i=1}^{k} Q_{iG} \right)$$

$$\times (-T_{G_1})\tilde{A}_{G_1} \cdots (-T_{G_n})\tilde{A}_{G_n} I_{G_1,\ldots,n} I'_{G_0}:\Phi^f_{1G}(Q_{1G}) \cdots \Phi^f_{kG}(Q_{kG}): , \tag{A.18}$$

[3] It is, of course, expected that the reader is familiar with the left-hand side of this expression (cf. Bjorken and Drell, 1965).

[4] The conditions $N \geq 2, n \geq 2$ correspond to contractions between different Wick-ordered products.

where the overall momentum conserving delta-function arises as a consequence of the fact that the (proper) subdiagram G, defined by

$$I_G = I_{G_1,\ldots,n}I'_{G_0},\tag{A.19}$$

must be connected, by definition. Alternatively, I'_{G_0} may be defined by

$$I'_{G_0} = I_{G/G_1,\ldots,n}\tag{A.20}$$

i.e., by replacing each of the $I_{G_1},\ldots,I_{G_n}$ in the expression I_G by unity. We recall that the $G_1,\ldots,G_n$ are connected, having no vertices in common. K_G in (A.18) includes, in addition to the components in $K_{G_1},\ldots,K_{G_n}$, the internal momentum components carried by the lines in $\hat{I}_{G_0}$. Also,

$$Q_G = (Q^0_{1G},\ldots,Q^3_{kG}).\tag{A.21}$$

We note that, as contractions between fields occur in pairs, $\hat{I}_{G_0}$ [see (A.17)] "contains" no vertices, i.e.,

$$\text{number of vertices of } G = \sum_{i=1}^{n} (\text{number of vertices of } G_i)$$

$$= \sum_{i=1}^{n} m_i = N,\tag{A.22}$$

and hence G has precisely N vertices.

We examine the expression (A.18). For $n = 2$, $m_1 + m_2 = N$, we have

$$\sum_{m_1+m_2=N} \sum_{\tilde{G}_1\in\mathscr{G}(m_1)} \sum_{\tilde{G}_2\in\mathscr{G}(m_2)} \sum_{\tilde{G}_0\in\mathscr{G}(2)} \int dQ_G\, dK_G\, \delta\left(\sum_{i=1}^{k} Q_{iG}\right)$$

$$\times (-T_{G_1})\tilde{A}_{G_1}(-T_{G_2})\tilde{A}_{G_2}I_{G_{12}}I'_{G_0}:\prod_{i=1}^{k} \Phi^f_{iG}(Q_{iG}):.\tag{A.23}$$

By using the symmetry over the summations $\sum_{\tilde{G}_1\in\mathscr{G}(m_1)}\sum_{\tilde{G}_2\in\mathscr{G}(m_2)}$, the fact that $m_1 + m_2 = N$, and using the identity in (A.10), we may rewrite (A.23) as

$$2! \sum_{G\in\mathscr{G}(N)} \int dQ_G\, dK_G\, \delta\left(\sum_{i=1}^{k} Q_{iG}\right) \sum_{\varnothing\subset G'\nsubseteq G}^{2} (-T_{G'})A_G:\prod_{i=1}^{k} \Phi^f_{iG}(Q_{iG}):.\tag{A.24}$$

The summation $\sum_{\varnothing\subset G'\nsubseteq G}^{2}$ is over all subdiagrams $G' \nsubseteq G$ such that, for $G' \neq \varnothing$, each G' is a union of *at most two* (i.e., 1 or 2) vertex-disjoint (connected) subgraphs and $I_{G/G'}$ corresponds to a subdiagram with exactly two vertices. The situation $G' = \varnothing$ occurs when $N = 2$. In this latter case $m_1 = m_2 = 1$ and the $-T_{G_i}\tilde{A}_{G_i}I_{G_i}$ are simply replaced by c_{G_i}, respectively; I_G then has also associated with it exactly two vertices. In general, if, say, $m_1 = 1$, then $m_2 = N - 1$ and $-T_{G_1}\tilde{A}_{G_1}I_{G_1}$ is replaced by c_{G_1}.

For $n = 3, m_1 + m_2 + m_3 = N$, we have

$$
\sum_{m_1+m_2+m_3=N} \sum_{\tilde{G}\in\mathscr{G}(m_1)} \sum_{\tilde{G}\in\mathscr{G}(m_2)} \sum_{\tilde{G}\in\mathscr{G}(m_3)} \sum_{\tilde{G}_0\in\mathscr{G}(3)} \int dQ_G\, dK_G
$$
$$
\times \delta\left(\sum_{i=1}^{k} Q_{iG}\right) (-T_{G_1})\tilde{A}_{G_1}(-T_{G_2})\tilde{A}_{G_2}(-T_{G_3})\tilde{A}_{G_3} I_{G_{123}} I'_{G_0} : \prod_{1=1}^{k} \Phi^{\mathrm{f}}_{iG}(Q_{iG}) :.
$$

$$(A.25)$$

As in (A.24) we may write this expression

$$
3!\sum_{\tilde{G}\in\mathscr{G}(N)} \int dQ_G\, dK_G\, \delta\left(\sum_{i=1}^{k} Q_{iG}\right) \sum_{\varnothing\subset G'\nsubseteq G}^{3} (-T_{G'})\tilde{A}_{G'} I_G : \prod_{i=1}^{k} \Phi^{\mathrm{f}}_{iG}(Q_{iG}) :, \quad (A.26)
$$

The summation $\sum_{\varnothing\subset G'\nsubseteq G}^{3}$, for $G' \neq \varnothing$, is over all subdiagrams $G' \nsubseteq G$ such that each G' is a union of *at most three* (i.e., 1 or 2 or 3) vertex-disjoint (connected) subgraphs. Again the situation $G' = \varnothing$ arises when $N = 3$. $I_{G/G'}$ corresponds to a subdiagram with exactly three vertices.

In general, we may then write for (A.18)

$$
n!\sum_{\tilde{G}\in\mathscr{G}(N)} \int dQ_G\, dK_G\, \delta\left(\sum_{i=1}^{k} Q_{iG}\right) \sum_{\varnothing\subset G'\nsubseteq G}^{n} (-T_{G'})\tilde{A}_{G'} I_G : \prod_{i=1}^{k} \Phi^{\mathrm{f}}_{iG}(Q_{iG}) :, \quad (A.27)
$$

where $\sum_{\varnothing\subset G'\nsubseteq G}^{n}$, for $G' \neq \varnothing$, denotes a sum over all $G' \nsubseteq G$ such that each G' is a union of *at most n* (i.e., 1 or 2 or ... or n) vertex-disjoint subgraphs. The situation $G' = \varnothing$ occurs when $N = n$. $I_{G/G'}$ corresponds to a sub-diagram with precisely n vertices.[5] For $G' = \varnothing$, i.e., when $N = n$, $m_1 = \cdots = m_n = 1$, I_G, again, corresponds to a subdiagram with $N = n$ vertices.

Upon summing over n and N we obtain for (A.15), from (A.27), the expression

$$
1 + \sum_{N=1}^{\infty} (-i)^N \sum_{G\in\mathscr{G}(N)} \int dQ_G\, dK_G\, \delta\left(\sum_{i=1}^{k} Q_{iG}\right) \sum_{\varnothing\subset G'\subset G}' (-T_{G'})\tilde{A}_{G'} I_G
$$
$$
\times : \prod_{i=1}^{k} \Phi^{\mathrm{f}}_{iG}(Q_{iG}) :.
$$

$$(A.28)$$

For $N = 1$, $\sum_{\varnothing\subset G'\subset G}'$ is to be replaced by one. For $G' \neq \varnothing$ and $G' \neq G$, the summation $\sum_{\varnothing\neq G'\nsubseteq G}'$ is over all subdiagrams G' such that each G' is a union of *at most N* (i.e., 2 or 3 or ... or N) vertex-disjoint (connected) subgraphs and $I_{G/G'}$ corresponds to a subdiagram with *at most N* (i.e., 2 or 3 or ... or N) vertices. The analytical expression for these latter vertices in $I_{G/G'}$ is unity. By definition $-T_{G'}\tilde{A}_{G'}$ is to be replaced by one if $G' = \varnothing$; and for $G' \neq \varnothing$, $(-T_{G'}) = 0$ if $d(G') < 0$. The $G \in \mathscr{G}(N)$ have exactly N vertices, by definition.

---

[5] Note that the analytical expression for these n vertices in $I_{G/G'}$ is unity, by definition.

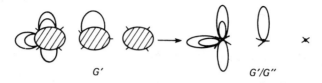

G' G'/G''

Fig. A.1 A subdiagram G' such that $G'' \nsubseteq G'$ with $d(G'') \geq 0$. The subdiagram G'' is depicted by the shaded area in the figure. G'' has the same number of vertices and the same number of connected parts as G', but the latter has one or more lines each of which joining two vertices within the same connected parts of G'. The subdiagram on the right-hand-side depicts G'/G'', showing, in particular, that the number of vertices of G'/G'' coincides with the number of connected parts of G' (or of G'').

Consider a subdiagram G' in (A.28). Suppose that G' contains a sub-diagram G'': $G'' \nsubseteq G'$ such that G'' has the same number of vertices as G' and the latter has one or more lines each of which joining two vertices within the same connected parts of G' (see Fig. A.1) and $d(G'') \geq 0$. Note that G'' has the same number of connected parts as G'. Consider the following expression:

$$[(-T_{G'})(-T_{G''}) + (-T_{G''})] \cdots I_{G'} = (1 - T_{G'})(-T_{G''}) \cdots I_{G'}. \quad (A.29)$$

If G' is a union of, say, n vertex-disjoint subgraphs, then G'' is not a union of vertex-disjoint subgraphs, and hence both terms in (A.29) do not occur, in the summation in $\sum'$ in (A.28). Integrating (A.29) over internal momenta of G'' we obtain an expression

$$(1 - T_{G'})I_{G'/G''}P_{G''}[Q_{G''}], \quad (A.30)$$

where $P_{G''}[Q_{G''}]$ is a polynomial of degree $\leq d(G'')$ in the extremal momenta of G'' by the definition of the $(-T_{G''})$ operation. The external momenta of G'' are now to be written as linear combinations of the external variables of G' and the internal variables in G'/G''. Integrating (A.30) over the internal variables of G'/G'' we obtain an expression of the form.

$$(1 - T_{G'})P_{G'}[Q_{G'}], \quad (A.31)$$

where $P_{G'}[Q_{G'}]$ is a polynomial of degree $\leq d(G')$ in the external momenta of G'. Using a well-known integral representation of the remainder of a Taylor operation[6] we obtain for (A.31).

$$\frac{1}{d(G')!} \int_0^1 dx(1-x)^{d(G')} \left(\frac{\partial}{\partial x}\right)^{d(G')+1} P_{G'}[xQ_{G'}] = 0.^{[7]} \quad (A.32)$$

---

[6] This integral representation is easily proved by induction in $d(G')$.

[7] The vanishing property of expressions like in (A.31) are well known (cf. Hepp, 1971, p. 483).

Conversely suppose that G'' is any proper subdiagram, with $d(G'') \geq 0$. We add to it some lines, each of which joins two vertices *within* the same connected parts of G'', and we form a subdiagram G' with the same number of vertices as G''. Then, by definition of the dimensionality of a subdiagram, we note that $d(G') \geq d(G'')$. Accordingly for *any* proper subdiagram G'' with $d(G'') \geq 0$, with n connected parts, but which is not a union of vertex-disjoint (n connected) subgraphs we can *always* find a proper subdiagram G' with $d(G') \geq 0$, with n connected parts, having the same vertices as G'' but with one or more lines, and a similar expression as the one in (A.29) involving a pair of terms may be written down which cancel out, as seen in (A.32). [Note that such a subdiagram G' may or may not itself be the union of vertex-disjoint (n connected) subgraphs.] As a result of cancellation between such pairs, we may then rigorously include subtractions over proper subdiagrams in the sum $\sum'_{\varnothing \subset G' \subset G}$ not necessarily being vertex-disjoint subgraphs. Hence we may simply replace the sum $\sum'_{\varnothing \subset G' \subset G}$ in (A.28) by a sum over *all* proper subdiagrams G' with $d(G') \geq 0$, and we may replace the $\tilde{A}_{G'}$, defined through (A.11), by the $A_{G'}$ introduced in Chapter 5 involving general proper subdiagrams. The additional sets of subdiagrams thus introduced involving one or more subdiagrams which are not the union of vertex-disjoint subgraphs *will not contribute* to the final expression for $\sum_{\varnothing \subset G' \nsubseteq G} (-T_{G'}) A_{G'} I_G$.

Accordingly we may rewrite (A.28)

$$1 + \sum_{N=1}^{\infty} (-i)^N \sum_{\tilde{G} \in \mathscr{G}(N)} \int dQ_G \, dK_G \, \delta\left(\sum_{i=1}^{k} Q_{iG}\right) \sum_{\phi \subset G' \subset G} (-T_{G'}) A_{G'} I_G$$

$$\times : \prod_{i=1}^{k} \Phi_{iG}^{\mathrm{f}}(Q_{iG}):, \tag{A.33}$$

$\sum_{\varnothing \nsubseteq G' \nsubseteq G}$ represents a sum over all proper, but not necessarily connected, subdiagrams of G with $(-T_{G'}) = 0$ if $d(G') < 0$. Upon using the basic property [see (5.2.22) and (5.4.2)].

$$R = \sum_{\varnothing \subset G' \subset G} (-T_{G'}) A_{G'} I_G = (1 - T_G) A_G I_G \equiv R_G, \tag{A.34}$$

we obtain for (A.33)

$$1 + \sum_{N=1}^{\infty} (-i)^N \sum_{\tilde{G} \in \mathscr{G}(N)} \int dQ_G \, dK_G \, \delta\left(\sum_{i=1}^{k} Q_{iG}\right) R_G : \prod_{i=1}^{k} \Phi_{iG}^{\mathrm{f}}(Q_{iG}):, \tag{A.35}$$

where for $N = 1$, R_G is to be replaced by c_G. Expression (A.35) is the proper and connected part of the renormalized perturbation expansion with R_G

(the renormalized Feynman integrands) introduced in Chapter 5 replacing I_G (the unrenormalized Feynman integrands).[8]

This completes the demonstration of the equivalence of the subtractions and the counterterms formalisms. We shall not, however, go into the philosophical implications of counterterms.

The formal local character of the interaction Lagrangian density $\mathscr{L}_c$ including the counterterms is readily seen. We may rewrite (A.12) and (A.13) as

$$\mathscr{L}_c(x) = \sum_{m=1}^{\infty} (-i)^{m-1} \sum_{\tilde{G} \in \mathscr{G}(m)} \int dQ_G \exp[i \sum_{j=1}^{e(G)} Q_{m'j}x] P_G[Q_G]$$
$$\times \, : \prod_{i=1}^{e(G)} \Phi_{mi}(Q_{mi}): , \tag{A.36}$$

where for $m > 1$, $P_G[Q_G]$ is a polynomial [of degree $\leq d(G)$] arising as a result of the Taylor operations in $(-T_G)A_G$. The terms in the expression (A.36) are well defined with cutoffs. Finally, upon integrating over Q_G we obtain for $\mathscr{L}_c(x)$

$$\mathscr{L}_c(x) = \sum_{m=1}^{\infty} (-i)^{m-1} \sum_{\tilde{G} \in \mathscr{G}(m)} P_G\left[\cdots \frac{\partial}{\partial x_i} \cdots\right] \tag{A.37}$$
$$\times \, :\Phi_{m1}(x_1) \cdots \Phi_{me(G)}(x_{e(G)}): |_{x_1 = \cdots = x_{e(G)} = x},$$

exhibiting the formal local character of $\mathscr{L}_c(x)$.

NOTES

There is a long history associated with the equivalence of subtractions and counterterms. For a partial list of references see Schwinger (1958), Gupta (1951), Takeda (1952), Matthews and Salam (1954), Bogoliubov and Shirkov (1959), Bjorken and Drell (1965), Jauch and Rohrlich (1976), Speer (1969), Zav'yalov and Anikin (1976), and Manoukian (1979b). The appendix is based on Manoukian (1979b).

---

[8] We note that with cutoffs adopted, as mentioned earlier, to make all the intermediate steps leading to (A.35) justifiable, R_G will depend on the cutoffs. More precisely, then, the renormalized Feynman integrand will coincide with the cutoff-independent part of R_G. In any case, if mass regulators are suitably introduced as cutoffs, for example, then we may apply our generalized decoupling theorem (Theorem 6.8.1) to conclude formally that the part of the amplitude, depending on these mass regulators, will vanish when the limit of infinite mass regulators is taken.

REFERENCES

Abers, E. S., and Lee, B. W. (1973). Gauge theories. *Phys. Rep.* **9C**, 1.

Adler, S. (1969). Axial-vector in spinor electrodynamics. *Phys. Rev.* **177**, 2426.

Akulov, V. P., Volkov, D. V., and Soroka, V. A. (1975). Gauge fields on superspaces with different holonomy groups. *JETP Lett.* **22**, 187.

Ambjørn, J. (1979). On the decoupling of massive particles in field theory. *Comm. Math. Phys.* **67**, 109.

Anderson, P. W. (1963). Plasmons, gauge invariance and mass. *Phys. Rev.* **130**, 439.

Anikin, S. A., and Polivanov, M. K. (1974). Proof of the Bogoliubov-Parasiuk theorem for nonscalar case. *Theoret. Math. and Phys.* **21**, 1058.

Anikin, S. A., Zav'yalov, O. I., and Polivanov, M. K. (1973). Simple proof of the Bogoliubov–Parasiuk theorem. *Theoret. Math. and Phys.* **17**, 1082.

Appelquist, T., and Carrazzonne, J. (1975). Infrared singularities and massive fields. *Phys. Rev.* **D 11**, 2856.

Arnowitt, R., Nath, P., and Zumino, B. (1975). Superfield densities and action principle in curved superspace. *Phys. Lett B* **56**, 81.

Atiyah, M. (1970). Resolution of singularities and division of distributions. *Comm. Pure Appl. Math.* **23**, 145.

Becchi, C., Rouet, A., and Stora, R. (1975). Renormalization of the Abelian Higgs-Kibble model. *Comm. Math. Phys.* **42**, 127.

Becchi, C., Rouet, A., and Stora, R. (1976). Renormalization of gauge theories. *Ann. Phys.* **98**, 287.

Bég, M. A., and Sirlin, A. (1974). Gauge theories of weak interactions. *Annual Rev. Nuclear Sci.* **24**, 379.

Bell, J. S., and Jackiw, R. (1969). A PCAC puzzle: $\pi^0 \to \gamma\gamma$ in the σ-model. *Nuovo Cimento A* **60A**, 47.

Berberian, S. K. (1965). "Measure and Integration." Macmillan, New York.

Bergère, M. C., and Zuber, J. B. (1974). Renormalization of Feynman amplitudes and parametric integral representations. *Comm. Math. Phys.* **35**, 113.

Bergère, M. C., de Calan, C., and Malbouisson, A. P. C. (1978). A theorem on asymptotic expansion of Feynman amplitudes. *Comm. Math. Phys.* **62**, 137.

Bernard, C., and Duncan, A. (1975). Lorentz covariance and Matthew's theorem for derivative-coupled field theories. *Phys. Rev. D* **11**, 848.

Bernstein, I. N., and Gel'fand, S. I. (1969). Meromorphic property of the functions P^λ. *Functional Anal. Appl.* **3**, 68.

Bernstein, J. (1974). Spontaneous symmetry breaking, gauge theories, the Higgs mechanism and all that. *Rev. Modern Phys.* **46**, 7.

Bethe, H. A. (1947). The electromagnetic shift of energy levels. *Phys. Rev.* **72**, 339.

Bialynicki-Birula, I. (1962). On the gauge covariance of quantum electrodynamics. *J. Math. Phys.* **3**, 1094.

Bjorken, J. D. (1972). *In* "The Proceedings of the XVI International Conference on High-Energy Physics at Chicago-Batavia" (J. D. Jackson and A. Roberts, eds.), Vol. 2, p. 304. Natl. Accelerator Lab., Batavia, Illinois.

Bjorken, J. D., and Drell, S. (1965). "Relativistic Quantum Fields." McGraw-Hill, New York.

Bogoliubov, N. N., and Parasiuk, O. S. (1957). Über die Multiplikation der Kansalfunktionen in der Quantentheorie der Felder. *Acta Math.* **97**, 227.

Bogoliubov, N. N., and Shirkov, D. V. (1959). "Introduction to the Theory of Quantized Fields." Wiley (Interscience), New York.

Born, M., Heisenberg, W., and Jordan, P. (1926). Zur Quantenmechanik II. *Z. Phys.* **35**, 557.

Bouchiat, C., Iliopoulos, J., and Mayer, P. (1972). An anomaly-free version of Weinberg's model. *Phys. Lett. B* **38B**, 519.

Brandt, R., and Preparata, G. (1971). Operator product expansions near the light cone. *Nuclear Phys. B* **27**, 541.

Brink, L., Gell-Mann, M., Ramond, P., and Schwarz, J. H. (1978). Supergravity as geometry of superspace. *Phys. Lett. B* **74**, 336.

Brodsky, S. J., and Drell, S. D. (1970). The present status of quantum electrodynamics. *Annual Rev. Nuclear Sci.* **20**, 147.

Buras, A. J. (1980). Asymptotic freedom in deep inelastic processes in the leading order and beyond. *Rev. Modern Phys.* **52**, No. 1, 199.

Buras, J., Ellis, J., Gaillard, M. K., and Nanopoulos, D. V. (1978). Aspects of the grand unification of strong, weak and electromagnetic interactions. *Nuclear Phys. B* **135**, 66.

Caianiello, L. R., Guerra, R. F., and Marinaro, M. (1969). Form-invariant renormalization. *Nuovo Cimento A* **60A**, 713.

Callan, G. C. (1970). Broken scale invariance in scalar field theory. *Phys. Rev. D* **2**, 1541.

Collins, E. G. J., Wilczek, F., and Zee, A. (1978). Low-energy manifestations of heavy particles: Application to the neutral current. *Phys. Rev. D* **18**, 242.

Costa, G., and Tonin, M. (1975). Renormalization of non-Abelian gauge theories. *Riv. Nuovo Cimento* **5**, No. 1, 29.

Deser, S., and Zumino, B. (1976). Consistent supergravity. *Phys. Lett. B* **62**, 335.

deWitt, B. (1964). Theory of radiative corrections for non-Abelian gauge fields. *Phys. Rev. Lett.* **12**, 742.

deWitt, B. (1967). Quantum theory of gravity. II. The manifestly covariant gauge. *Phys. Rev.* **162**, 1195.

Dirac, P. A. M. (1927). The quantum theory of the emission and absorption of radiation. *Proc. Roy. Soc. London Ser. A* **114**, 243.

Dyson, F. J. (1949a). The radiation theories of Tomonaga, Schwinger and Feynman. *Phys. Rev.* **75**, 486.

Dyson, F. J. (1949b). The *S*-matrix in quantum electrodynamics. *Phys. Rev.* **75**, 1736.

Englert, F., and Brout, R. (1964). Broken symmetry and the mass of gauge vector mesons. *Phys. Rev. Lett.* **13**, 321.

Fadeev, L. D., and Popov, V. N. (1967). Feynman diagrams for the Yang–Mills field. *Phys. Lett. B* **25B**, 29.

Feynman, R. P. (1949a). The theory of positrons. *Phys. Rev.* **76**, 749.

Feynman, R. P. (1949b). Space–time approach to quantum electrodynamics. *Phys. Rev.* **76**, 769.

Feynman, R. P. (1950). Mathematical formulation of the quantum theory of electromagnetic interaction. *Phys. Rev.* **80**, 440.

Feynman, R. P. (1963). Quantum theory of gravitation. *Acta Phys. Polon.* **24**, 697.

Feynman, R. P. (1972). "The Development of the Space–Time View of Quantum Electrodynamics," Nobel Lect. Phys. 1965, p. 155. Elsevier, Amsterdam.

Fink, J. P. (1967). Asymptotic behavior of Feynman integrals, Doctoral Dissertation, Stanford University, Stanford, California.

Fink, J. P. (1968). Asymptotic estimates of Feynman integrals. *J. Math. Phys.* **9**, 1389.

Fradkin, E. S., and Tyutin, I. V. (1969). Feynman rules for the massless Yang–Mills field, renormalizability of the theory of the massive Yang–Mills field. *Phys. Lett. B* **30B**, 562.

Fradkin, E. S., and Tyutin, I. V. (1970). *S*-matrix for Yang–Mills and gravitational fields. *Phys. Rev. D* **2**, 2841.

Fradkin, E. S., and Tyutin, I. V. (1974). Renormalizable theory of massive vector particles. *Riv. Nuovo Cimento* **4**, No. 1, 1.

Freedman, D. Z., van Nieuwenhuizen, P., and Ferrara, S. (1976). Progress toward a theory of supergravity. *Phys. Rev. D* **13**, 3214.

Frishman, Y. (1970). Scale invariance and current commutators near the light cone. *Phys. Rev. Lett.* **25**, 966.

Gell-Mann, M., and Low, F. E. (1954). Quantum electrodynamics at small distances. *Phys. Rev.* **95**, 1300.

Georgi, H., and Glashow, S. L. (1974). Unity of all elementary particle forces. *Phys. Rev. Lett.* **32**, 438.

Georgi, H., Quinn, H. R., and Weinberg, S. (1974). Hierarchy of interactions in unified gauge theories. *Phys. Rev. Lett.* **33**, 451.

Glashow, S. L. (1959). *The vector meson in elementary particle decays.* Ph.D. Thesis, Harvard University, Cambridge, Massachusetts.

Glashow, S. L. (1961). Partial-symmetries of weak interactions. *Nuclear Phys.* **22**, 579.

Glashow, S. L. (1980). Toward a unified theory: Threads in a tapestry, Novel Lectures in Physics 1979. *Rev. Mod. Phys.* **52**, No. 3, 539.

Goldstone, J. (1961). Field theories with "superconductor" solutions. *Nuovo Cimento* **19**, 154.

Goldstone, J., Salam, A., and Weinberg, S. (1962). Broken symmetries. *Phys. Rev.* **127**, 965.

Gross, D. J., and Jackiw, R. (1972). Effect of anomalies on quasi-renormalizable theories. *Phys. Rev. D* **6**, 477.

Gross, D. J., and Wilczek, F. (1973). Ultraviolet behavior of non-Abelian gauge theories. *Phys. Rev. Lett.* **30**, 1343.

Gupta, N. (1951). On the elimination of divergencies from quantum electrodynamics. *Proc. Phys. Soc. London Ser. A* **64**, 426.

Guralnik, G. S., Hagen, C. R., and Kibble, T. W. (1964). Broken symmetry and the mass of gauge vector mesons. *Phys. Rev. Lett.* **13**, 321.

Hänsch, T. W., Schawlow, A. L., and Series, G. W. (1979). The spectrum of atomic hydrogen. *Sci. Am.* **240**, No. 3, 94.

Hagiwara, T., and Nakazawa, N. (1981). Perturbative effect of heavy particles in an effective-Lagrangian approach. *Phys. Rev. D* **23**, 959.

Hahn, Y., and Zimmermann, W. (1968). An elementary proof of Dyson's power counting theorem. *Comm. Math. Phys.* **10**, 330.

Halmos, P. R. (1974a). "Measure Theory." Springer-Verlag, Berlin and New York.

Halmos, P. R. (1947b). "Finite Dimensional Vector Spaces." Springer-Verlag, Berlin and New York.

Heisenberg, W. (1938). Über die in der Theorie der Elementariteilchen auftretende Universelle Länge. *Ann. Phys.* (5) **32**, 20.

Heisenberg, W., and Pauli, W. (1929). Zur Quantendynamik der Wellenfelder. *Z. Phys.* **56**, 1.

Heisenberg, W., and Pauli, W. (1930). Zur Quantentheorie der Wallenfelder. II. *Z. Phys.* **59**, 168.

Hepp, K. (1966). Proof of the Bogoliubov–Parasiuk theorem of renormalization. *Comm. Math. Phys.* **2**, 301.

Hepp, K. (1971). *In* "Statistical Mechanics and Quantum Field Theory, 1970 Les Houches Lectures" (C. deWitt and R. Stora, eds.), p. 429. Gordon & Breach, New York.

Higgs, P. W. (1964). Broken symmetries, massless particles and gauge fields. *Phys. Lett.* **12**, 132.

Higgs, P. W. (1966). Spontaneous symmetry breakdown without massless bosons. *Phys. Rev.* **145**, 1156.

Hironaka, H. (1964). Resolution of singularities of an algebraic variety over a field of characteristic zero. *Ann. Math.* **79**, 109.

Hoffman, K. (1975). "Analysis in Euclidean Space." Prentice-Hall, Englewood Cliffs, New Jersey.

Isaacson, E., and Keller, H. B. (1966). "Analysis of Numerical Methods." Wiley, New York.

Jauch, J. M., and Rohrlich, F. (1976). "Theory of Photons and Electrons." Springer-Verlag, Berlin and New York.

Kanazawa, S., and Tomonaga, S. (1948). Corrections due to the reaction of "Cohesive force field" to the elastic scattering of an electron. I. *Progr. Theoret. Phys.* **3**, 276.

Kazama, Y., and Yao, Y.-P. (1979). Effects of heavy particles through factorization and renormalization group. *Phys. Rev. Lett.* **43**, 1562.

Kibble, T. W. B. (1967). Symmetry breaking in non-Abelian gauge theories. *Phys. Rev.* **155**, 1554.

Koba, Z., and Tomonaga, S. (1948). On the reaction of collision processes. I. *Progr. Theoret. Phys.* **3**, 290.

Koba, Z., Tati, T., and Tomonaga, S. (1947). On a relativistically invariant formulation of the quantum theory of wave fields. II. *Prog. Theoret. Phys.* **2**, 101 (1947).

Kuo, P. K. and Yennie, D. R. (1969). Renormalization theory. *Ann. Physics* **51**, No. 3, 496.

Lam, C. S. (1965). Feynman rules and Feynman integrals for system with higher spin fields. *Nuovo Cimento* **38**, 1755.

Lamb, W. E., and Retherford, R. C. (1947). Fine structure of the hydrogen atom by a microwave method. *Phys. Rev.* **72**, 241.

Lautrup, B. E., Peterman, A., and deRafael, E. (1972). Recent developments in the comparison between theory and experiments in quantum electrodynamics. *Phys. Rep.* **3C**, 193.

Lee, B. W. (1972). Renormalizable massive vector-meson theory, perturbation theory of the Higgs phenomenon. *Phys. Rev. D* **5**, 823.

Lee, B. W. (1973). Transformation properties of proper vertices in gauge theories. *Phys. Lett. B* **46B**, 214.

Lee, B. W., and Zinn-Justin, J. (1972a). Spontaneously broken gauge symmetries. I. Preliminaries. *Phys. Rev. D* **5**, 3121.

Lee, B. W., and Zinn-Justin, J. (1972b). Spontaneously broken gauge symmetries. II. Perturbation theory and renormalization. *Phys. Rev. D* **5**, 3137.

Lee, B. W., and Zinn-Justin, J. (1972c). Spontaneously broken gauge symmetries. III. Equivalence. *Phys. Rev. D* **5**, 3155.

Lee, T. D., and Yang, C. N. (1962). Theory of charged vector mesons interacting with the electromagnetic field. *Phys. Rev.* **128**, 885.

Lowenstein, J. H., and Speer, E. (1976). Distributional limits of renormalized Feynman integrals with zero-mass denominators. *Comm. Math. Phys.* **47**, 43.

Lukierski, J. (1963). Gauge transformations in quantum field theory. *Nuovo Cimento* **24**, 561.

Mandelstam, S. (1968a). Feynman rules for electromagnetic and Yang–Mills fields from the gauge-independent field theoretic formalism. *Phys. Rev.* **175**, 1580.

Mandelstam, S. (1968b). Feynman rules for the gravitational field from the coordinate-independent field-theoretic formalism. *Phys. Rev.* **175**, 1604.

Manoukian, E. B. (1976). Generalization and improvement of the Dyson–Salam renormalization scheme and equivalence with other schemes. *Phys. Rev. D* **14**, 966, 2202 (E).

Manoukian, E. B. (1977). Convergence of the generalized and improved Dyson-Salam renormalization scheme. *Phys. Rev. D* **15**, 535; **25**, 1157(E) (1982).

Manoukian, E. B. (1978). High-energy behavior of renormalized Feynman amplitudes. *J. Math. Phys.* **19**, 917.

Manoukian, E. B. (1979a). Low-energy behavior of massive field theories. *Acta Phys. Austriaca* **51**, 311.

Manoukian, E. B. (1979b). Subtractions vs. counterterms. *Nuovo Cimento A* **53A**, 345.

Manoukian, E. B. (1980a). Dimensional analysis of subtracted out Feynman integrands. *J. Phys. A* **13**, 2903.

Manoukian, E. B. (1980b). Zero-mass behavior of Feynman amplitudes. *J. Math. Phys.* **21**, 1218.

Manoukian, E. B. (1980c). High-energy behavior of renormalized Feynman amplitudes. II. *J. Math. Phys.* **21**, 1662.

Manoukian, E. B. (1980d). Low-energy analysis of subtracted Feynman amplitudes. *Nuovo Cimento A* **57A**, 377.

Manoukian, E. B. (1981a). On the decoupling theorem in quantum field theory. *J. Math. Phys.* **22**, 572.

Manoukian, E. B. (1981b). Dimensional analysis of subtracted-out Feynman integrands. II. *J. Phys. G* **7**, 1149.

Manoukian, E. B. (1981c). General asymptotic behavior in quantum field theory. *J. Phys. G* **7**, 1159.

Manoukian, E. B. (1981d). Generalized decoupling theorem in quantum field theory. *J. Math. Phys.* **22**, 2258.

Manoukian, E. B. (1982a). Class B_n-functions: Convergence of subtractions. *Nuovo Cimento A* **67A**, 101.

Manoukian, E. B. (1982b). Class B_n-function property of Feynman integrals. *J. Phys. G* **8**, 599.

Marciano, W., and Pagels, H. (1978). Quantum chromodynamics. *Phys. Rep.* **36C**, No. 3, 136.

Matthews, P. T. (1949). The application of Dyson's methods to meson interactions. *Phys. Rev.* **76**, 684.

Matthews, P. T., and Salam, A. (1954). Renormalization. *Phys. Rev.* **94**, 185.

Munroe, M. E. (1959). "Introduction to Measure and Integration." Addison-Wesley, Reading, Massachusetts.

Nishijima, K. (1969). "Fields and Particles: Field Theory and Dispersion Relations." Benjamin, New York.

Oppenheimer, J. R. (1930). Note on the theory of the interaction of field and matter. *Phys. Rev.* **35**, 461.

Ovrut, B. A., and Schnitzer, H. J. (1981). Gauge theories with minimal subtractions and the decoupling theorem. *Nuclear Phys. B* **179**, 381.

Parasiuk, O. S. (1960). On Bogoliubov's theory of R-operation. *Ukrainian Math. J.* **12**, 287.

Pati, J. C., and Salam, A. (1973). Unified Lepton–Hadron symmetry and a gauge theory of the basic interactions. *Phys. Rev. D* **8**, 1240.

Peierls, R. E. (1973). *The development of quantum field theory*. In "The Physicist's Conception of Nature" (J. Mehra, ed.), p. 370. Reidel Publ., Dordrecht, Netherlands.

Poggio, E. C., Quinn, H. R., and Zuber, J. B. (1977). Renormalization group and the infrared behavior of gauge theories. *Phys. Rev. D* **15**, 1630.

Politzer, H. D. (1973). Reliable perturbation results for strong interactions. *Phys. Rev. Lett.* **30**, 1346.

Politzer, H. D. (1974). Asymptotic freedom: An approach to strong interactions. *Phys. Rep.* **14C**, 130.

Royden, H. L. (1963). "Real Analysis." Macmillan, New York.

Rudin, W. (1964). "Principles of Mathematical Analysis." McGraw-Hill, New York.

Rudin, W. (1966). "Real and Complex Analysis." McGraw-Hill, New York.

Salam, A. (1951a). Overlapping divergences and the S-matrix. *Phys. Rev.* **82**, 217.

Salam, A. (1951b). Divergent integrals in renormalizable field theories. *Phys. Rev.* **84**, 426.

Salam, A. (1952). Renormalized S-matrix for scalar electrodynamics. *Phys. Rev.* **86**, 731.

Salam, A. (1968). *Weak and electromagnetic interactions*. In "Elementary Particle Physics" (N. Svartholm, ed.), Nobel Symp. 8, p. 367. Wiley, New York.

Salam, A. (1980). Gauge unification of fundamental forces, Nobel Lectures in Physics 1979. *Rev. Modern Phys.* **52**, No. 3, 525.

Salam, A., and Strathdee, J. (1972). A renormalizable gauge model of Lepton interactions. *Nuovo Cimento A* **11A**, 397.

Salam, A., and Strathdee, J. (1978). Supersymmetry and superfields. *Fortschr. Phys.* **26**, 57.

Schwartz, L. (1978). "Theorie des distributions." Hermann, Paris.

Schweber, S. (1961). "An Introduction to Relativistic Quantum Field Theory." Harper & Row, New York.

Schwinger, J. (1948a). On quantum electrodynamics and the magnetic moment of the electron. *Phys. Rev.* **73**, 416.

Schwinger, J. (1948b). Quantum electrodynamics. I. A covariant formulation. *Phys. Rev.* **74**, 1439.

Schwinger, J. (1949a). Quantum electrodynamics. II. Vacuum polarization and self-energy. *Phys. Rev.* **75**, 651.

Schwinger, J. (1949b). Quantum electrodynamics. III. The electromagnetic properties of the electron-radiative corrections to scattering. *Phys. Rev.* **76**, 790.

Schwinger, J. (1951a). On gauge invariance and vacuum polarization. *Phys. Rev.* **82**, 664.

Schwinger, J. (1951b). The theory of quantized fields. I. *Phys. Rev.* **82**, 914.

Schwinger, J. (1957). A theory of the fundamental interactions. *Ann. Physics.* **2**, 407.

Schwinger, J. (1958). "Quantum Electrodynamics." Dover, New York.

Schwinger, J. (1972). "Relativistic Quantum Field Theory." Nobel Lect. Phys. 1965, p. 140. Elsevier, Amsterdam.

Shaw, R. (1955). The problem of particle types and other contributions to the theory of elementary particles, Ph.D. Thesis, Cambridge University.

Slavnov, A. A. (1972). Ward identities in gauge theories. *Theoret. and Math. Phys.* **10**, 99.

Slavnov, A. A. (1974). Generalized Pauli–Villars regularization in the presence of zero-mass particles. *Theoret. and Math. Phys.* **19**, 3.

Slavnov, A. A. (1975). Renormalization of gauge invariant theories. *Soviet. J. Particles and Nuclei* **5**, No. 3, 303.

Speer, E. (1969). "Generalized Feynman Amplitudes." Princeton Univ. Press, Princeton, New Jersey.

Steinmann, O. (1966). Renormalizability in LSZ perturbation theory. *Ann. Physics.* **36**, 267.

Stückelberg, E. C. G., and Petermann, A. (1953). Normalization of constants in quantum theory. *Helv. Phys. Acta* **26**, 499.

Symanzik, K. (1970). Small distance behavior in field theory and power counting. *Comm. Math. Phys.* **18**, 227.

Symanzik, K. (1971). Small distance behavior in field theory. *Springer Tracts Modern Phys.* **57**, 222.

Takeda, G. (1952). On the renormalization theory of the interaction of electrons and photons. *Progr. Theoret. Phys.* **7**, 359.

Taylor, J. C. (1971). Ward identities and charge renormalization of the Yang-Mills field. *Nuclear Phys. B* **33**, 436.

Taylor, J. C. (1976). "Gauge Theories of Weak Interactions." Cambridge Univ. Press, London and New York.

't Hooft, G. (1971a). Renormalization of Massless Yang-Mills fields. *Nuclear Phys. B* **33**, 173.

't Hooft, G. (1971b). Renormalizable Legrangians for massive Yang-Mills. *Nuclear Phys. B* **35**, 167.

't Hooft, G. (1973). Dimensional regularization and the renormalization group. *Nuclear Phys. B* **61**, 455.

't Hooft, G., and Veltman, M. (1972a). Regularization and renormalization of gauge fields. *Nuclear Phys. B* **44**, 188.

't Hooft, G., and Veltman, M. (1972b). Combinatorics of gauge fields. *Nuclear Phys. B* **50**, 318.

Todorov, I. T. (1971). "Analytical Properties of Feynman Diagrams in Quantum Field Theory." Pergamon, Oxford.

Tomonaga, S. (1946). On a relativistically invariant formulation of the quantum theory of wave fields. *Progr. Theoret. Phys.* **2**, 27.

Tomonaga, S. (1972). "Development of Quantum Electrodynamics," Nobel Lect. Phys. 1965, p. 126. Elsevier, Amsterdam.

Toussaint, D. (1978). Renormalization effects from superheavy Higgs particles. *Phys. Rev. D* **18**, 1626.

Utiyama, R. (1956). Invariant theoretical interpretation of interaction. *Phys. Rev.* **101**, 1597.

Vaĭnshtein, A. I., and Khriplovich, I. B. (1974). Renormalizable models of the electromagnetic and weak interactions. *Soviet. Phys. Uspekhi* **17**, No. 2, 263.

Velo, G., and Wightman, A. S., eds. (1976). "Renormalization Theory, Proceedings of the NATO Advanced Study Institute held at the International School of Mathematical Physics at the "Ettore Majorana." Reidel, Dordrecht, Netherlands.

Waller, I. (1930). Bemerkungen über die Rolle der Eigenenergie des Elektrons in der Quantentheorie der Stranhlung. *Z. Phys.* **62**, 673.

Weinberg, S. (1960). High-energy behavior in quantum field theory. *Phys. Rev.* **118**, 838.

Weinberg, S. (1964a). Feynman rules for any spin. *Phys. Rev. B* **133B**, 1318.

Weinberg, S. (1964b). Feynman rules for any spin. II. Massless particles. *Phys. Rev. B* **134B**, 882.

Weinberg, S. (1967). A model of leptons. *Phys. Rev. Lett.* **19**, 1264.

Weinberg, S. (1969). Feynman rules for any spin. III. *Phys. Rev.* **181**, 1893.

Weinberg, S. (1970). Physical processes in a convergent theory of the weak and electromagnetic interactions. *Phys. Rev. Lett.* **27**, 1688.

Weinberg, S. (1973). New approach to renormalization group. *Phys. Rev. D* **8**, 3497.

Weinberg, S. (1974). Recent progress in gauge theories of the weak, electromagnetic and strong interactions. *Rev. Modern. Phys.* **46**, 255.

Weinberg, S. (1979). Ultraviolet divergences in quantum theories of gravitation. In "General Relativity—An Einstein Centenary Survey" (S. W. Hawking and W. Israel, eds.), Cambridge Univ. Press, London and New York.

Weinberg, S. (1980). Conceptual foundations of the unified theory of weak and electromagnetic interactions, Nobel Lectures in Physics 1979. *Rev. Modern Phys.* **52**, No. 3, 515.

Weisskopf, V. (1934a). Über die Selbstenergie des Elektrons. *Z. Phys.* **89**, 27.

Weisskopf, V. (1934b). Berichtigung zu der Arbeit: Über die Selbstenergie des Elektrons. *Z. Phys.* **90**, 817.

Weisskopf, V. S. (1936). Über die Elektrodynamik des Vakuums auf Grund der Quantentheorie des Elektrons. *K. Dan. Vidensk. Selsk. Mat.-Fys. Medd.* **15**, No. 6, 3.

Weisskopf, V. S. (1939). On the self-energy and the electromagnetic field of the electron. *Phys. Rev.* **56**, 72.

Weisskopf, V., and Wigner, E. (1930a). Berechnung der natürlichen Linienbreite auf Grund der Diracschen Lichttheorie. *Z. Phys.* **63**, 54.

Weisskopf, V., and Wigner, E. (1930b). Über die natürliche Linienbreite in der Strahlung des harmonischen Oszillators. *Z. Phys.* **65**, 18.

Wentzel, G. (1973). Quantum theory of fields (until 1947). *In* "The Physicist's Conception of Nature" (J. Mehra, ed.), p. 380. Reidel, Dordrecht, Netherlands.

Wess, J., and Zumino, B. (1977). Superspace formulation of supergravity. *Phys. Lett. B* **66**, 361.

Wick, G. C. (1950). The evaluation of the collision matrix. *Phys. Rev.* **80**, 268.

Wilson, K. (1969). Non-Lagrangian models of current algebra. *Phys. Rev.* **179**, 1499.

Yang, C. N., and Mills, R. (1954). Conservation of isotropic spin and isotropic gauge invariance. *Phys. Rev.* **96**, 191.

Zav'yalov, O. I., and Anikin, A. S. (1976). Counterterms in the formalism of normal product. *Theoret. and Math. Phys.* **26**, 105.

Zimmermann, W. (1968). The power counting theorem for Minkowski metric. *Comm. Math. Phys.* **11**, 1.

Zimmermann, W. (1969). Convergence of Bogoliubov's method of renormalization in momentum space. *Comm. Math. Phys.* **15**, 208.

Zimmermann, W. (1973a). Composite operators in the perturbation theory of renormalizable interactions. *Ann. Physics* **77**, 536.

Zimmermann, W. (1973b). Normal products and the short distance expansion in the perturbation theory of renormalizable interactions. *Ann. Physics* **77**, 570.

Zumino, B. (1960). Gauge properties of propagators in quantum electrodynamics. *J. Math. Phys.* **1**, 1.

Zumino, B. (1975). Supersymmetry. *In* "Gauge Theories and Modern Field Theory, Proceedings of a Conference Held at Northeastern University, Boston" (R. Arnowitt and P. Nath, eds.), MIT Press, Cambridge, Massachusetts.

LIST OF SYMBOLS

INDEX

Pure and Applied Mathematics

A Series of Monographs and Textbooks

Editors **Samuel Eilenberg and Hyman Bass**

Columbia University, New York

RECENT TITLES

CARL L. DeVITO. Functional Analysis

MICHIEL HAZEWINKEL. Formal Groups and Applications

SIGURDUR HELGASON. Differential Geometry, Lie Groups, and Symmetric Spaces

ROBERT B. BURCKEL. An Introduction to Classical Complex Analysis: Volume 1

JOSEPH J. ROTMAN. An Introduction to Homological Algebra

C. TRUESDELL AND R. G. MUNCASTER. Fundamentals of Maxwell's Kinetic Theory of a Simple Monatomic Gas: Treated as a Branch of Rational Mechanics

BARRY SIMON. Functional Integration and Quantum Physics

GRZEGORZ ROZENBERG AND ARTO SALOMAA. The Mathematical Theory of L Systems.

DAVID KINDERLEHRER and GUIDO STAMPACCHIA. An Introduction to Variational Inequalities and Their Applications.

H. SEIFERT AND W. THRELFALL. A Textbook of Topology; H. SEIFERT. Topology of 3-Dimensional Fibered Spaces

LOUIS HALLE ROWEN. Polynominal Identities in Ring Theory

DONALD W. KAHN. Introduction to Global Analysis

DRAGOS M. CVETKOVIC, MICHAEL DOOB, AND HORST SACHS. Spectra of Graphs

ROBERT M. YOUNG. An Introduction to Nonharmonic Fourier Series

MICHAEL C. IRWIN. Smooth Dynamical Systems

JOHN B. GARNETT. Bounded Analytic Functions

EDUARD PRUGOVEČKI. Quantum Mechanics in Hilbert Space, Second Edition

M. SCOTT OSBORNE AND GARTH WARNER. The Theory of Eisenstein Systems

K. A. ZHEVLAKOV, A. M. SLIN'KO, I. P. SHESTAKOV, AND A. I. SHIRSHOV. Translated by HARRY SMITH. Rings That Are Nearly Associative

JEAN DIEUDONNÉ. A Panorama of Pure Mathematics; Translated by I. Macdonald

JOSEPH G. ROSENSTEIN. Linear Orderings

AVRAHAM FEINTUCH AND RICHARD SAEKS. System Theory: A Hilbert Space Approach

ULF GRENANDER. Mathematical Experiments on the Computer

HOWARD OSBORN. Vector Bundles: Volume 1, Foundations and Stiefel-Whitney Classes

K. P. S. BHASKARA RAO AND M. BHASKARA RAO. Theory of Charges

RICHARD V. KADISON AND JOHN R. RINGROSE. Fundamentals of the Theory of Operator Algebras, Volume I

EDWARD B. MANOUKIAN. Renormalization

IN PREPARATION

ROBERT B. BURCKEL. An Introduction to Classical Complex Analysis: Volume 2

RICHARD V. KADISON AND JOHN R. RINGROSE. Fundamentals of the Theory of Operator Algebras, Volume II

BARRETT O'NEILL. Semi-Riemannian Geometry: With Applications to Relativity

E. J. McSHANE. Unified Integration

A. P. MORSE. A Theory of Sets, Revised and Enlarged Edition